Material
Forming
Processes

Material Forming Processes

edited by
Anne Marie Habraken

First published in 2001 by Hermes Science Publications, Paris
First published in Great Britain and the United States in 2003 by Kogan Page Science, an
imprint of Kogan Page Limited
Derived from *International Journal of Forming Processes, Material Forming,*
Vol. 4, No. 3-4.

First South Asian Edition 2007

Kogan Page Limited   Kogan Page India
120 Pentonville Road   4737/23 Ansari Road
London N1 9JN   New Delhi- 110002
UK
www.koganpagescience.com

© Hermes Science Publishing Limited

© Kogan Page Limited

ISBN 1-9039-9633-3

British Library Cataloguing-in-Publication Data
A CIP record for this book is available from the British Library.

Library of Congress Cataloging-in-Publication Data

Material forming processes / edited by Anne Marie Habraken.
    p. cm.
ISBN 1-9039-9633-3
    1. Materials. [1. Materials.] I. Habraken, Anne Marie, 1961-
TA403.8 .M369   2003
620.1'1--dc21
                                            2002152251

Printed in Brijbasi Art Press Ltd., I-72, Sector-9, Noida, U.P. India.

Contents

Foreword
Anne Marie Habraken   ix

1. **Analysis of Phase Segregation Effects Arising in Fluid-particle Flows
    During Metal Injection Molding**
    Thierry Barriere, Boasheng Liu and Jean-Claude Gelin   1

2. **A Meshless Simulation of Injection Processes Involving Short Fiber
    Molten Composites**
    Miguel A. Martínez, Elias Cueto, Manuel Doblaré and Francisco Chinesta   19

3. **Semi-solid Processing of Engineering Alloys by a Twin-screw
    Rheomolding Process**
    Z. Fan, S. Ji and M. J. Bevis   39

4. **Micro Injection Molding**
    Christian G. Kukla   55

5. **Materials Characterization Methods and Material Models for
    Stamping of Plain Woven Composites**
    J. Chen, D. S. Lussier , J. Cao and X. Q. Peng   71

6. **Characterization and Modeling of Fabric Deformation During
    Forming of Textile Composites**
    Andrew Long   87

7. **Comparison of Ductile Damage Models**
    Cyril Bordreuil, Emmanuelle Vidal-Sallé and Jean-Claude Boyer   105

8. **Experimental and Numerical Analysis of Blanking for Thin Sheet
    Metal Parts**
    Vincent Lemiale, Philippe Picart and Sébastien Meunier   123

9. **Development in Finite Element Simulations of Aluminum Extrusion**
    J. Lof, K. Valkering and J. Huétink   139

10. **Investigation of Springback Using Two Different Testing Methods**
    Martin Rohleder, Karl Roll, Alexander Brosius and Matthias Kleiner   151

11. **Modeling of AlMg Sheet Forming at Elevated Temperatures**
    A. H. van den Boogaard, P. J. Bolt and R. J. Werkhoven   165

12. **Hydroforming Processes for Tubular Parts: Optimization by means of
    Adaptive and Iterative FEM Simulation**
    Wim H. Sillekens and Robert J. Werkhoven   179

13. **Identification of Non-linear Kinematic Hardening with Bend-reverse
    Bend Experiments in Anisotropic Sheet-metals**
    Michel Brunet, Fabrice Morestin, Françis Sabourin and Stephane Godereaux   197

14. **On the Simulation of Microhardness at Large Strains Using a
    Gradient Theory of Plasticity**
    Thomas Svedberg and Kenneth Runesson   215

15. **A Simplified Model of Thermal Contact Resistance Adapted to
    Forging Simulation**
    Brahim Bourouga, Vincent Goizet and Jean-Pierre Bardon   235

16. **Validation of the Cockcroft and Latham Fracture Criterion for Cold
    Heading of Steel Fasteners Using Drop Weight Compression Testing
    and Finite Element Modeling**
    Nicolas Nickoletopoulos, Michel Hone, Yves Verreman and
    James A. Nemes   249

17. **Industrial Forging Design Using an Inverse Technique**
    Luísa C. Sousa, Catarina F. Castro, Carlos C. António and Abel D. Santos   267

18. **Simulation of Face Milling and Turning with the Finite Element
    Method**
    Luc Masset   285

19. **Analysis of Material Behaviour at High Strain Rates for Modeling
    Machining Processes**
    Paolo Bariani, Guido Berti and Stefano Corazza   303

20. **A Comparative Study of Crystallization of iPP and PA6 Under Pressure**
Vicenzo La Carrubba, Valerio Brucato and S. Piccarolo   315

21. **Arbitrary Lagrangian-Eulerian Simulation of Powder Compaction Processes**
A. Pérez-Foguet, A. Rodríguez-Ferran and A. Huerta   341

22. **Austenite-to-Ferrite Phase Transformation During Continuous Casting of Steels**
Ernst Gamsjäger, Franz Dieter Fischer, Christian M. Chimani and
Jiri Svoboda   357

Index   379

Foreword

This publication consists of extended and edited versions of selected papers presented at the 4th international ESAFORM conference on Material Forming held at the University of Liège in Belgium during April 2001. The purpose of this conference was to bring together scientists working in different scientific disciplines on material forming processes with all types of materials.

The contributions gathered together in this publication cover a wide range of topics including:

Advection equation in forming processes: towards an efficient numerical treatment

Advanced injection molding technologies

Composites forming processes

Damage modeling in metal forming processes

Die extrusion simulation and optimization

Further study of NUMISHEET 99 Benchmark B1 (Cup drawing simulation)

Hydro forming-blowing

Inverse modeling dedicated to rheology parameter identification

Micro-macro modeling in forming processes

No local and gradient models in forming

Numerical and physical investigations of forging

Simulation of machining and other cutting processes

Structural development during processing from molten materials

Super plastic forming

Specific topics like casting, rolling, optimization, cracks and powder compaction were also covered.

Forming processes were thus analyzed at different levels: the physics, experimental investigation and modeling.

The scientific community is both focused on a better understanding of phenomena and on working on process optimization.

The subjects considered here therefore include mathematical treatments and specific topics of finite element simulation but experimental devices to provide data are also discussed, creating an important link with studies of material forming processes. Materials as varied as metals, polymers, composites and also wood and bones are also considered in some of the contributions.

Anne Marie Habraken
National Fund for Scientific Research
University of Liège

Chapter 1

Analysis of Phase Segregation Effects Arising in Fluid-particle Flows During Metal Injection Molding

Thierry Barriere and Jean-Claude Gelin
University of Franche-Comté and CNRS, Besançon, France

Boasheng Liu
Institute of Engineering Science, Southwest Jiatong University, Sichuan, P R China

1. Introduction

Metal injection molding (MIM) technology has become a very active research field in the last ten years. MIM combines the well-known polymer injection molding and powder metallurgy technology, enabling parts with very complex geometries to be shaped as in traditional polymer injection molding. This technology is especially suitable for the production of small and intricate metal parts in large batches at low cost. Compared with traditional powder metallurgy methods, one can produce much more complex parts with better surface quality and high mechanical properties using this technology.

The process consists mainly of four steps [GER 97; KIM 98; BAR99]: 1) The mixting of metallic powders with polymer binders to get the feedstock, 2) The injection molding of the feedstock to get the desired shape, 3) The elimination of polymer binder, normally performed in a debinding cell, 4) The sintering stage to obtain the final parts in nearly pure metallic form. The simulation begins to play a very important role for the determination of the different process parameters and optimization of component properties.

In MIM, the injection stage is perhaps the most important one. All possible appearances of the different defects, such as jetting, bubble and phase segregation are generally amplified during the next stages [GER 97]. Several authors have tried to model and simulate the injection stage in MIM by an equivalent viscosity for the feedstock mixture [BIL 00], [ILL 00]. This approach may predict accurately the filling states but is not able to provide data concerning the phase separation effects. Depending on the feedstock, mold geometry and injection parameters, the proportion between polymer binder and metallic powder may be slightly or even considerably changed during a complex shape injection. This so-called *segregation effect* may lead to undesired distortions in the final parts after sintering. To predict and simulate such an important effect in the mixture flows during MIM, several authors have adopted the mixture theory [LAN 96; DUT 98a; GEL 99; BAR 00a]. This theory uses a bi-phasic model to describe the flows of a mixture composed of a fluid phase charged with metallic powder at high concentration. It has been shown in previous work that such an approach is suitable for simulation of the injection stage in MIM, especially for prediction of the segregation effects. In such bi-phasic modeling, the flows of polymer binder and metallic powder are described distinctly by two different phases considered both as a viscous fluid, the fluid phase, and the solid phase respectively. The volume fractions of each phase, under the constraint of saturation condition, are taken as the key variables to describe directly the evolution of phase segregation, *i.e.* the change of their proportions during mold filling.

The solution of such an equation system resulting from the bi-phasic approach is very expensive in computational cost using a classical algorithm, because two Stokes equations coupled by a momentum exchange term should be solved simultaneously to obtain the pressure field and the velocity fields for each phase. With reference to the work on explicit models for 2D casting simulation [USM 92][LEW 95] for a viscous fluid the authors have proposed and developed a new explicit algorithm to carry out the simulation of incompressible bi-phasic flows [BAR 00b; GEL 00; BAR 01]. This fractional-step scheme runs without any iterations, while the physical phenomena are closely coupled by the sequential steps. Most operations may be performed locally without any global resolution, except the solution of the pressure field. One still needs a global solution for the pressure field to maintain incompressibility of the mixture. This phenomenon should be well recognized as each phase may change their proportion in the course of injection while the mixture remains in incompressible state. The solid and fluid phases behave like different non-Newtonian viscous fluids. The interaction between both phases is taken into account by a momentum exchange term.

In this paper, the developments are particularly focussed on the way to solve advection-diffusion equations related to the filling state and phase segregation. New methods are proposed for the calculation of phase volume fraction. Their applications in MIM simulation are then presented and discussed.

2. Advection-diffusion equations

As we deal with a fluid mechanics problem, the Eulerian description is the most suitable choice for injection molding of a powder-binder mixture. The only boundary unknown is the filling front in the mold at each instant during the injection course. In the filled portion, the mixture theory is applied. The flows of each phase are represented by two coupled Stokes equations. The interaction between the flows of both phases is taken into account by a momentum exchange term.

The volume fractions of each phase, which represent the distribution of their volume proportions at instant t, are assigned as two field variables ϕ_s and ϕ_f. The mass conservation equations for each phase in the mixture permit access to their density changes and description of segregation effects. The flows of solid and fluid phases possess two co-existent velocity fields v_s and v_f. The effective velocity v_{ef} is sometimes called the mixture's velocity and is defined as:

$$v_{ef} = \phi_s v_s + \phi_f v_f \qquad [1]$$

At each instant t during the injection course, one has to know which part of the mold domain is filled by the mixture. The filled domain is modeled using the mixture theory while a fictive flow model is used for modeling the air in the remaining part of the mold. A space and time variable F, the *filling state*, is used to represent the evolution of filled domain. This filling state variable is sometimes identified as a pseudo-concentration, or fictitious concentration, by some authors [USM 92; LEW 95]. This filling state variable is governed by an advection equation:

$$\frac{\partial F}{\partial t} + \nabla \bullet (v_{ef} F) = 0 \qquad [2]$$

In our definition, the filling state variable assumes a value of 1 in the filled portion of the mold while the void part in the mold is assigned the value 0 at each instant t during the injection course. In the mold domain Ω, one has then the filled portion Ω^F and the remained void portion Ω^V defined as:

$$\Omega^F = \left\{ X \in \Omega^F(t) \mid F(X,t) = 1, \quad \Omega^F(t) \cup \Omega^V(t) = \Omega \right\},$$

$$\Omega^V = \left\{ X \in \Omega^V(t) \mid F(X,t) = 0, \quad \Omega^F(t) \cup \Omega^V(t) = \Omega \right\}. \qquad [3]$$

This domain determination allows us to apply different models respectively in filled and void subdomains according to what really happens in the mold during injection.

For each spatial position in the filled part of the mold $X \in \Omega^r(t)$, two field variables (ϕ_s, ϕ_f) are used to define the volume fraction of solid and fluid phase in mixture flow. These two fields should certainly verify the so-called saturation conditions:

$$\phi_s + \phi_f = 1 \text{ and } \frac{\partial \phi_s}{\partial t} + \frac{\partial \phi_f}{\partial t} = 0. \qquad [4]$$

Each phase is considered to be intrinsically incompressible, so that the density of each phase remains constant (ρ_{s_0}, ρ_{f_0}). The apparent density for each phase in the mixture is then related to its volume fraction by the following relationship:

$$\rho_s = \phi_s \rho_{s_0} \text{ and } \rho_f = \phi_f \rho_{f_0}, \tag{5}$$

where ρ_s and ρ_f, are respectively the apparent density values in the mixture for solid and fluid phase. The mass conservation for the flow of each phase should be verified:

$$\frac{\partial \rho_s}{\partial t} + \nabla \bullet (\rho_s v_s) = 0 \text{ and } \quad \frac{\partial \rho_f}{\partial t} + \nabla \bullet (\rho_f v_f) = 0. \tag{6}$$

Accounting for the intrinsic incompressibility of each phase, one can express their mass conservation as:

$$\frac{\partial \phi_s}{\partial t} + \nabla \bullet (\phi_s v_s) = 0 \quad \text{and} \quad \frac{\partial \phi_f}{\partial t} + \nabla \bullet (\phi_f v_f) = 0. \tag{7}$$

These two equations are coupled by the saturation condition [BOW 76], [BAR 00a]. They are very important especially in the prediction of segregation effects during the injection course. The variation of both volume fractions represents directly the change of their proportions during the mixture's flow.

The volume saturation and mass conservation for solid and fluid phase result directly in the incompressibility condition for the mixture flow is expressed as:

$$\nabla \bullet v_{ef} = 0. \tag{8}$$

The boundary conditions for solid and fluid volume fractions, as well as the filling state variable should be imposed on the inlet Γ^I. These conditions are written as:

$$\forall X \in \Gamma^I, \quad \phi_s = \phi_{s_0}, \quad \phi_f = \phi_{f_0}, \quad F = 1. \tag{9}$$

Of course both volume fractions should satisfy the saturation condition $\phi_{s_0} + \phi_{f_0} = 1$. In the void part of the mold Ω^V, at each instant both fluid and solid volume fractions take a value equal to 0:

$$\forall X \in \Omega^V(t), \quad \phi_s = 0, \quad \phi_f = 0. \tag{10}$$

The void part of the mold is determined at each instant by the calculation of filling state variable F which indicates the portion not filled by the mixture.

3. Solution of advection equations for filling state

The stability of the filling state dominated by equation [2] had been discussed by several authors [DON 84], [BAR 00a]. The Taylor-Galerkin method is considered as a suitable choice for solving such an equation. Some formulations were proposed in 2D cases mainly for solving mold filling problems in casting [GAO 91], [USM 92], [LEW 95]. To get a global and precise vision of such a method, a formulation in 3D-operators form is proposed. The boundary condition integration is also discussed precisely. The void part of the mold is supposed to be filled by a fictitious fluid to simplify the simulation procedures. As fractional steps are performed to take different effects into account, the determination of filling state is the premise for other operations.

The time derivative of variable F in discretized form can be expressed by a second-order Taylor series development as follows:

$$\frac{F^{n+1} - F^n}{\Delta t} = \frac{\partial F^n}{\partial t} + \frac{\Delta t}{2} \frac{\partial^2 F^n}{\partial t^2} + O\,(\Delta t)^2 \qquad [11]$$

The advection equation [2] that controls the fluid front can lead to the following relation by using the approximation $\partial v_{ef} / \partial t = (v_{ef}^{n+1} - v_{ef}^n)/\Delta t$ and assuming that $\nabla \bullet v_{ef} = 0$:

$$\frac{\partial F^n}{\partial t} = -\, v_{ef}^n \bullet \nabla F^n$$

$$\frac{\partial^2 F^n}{\partial t^2} = v_{ef}^n \bullet \nabla\,(\,v_{ef}^n \bullet \nabla F^n) - \frac{v_{ef}^{n+1} - v_{ef}^n}{\Delta t} \bullet \nabla F^n \qquad [12]$$

The weak form of equation [11] can then be written as:

$$\int_\Omega f^* \frac{F^{n+1} - F^n}{\Delta t}\, d\Omega = -\int_\Omega f^* v_{ef}^{n+1/2} \bullet \nabla F^n\, d\Omega + \frac{\Delta t}{2} \int_\Omega f^* v_{ef}^n \bullet \nabla\,(\,v_{ef}^n \bullet \nabla F^n)\, d\Omega \qquad [13]$$

where $v_{ef}^{n+1/2} = (v_{ef}^{n+1} + v_{ef}^n)/2$, f^* is a homogeneous field associated to variable F. As the effective velocity v_{ef}^n is an incompressible field and using the identity $\nabla \bullet (a\,\vec{b}) = a\nabla \bullet \vec{b} + \vec{b} \bullet \nabla a$, the second term in the right-hand side of [13] can be written as:

$$\frac{\Delta t}{2} \int_\Omega \nabla \bullet [\, f^* v_{ef}^n \bullet (\, v_{ef}^n \bullet \nabla F^n)\,]\, d\Omega - \frac{\Delta t}{2} \int_\Omega (\, v_{ef}^n \bullet \nabla F^n) \bullet v_{ef}^n \bullet \nabla f^*\,]\, d\Omega \qquad [14]$$

Using the divergence theorem, the first term of the [14] can be transformed to a surface integral as:

$$\frac{\Delta t}{2} \int_{\Omega} \nabla \bullet [\, f^* v_{ef}^n \bullet (\, v_{ef}^n \bullet \nabla F^n) \,] \, d\Omega = \frac{\Delta t}{2} \oint_{\Gamma} f^* v_{ef}^n \bullet (\, v_{ef}^n \bullet \nabla F^n) \bullet \bar{n} \, dS \qquad [15]$$

Except for the boundary conditions on the outlet $X \in \Gamma^O$, the above integral vanishes for other parts of the mold surface. The integral on the outlet surface takes also a zero value until the filling front is near the outlet. When the mold filling is being completed this term must be taken into account to correctly simulate the filling process at the final stage.

The discretized form of equation [13] by the finite element method can then be written as:

$$\mathbf{M} \frac{F^{n+1} - F^n}{\Delta t} = -\left[\mathbf{C_{ad}}(v_{ef}) + \mathbf{K_{df}}(v_{ef}) \right] F^n + \mathbf{K_O}(v_{ef}) \, F^n \qquad [16]$$

where $\mathbf{M}$ is a lumped pseudo matrix, $\mathbf{C_{ad}}(v_{eff})$ is the matrix that represents the advection effect, $\mathbf{K_{df}}(v_{eff})$ is a stiffness matrix that stands for diffusion effects and $\mathbf{K_O}(v_{ef})$ is the operator for surface integration terms, F^{n+1} and F^n are the filling states at time step t_{n+1} and t_n. Denoting as N_{out} the nodes associated to outlet mold surface and with unit normal vector $\bar{n}_{out}$, these matrices take respectively the following forms:

$$\mathbf{M} = \mathop{A}_{\Omega} \int_{\Omega^e} \mathbf{N}^T \, \mathbf{N} \, d\Omega \qquad [17]$$

$$\mathbf{C_{ad}}(v_{ef}) = \mathop{A}_{\Omega} \int_{\Omega^e} \mathbf{N}^T (v_{ef}^{n+1/2})^T \mathbf{G} \, d\Omega \qquad [18]$$

$$\mathbf{K_{df}}(v_{ef}) = \frac{\Delta t}{2} \mathop{A}_{\Omega} \int_{\Omega^e} \mathbf{G}^T (v_{ef}^n)^T v_{ef}^n \, \mathbf{G} \, d\Omega \qquad [19]$$

$$\mathbf{K_O}(v_{ef}) = \frac{\Delta t}{2} \mathop{A}_{N_{out}} N_{out} \cap \int_{\Omega_{out}^e} \mathbf{N}^T (v_{ef}^n)^T v_{ef}^n \bullet \bar{n}_{out}^T \mathbf{G} \, d\Omega \qquad [20]$$

In the above equations, N is the shape function in each element for filling state variables, $\mathbf{G}$ is a gradient operator built from the derivatives of these shape functions, $v_{ef}^{n+1/2}$ and v_{ef}^n are the velocity values within each element. Ω_{out}^e stands for the elements linked with mold outlet nodes N_{out}:

$$\Omega_{out}^e = \left\{ \Omega^e \mid N_{out} \subset \Omega^e \right\} \qquad [21]$$

The solution scheme consists of only local and explicit operations, so the prediction stage is very efficient.

4. Numerical schemes for segregation predictions

The variation of volume fraction fields for each phase represents directly the segregation effects during the injection course. From the velocity fields for solid and fluid phase respectively, both volume fractions are governed by mass conservation and saturation conditions, as in equations [4]–[7]. The solution of such problems needs only to be performed in the filled domain. In the void part of mold, these variables always take values equal to zero. The following numerical methods may be chosen to calculate the volume fraction fields.

4.1. *Simple forward explicit method*

Equation [7] can be written in the following form:

$$\frac{\partial \phi_s}{\partial t} = -\phi_s \nabla \bullet \mathbf{v}_s - \mathbf{v}_s \bullet \nabla \phi_s \quad \text{and} \quad \frac{\partial \phi_f}{\partial t} = -\phi_f \nabla \bullet \mathbf{v}_f - \mathbf{v}_f \bullet \nabla \phi_f \qquad [22]$$

Both equations can be discretized in space by the standard Galerkin method [DUT 98a; 98b]. The integration with respect to time can be performed by the standard finite difference method. Once a forward difference method is applied, one can solve the following equation for each phase:

$$\mathbf{M} \frac{\Phi_{p*}^{n+1} - \Phi_p^n}{\Delta t} = -[\mathbf{K}_{dv}(\mathbf{v}_p^{n+1}) + \mathbf{C}_{ad}(\mathbf{v}_p^{n+1})]\, \Phi_p^n \qquad [23]$$

where p represents respectively the solid and fluid phase, such as $p \in [s, f]$, Φ_p^n are the solid or fluid volume fractions discretized in vector form at instant t_n, Φ_{p*}^{n+1} are the predicted values for solid or fluid volume fractions at instant t_{n+1}. The different operators are defined as follows:

$$\mathbf{M} = \underset{\Omega^F}{A} \int_{\Omega^e} \mathbf{N}^T \mathbf{N} \, d\Omega, \qquad [24]$$

$$\mathbf{K}_{dv}(\mathbf{v}_p^{n+1}) = \underset{\Omega^F}{A} \int_{\Omega^e} \mathbf{N}^T (\nabla \bullet \mathbf{v}_p^{n+1}) \mathbf{N} \, d\Omega, \qquad [25]$$

$$\mathbf{C_{ad}}(\mathbf{v_p^{n+1}}) = \underset{\Omega^F}{A} \int_{\Omega^e} \mathbf{N}^T (\mathbf{v_p^{n+1}})^T \mathbf{G}\, d\Omega, \qquad\qquad\qquad [26]$$

where vector $\mathbf{v_p^{n+1}}$ stands for the velocity field of solid or fluid phases obtained at time t_{n+1}, $\mathbf{N}$ and $\mathbf{G}$ have the same definition as in equations [17]–[20], the pseudo mass matrix $\mathbf{M}$ can be lumped in a diagonal form. In that case, the proposed method becomes an explicit and very efficient one.

The saturation condition cannot be satisfied automatically in such a scheme. A numerical regularization at each node is needed for both predicted volume fractions to satisfy the saturation condition:

$$\phi_s^{n+1} = \frac{\phi_{s*}^{n+1}}{\phi_{s*}^{n+1} + \phi_{f*}^{n+1}} \quad \text{and} \quad \phi_f^{n+1} = \frac{\phi_{f*}^{n+1}}{\phi_{s*}^{n+1} + \phi_{f*}^{n+1}}. \qquad\qquad [27]$$

4.2. Backward global solution scheme

If the integration of equation [22] with respect to time is performed by a backward difference method, with Galerkin discretization in space, one obtains another method for solution of the volume fractions:

$$[\, \mathbf{M} + \Delta t\, \mathbf{K_{dv}}(\mathbf{v_p^{n+1}}) + \Delta t\, \mathbf{C_{ad}}(\mathbf{v_p^{n+1}}) \,]\, \Phi_p^{n+1} = \mathbf{M}\, \Phi_p^n \qquad\qquad [28]$$

The variables and operators in equation [28] have the same definitions as for the case related to equation [23]. These equations for each phase lead to two independent linear systems to be solved globally. The boundary conditions should be applied in a standard manner. Even if this solution scheme is very straightforward without any iteration, it is much more expensive than the explicit one. The advantage of such a method remains in its stability and accuracy. As in the case of the forward explicit method, a numerical regularization is needed to verify the saturation condition and is identical to equation [27].

4.3. Taylor-Galerkin method

In order to get a compromise between the stability and computational cost, the Taylor-Galerkin method may be considered as a practical algorithm to solve the advection equations. Such a method has been used to calculate the filling state in the injection course. For volume fraction calculations, a similar method is developed to perform stable simulations with reasonable cost. To achieve such a goal, a Taylor development for each phase should be performed. The system to be solved is written in the following manner:

$$\frac{\phi_p^{n+1} - \phi_p^n}{\Delta t} = \frac{\partial \phi_p^n}{\partial t} + \frac{\Delta t}{2} \frac{\partial^2 \phi_p^n}{\partial t^2} + O(\Delta t)^2,$$ [29]

where from equation [7]:

$$\frac{\partial \phi_p^n}{\partial t} = -(\nabla \bullet \mathbf{v}_p^n + \mathbf{v}_p^n \bullet \nabla) \phi_p^n,$$ [30]

$$\frac{\partial^2 \phi_p^n}{\partial t^2} = -(\nabla \bullet \mathbf{v}_p^n + \mathbf{v}_p^n \bullet \nabla) \frac{\partial \bar{\phi}_p^n}{\partial t} - (\nabla \bullet \frac{\partial \mathbf{v}_p^n}{\partial t} + \frac{\partial \mathbf{v}_p^n}{\partial t} \bullet \nabla) \phi_p^n,$$ [31]

where $\partial \bar{\phi}_p^n / \partial t$ and $\partial \mathbf{v}_p^n / \partial t$ are expressed as:

$$\frac{\partial \bar{\phi}_p^n}{\partial t} = \frac{\phi_p^n - \phi_p^{n-1}}{\Delta t}, \qquad \frac{\partial \mathbf{v}_p^n}{\partial t} = \frac{\mathbf{v}_p^{n+1} - \mathbf{v}_p^n}{\Delta t}.$$ [32]

Substituting [32 a, b] in equation [29] results in the following equation:

$$\frac{\phi_p^{n+1} - \phi_p^n}{\Delta t} = -(\nabla \bullet \mathbf{v}_p^{n+1/2} + \mathbf{v}_p^{n+1/2} \bullet \nabla) \phi_p^n - \frac{\Delta t}{2} (\nabla \bullet \mathbf{v}_p^n + \mathbf{v}_p^n \bullet \nabla) \frac{\partial \bar{\phi}_p^n}{\partial t},$$ [33]

where $\mathbf{v}_p^{n+1/2} = (\mathbf{v}_p^{n+1} + \mathbf{v}_p^n)/2$. The discretized weak form of equation [33] is now expressed as:

$$\mathbf{M} \frac{\Phi_{p*}^{n+1} - \Phi_p^n}{\Delta t} = \mathbf{C_a}(\mathbf{v}_p^{n+1/2}) \Phi_p^n + \frac{1}{2} \mathbf{K_d}(\mathbf{v}_p^n)(\Phi_p^n - \Phi_p^{n-1})$$ [34]

where $\mathbf{M}$ is the lumped pseudo mass matrix. For the other matrices, they are written as:

$$\mathbf{C_a}(\mathbf{v}_p^{n+1/2}) = -\underset{\Omega^F}{A} \int_{\Omega^e} \mathbf{N}^T [(\nabla \bullet \mathbf{v}_p^{n+1/2}) \mathbf{N} + (\mathbf{v}_p^{n+1/2})^T \mathbf{G}] \, d\Omega,$$ [35]

$$\mathbf{K_d}(\mathbf{v}_p^n) \doteq -\underset{\Omega^F}{A} \int_{\Omega^e} \mathbf{N}^T [(\nabla \bullet \mathbf{v}_p^n) \mathbf{N} + (\mathbf{v}_p^n)^T \mathbf{G}] \, d\Omega.$$ [36]

Equation [34] can be solved explicitly with high efficiency. As the equations for both phases are solved independently, numerical regularization in equation [27] is needed to satisfy the saturation condition.

4.4. Self-regularization method

If the saturation condition is used, the fluid volume fraction ϕ_f can be replaced by $(1-\phi_s)$ in the mass conservation for the fluid phase. These two different mass conservation equations can be condensed into one when attention is paid to choose their signs. This choice is important to avoid the links between the two equations. One has then the following relation when the velocity fields for each phase are obtained previously in the same time step:

$$(\rho_{s_0} + \rho_{f_0}) \frac{\partial \phi_s}{\partial t} = -[\rho_{s_0} \nabla \bullet \mathbf{v}_s + \rho_{f_0} \nabla \bullet \mathbf{v}_f] \phi_s$$

$$-[\rho_{s_0} \mathbf{v}_s + \rho_{f_0} \mathbf{v}_f] \bullet \nabla \phi_s + \rho_{f_0} \nabla \bullet \mathbf{v}_f . \qquad [37]$$

After discretization by a Galerkin method, the solution for the solid phase volume fraction can be calculated from the following matrix equation:

$$\mathbf{M}_D \frac{\Phi_s^{n+1} - \Phi_s^{n+1}}{\Delta t} = -\left[\mathbf{K}_D^{div}(\mathbf{v}_s, \mathbf{v}_f) + \mathbf{C}_D^{adv}(\mathbf{v}_s, \mathbf{v}_f) \right] \Phi_s^n + \mathbf{F}_D(\mathbf{v}_s) , \qquad [38]$$

where ϕ_S^n, ϕ_S^{n+1} are the discretized solid volume fractions in vector form at instant t_n and t_{n+1}, $\mathbf{M}_D$ is a lumped mass matrix in diagonal form, $\mathbf{K}_D^{div}(\mathbf{v}_S, \mathbf{v}_f)$ and $\mathbf{C}_D^{adv}(\mathbf{v}_S, \mathbf{v}_f)$ are the diffusion and advection matrices, $\mathbf{F}_D$ stands for an effect provided by the fluid phase flow. These terms take, respectively, the following forms:

$$\mathbf{M}_D = \underset{\Omega^F}{A} \int_{\Omega^e} \mathbf{N}^T (\rho_{s_0} + \rho_{f_0}) \mathbf{N} \, d\Omega , \qquad [39]$$

$$\mathbf{K}_D^{div}(\mathbf{v}_s, \mathbf{v}_f) = \underset{\Omega^F}{A} \int_{\Omega^e} \mathbf{N}^T [\rho_{s_0} \nabla \bullet \mathbf{v}_s^e + \rho_{f_0} \nabla \bullet \mathbf{v}_f^e] \mathbf{N} \, d\Omega , \qquad [40]$$

$$\mathbf{C}_D^{adv}(\mathbf{v}_s, \mathbf{v}_f) = \underset{\Omega^F}{A} \int_{\Omega^e} \mathbf{N}^T [\rho_{s_0} (\mathbf{v}_s^e)^T + \rho_{f_0} (\mathbf{v}_f^e)^T] \mathbf{G} \, d\Omega , \qquad [41]$$

$$\mathbf{F}_D = \underset{\Omega^F}{A} \int_{\Omega^e} \mathbf{N}^T (\rho_{f_0} \nabla \bullet \mathbf{v}_f^e) \, d\Omega . \qquad [42]$$

This solution step can be performed locally as the mass matrix $\mathbf{M}_D$ a is lumped one. As the saturation condition is automatically satisfied, it becomes straightforward to get the evolution of segregation effects in the simulation of the injection molding. The biggest advantage of such a method is that it automatically regularizes the saturation condition and needs only one solution instead of two in other methods.

Control of time steps

The time step control during calculations of the solid and fluid volume fractions, as well as for the filling state variable is determined by a lower limit associated with the advection equations [LOH 84]. This time step limit can be written as:

$$\Delta t \leq \alpha \, C \min \left(\frac{h}{v_s^m}, \frac{h}{v_f^m}, \frac{h}{v_{eff}^m} \right) \tag{43}$$

where h is a characteristic value associated with the size of the smallest element. α is a safety factor, its value may be chosen as $0.5 \leq \alpha \leq 0.9$, $C \leq 1/\sqrt{3} = 0.5773$ is the so-called Courant number, v_s^m, v_f^m and v_{ef}^m are three characteristic values which represent the maximum norms of the solid, fluid and effective velocity fields at the time considered. This limit time step should be chosen so that it can be applied for both coupled Stokes equations. The minimum value satisfying the different control criteria should be taken into account.

5. Numerical examples

5.1. *Flow in a convergent die*

The flow of a powder-binder mixture through a convergent die illustrates well the phenomenon associated with powder accumulation in the convergent die.

The geometry of the die and process parameters, as well as viscosity coefficients, interaction parameter and boundary conditions are given in Table 1.

Table 1. *Material parameters, geometry and mesh for a convergent die*

Imposed inlet pressure:	5 MPa	
Sliding wall conditions		
Initial powder volume fraction: 50%		
Interaction term:	1.10^9 Pa.s.m^{-2}	
	Solid phase	Fluid phase
Density	7500 kg/m^3	1000 kg/m^3
Viscosity	10 Pa.s	1 Pa.s

318 elements, 508 nodes

Figure 1 illustrates well the important segregation effects as the powder volume fraction increases from 50% to 61.6% for the injection parameters and processing conditions indicated in Table 1.

The powder segregation in that case is mainly sensitive to the interaction parameter between both phases, binder and powder viscosities as well as geometry of the flow and particularly die reduction angle.

The results of the analysis are reported in Figures 2 (a, b and c) and they illustrate the variation of powder volume fraction $\Delta\phi_s$. It is demonstrated that $\Delta\phi_s$ increases with the interaction parameter k and the powder viscosity μ_s and there exists an optimum die angle α that minimizes it.

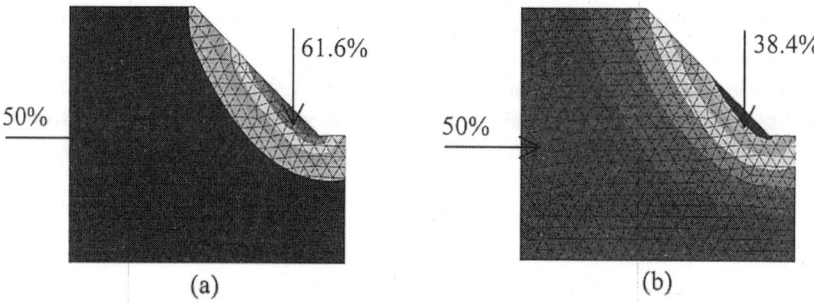

Figure 1. *Contours of powder volume fraction (a) and fluid volume fraction (b) for the flow through a convergent die*

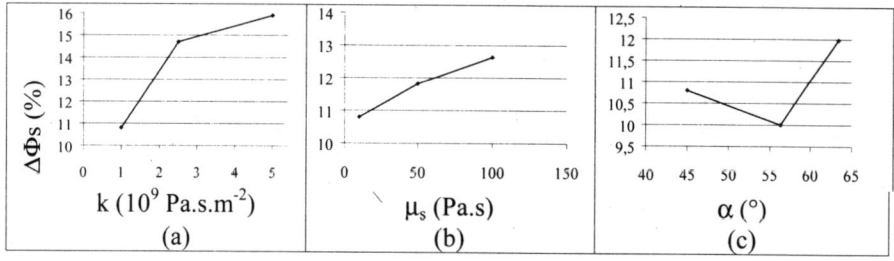

Figure 2. *Sensitivity of powder segregation vs. the interaction coefficient* k *(a), the powder viscosity* μ_s *(b), and convergent die angle* α *(c)*

5.2. *Mold filling of a complex shape*

The second example concerns a study of mold filling for a thin and complex mobile phone shell to reveal the possible segregation effects.

A forward explicit method is used in the simulation. The segregation evolution is studied with different initial powder volume fractions. The material parameters, component geometry and the finite element mesh are reported in Table 2. The pressure is imposed on the upper left corner of the mobile phone. The outlet is located on the underside.

The filling front prediction during injection is very important due to the complex mold cavity. The filling states at two different stages are represented in Figure 3. The width of the window edge at the right hand side is larger than the left hand one, but it is further from the inlet. This geometry provokes a significant difference in the distribution of the solid and fluid velocities at different injection stages. The numerical simulation provides obviously the dead zone and segregation locations. Two filling states during the injection are shown in Figure 3 corresponding respectively to 20% and to 50%.

Table 2. *Material parameters, geometry and mesh*

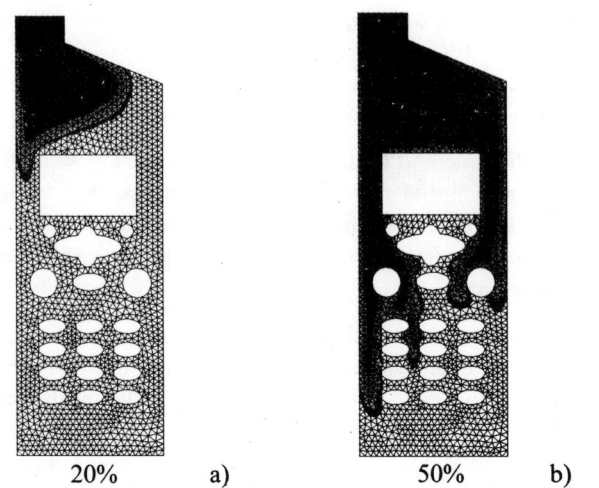

Figure 3. *Contours of mixture front during mobile phone mold filling, (a) 20% of mold filling, (b) 50% of mold filling*

Segregation zones develop near the wall mold boundaries. For a 50% filled state of the entire mold cavity, the solid and fluid volume fractions represent well such an effect, Figure 4.

Another simulation has been carried out with 75% as the initial powder volume fraction. The results corresponding to a filled state equal to 50% are shown in Figure 5. One can then compare the different segregation effects according to the initial powder volume fractions. The comparison between Figure 4 and Figure 5 shows analogous segregation zones. At a bigger initial powder volume fraction there corresponds the development of larger segregation zones during injection molding.

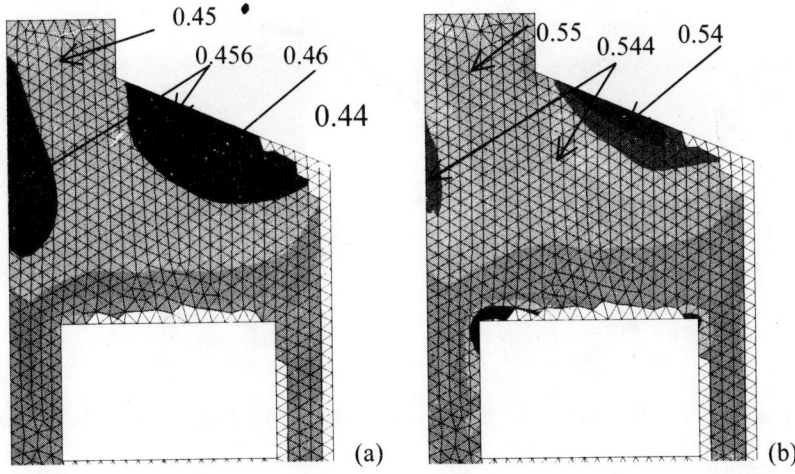

Figure 4. *Contours of solid (a) and fluid (b) volume fractions at 50% filled state (45% initial powder volume fraction)*

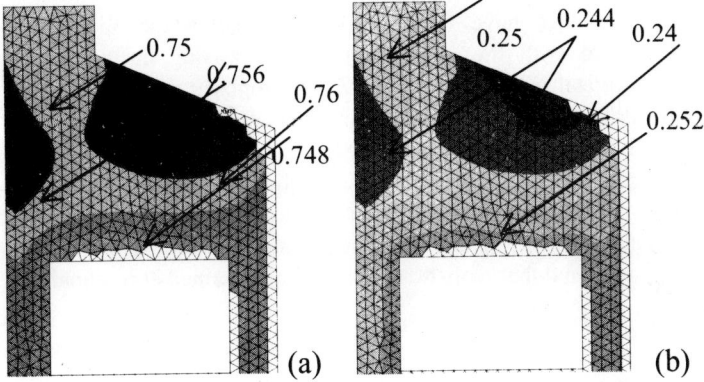

Figure 5. *Contours solid (a) and fluid (b) volume fractions at 75% filled state (45% initial powder volume fraction)*

Results plotted in Figure 6 show the influence of initial powder volume fraction on the mold filling time. The necessary filling time increases with the initial powder volume fraction.

The computational cost for different solution methods is about 20 minutes on a standard PC with a Pentium 3 processor. As the solution of volume fractions is only a part of the computational operation, it is difficult to have an exact measure of the computation cost related to volume fraction calculation. However, this work proves the feasibility to develop such software and its efficiency to study the evolution of segregation effects during powder injection molding.

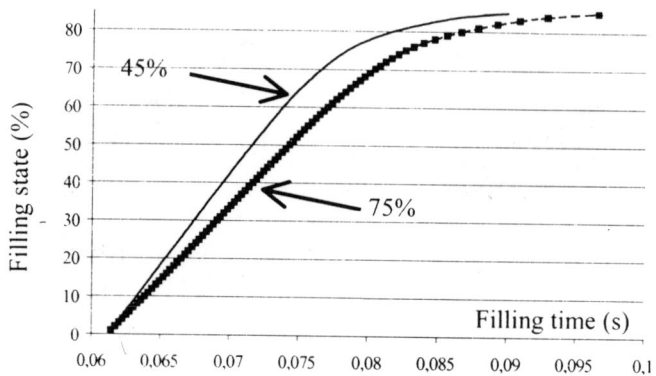

Figure 6. *Mold filled ratio vs. filling time for different initial powder volume fractions* $\phi_1 = 45\%$ *and* $\phi_2 = 75\%$

6. Conclusion

Different approaches have been proposed to solve the advection-diffusion equations associated with the segregation effects arising during powder-binder mixture flows in metal injection molding. Each proposed method contains its own specifications associated with different mathematical formulations and related computational costs. All the methods have led to feasible and accurate results. The simulations carried out for a mobile phone shell show that the segregation effects have been well predicted under different conditions. It is therefore concluded that the advection equations in bi-phasic injection simulation can be solved in a reliable and efficient manner and that realistic results can be obtained at reasonable cost.

Acknowledgments

This project has been supported by Part Time Invited Professor Program of the French Ministry of Education and the French-Chinese advanced research program.

References

[BAR 99] BARRIERE T., DUTILLY M., GELIN J.-C., "Analysis of metal injection molding as a near net shape forming process for microparts", *Proc. 6th Int. Conf. on Technology of Plasticity*, Nuremberg, Germany, pp. 19–23, 1999.

[BAR 00a] BARRIERE T., RENAULT D., GELIN J.-C., DUTILLY M., "Moulage par injection de poudres métalliques – Expérimentations, modélisation et simulation", *Mécanique et Industries*, 1, pp. 201–211, 2000.

[BAR 00b] BARRIERE T., "Expérimentations, modélisation et simulation numérique du moulage par injection de poudres métalliques", PhD Thesis, University of Franche-Comté, 2000.

[BAR 01] BARRIERE T., GELIN J.-C., LIU B., "Experimental and numerical investigations on the properties and quality of parts produced by MIM", *Powder Metallurgy*, 44, 3, pp. 228–234, 2001.

[BIL 00] BILOVOL V.V., KOWALSKI L. and DUSZCZYK J., "Application of fully 3-D simulation for studying of pressure development during powder injection molding process", *PIM 2000*, Ed. by German, R.M., Pennsylvania State University, USA, 2000.

[BOW 76] BOWEN R.M., *Theory of mixtures*, Ed. in Continuum Physics, Academic Press, New York, 3, 1976.

[DUT 98a] DUTILLY M, "Modélisation du moulage par injection de poudres métalliques", PhD Thesis, University of Franche-Comté, pp. 1–100, 1998.

[DUT 98b] DUTILLY M., GHOUATI O., GELIN J.-C., "Finite element analysis of the debinding and densification phenomena in the process of metal injection molding", *Journal of Material Processing Technology*, 83, pp. 170–175, 1998.

[DON 84] DONEA J., "A Taylor-Galerkin method for convective transport problems", *Inter. J. for Numer. Meth. in Engrg.*, 20, pp. 101–119, 1984.

[FOR 91] FORTIN M., BREZZI F., *Mixed and hybrid finite element method,* Springer Verlag, 1991.

[GAO 91] GAO D.M., "Modélisation numérique du remplissage des moules de fonderie par le méthode des éléments finis", PhD Thesis, Conpiegne University of Technology, 1991.

[GEL 00] GELIN J.-C., BARRIERE T., LIU B., "Modeling of the injection of loaded thermoplastic mixtures with application in metal injection molding", Met Soc, *Mathematical Modeling in Metals, Processing and Manufacturing*, Ed. by Martin P. *et al.*, Ottawa, Ontario, Canada, August 20–23 2000, CD-ROM, pp. 1–14, 2000.

[GEL 99] GELIN J.-C., BARRIERE T., DUTILLY T., "Experiments and computational modeling of metal injection molding for forming small parts", *Annals of the CIRP,* 48, 1, pp. 179–182, 1999.

[GER 97] GERMAN R.M. and BOSE A., *Injection molding of metals and ceramics*, MPI Press, Princeton, New Jersey, USA, 1997.

[ILI 00] ILINCA F., HETU J.F. HOLMES B., "Three dimensional modeling of metal powder injection molding", Mathematical Modeling in Metals, *Processing and Manufacturing*, Ed. by Martin P. *et al.*, Ottawa, Ontario, Canada, August 20–23 2000, CD-ROM, pp. 1–9, 2000.

[KIM 98] KIM K.T., JEON Y.C, "Densification behavior of 316L stainless steel powder under high temperature", *Materials Science and Engineering*, pp. 242–250, 1998.

[LAN 96] LANTERI B., BURLET H., POITOU A., CAMPION I., "Powder injection molding, an original simulation of paste flow", *European Journal of Mechanics A/ Solids*, 15, 1996.

[LEW 95] LEWIS R.W., USMANI A.S., CROSS T.J., "Efficient mould filling simulation in castings by an explicit finite element method", *International Journal for Numerical Methods in Fluids*, 20, pp. 493–506, 1995.

[LOH 84] LOHNER R., MORGAN K., ZIENKIEWIEZ O.C., "The solution of non-linear hyperbolic equation systems by the finite element method", *International Journal for Numerical Methods in Fluids*, 4, pp. 1043–1063, 1984.

[USM 92] USMANI A.S., CROSS J.T., LEWIS R.W., "A finite element model for simulations of mould filling in metal casting and the associated heat transfer", *International Journal for Numerical Methods in Engineering*, 35, pp. 787–806, 1992.

Chapter 2

A Meshless Simulation of Injection Processes Involving Short Fiber Molten Composites

Miguel A. Martínez, Elias Cueto and Manuel Doblaré
Mechanical Engineering Dept, University of Zaragoza, Spain

Francisco Chinesta
Conservatoire National des Arts et Métiers, Laboratoire des Matériaux Macromoléculaires, Paris, France

1. Introduction

There are two main kinds of composite materials, their choice depending on the field of application. The first concerns the composites widely used in many common industrial applications where no special mechanical properties are required and the main design criterion is low cost (car industry, packaging, home electrical appliances etc). Usually, their price does not exceed dollars per kg. The second group includes technical composites used for more specific applications (aerospace and naval industries, sport articles etc). These composites are mainly used because of their specific resistance in spite of their high price. In general, technical composites consist of continuous fibers in a thermo set matrix, whereas the others are usually composed of a thermoplastic matrix reinforced with short fibers. Although short fiber composites represent the essential part of the market, the number of studies concerning their mechanical properties or the industrial process control required to optimize these properties is very limited. The main aim of this work is to propose a new technique which allows accurate simulations of injection processes involving short fiber molten composites. Industrial process optimization needs accurate and fast simulation techniques, in order to define the mold geometry, the position of injection nozzles, the injection flow rate etc, producing suitabvle fiber orientation. Strength can be improved for example by orientating the fibers in the direction of stresses.

During the injection process the fluid flow induces large material deformation, which leads to composites exhibiting specific phenomena: the fibers affect the flow kinematics, and simultaneously the flow kinematics governs the fiber orientation. Thus, the injection process induces an anisotropy in the manufactured parts. This anisotropy is usually desired, but it needs to be controlled in order to optimize the mechanical properties of the final product (for example orientating the fibers in the principal stress directions). One possibility to control the induced anisotropy lies in the experimental determination of the fiber orientation in prototype pieces. However, this fully experimental trial and error procedure is known to be very expensive because it requires the design and manufacture of several molds. In this context the process simulation is very attractive for industry because it allows a significant reduction of the design cost.

The numerical modeling of short fiber molten composites forming processes requires firstly derivation of a constitutive equation for the suspension, from which the numerical simulation can be carried out by solving numerically the partial differential equations defining the mechanical model.

1.1. *Mechanical model*

Mechanical models for fiber suspensions involve averaging procedures because it is not possible to take into account the movement of each fiber in the suspension. These averaging procedures are of two kinds. The first one is a spatial averaging often referred to as homogenization. It allows one to derive macroscopic quantities as soon as the microscopic geometry is characterized. The second averaging procedure is statistical because the morphology can never be characterized exactly. It makes possible derivation of statistical averaged macroscopic quantities from a statistical description of the morphology [BAT 70], [HIN 75], [HIN 76], [MES 97].

This paper is focused on the description of a new numerical technique able to offer accurate solutions to complex flows simulations. We will consider a simple model for the description of the fiber suspension behaviour. However, this model retains all the mathematical particularities of more accurate mechanical models.

The mechanical model is defined by the momentum equations, which, neglecting mass and inertial effects are given by

$$Div\underline{\underline{\sigma}} = \underline{0} \tag{1}$$

where $\underline{\underline{\sigma}}$ is the stress tensor and Div the divergence operator. The suspension incompressibility is written in the usual form

$$Div\underline{v} = 0 \tag{2}$$

where $\underline{v}$ is the velocity field. We consider the following constitutive equation for the suspension

$$\underline{\underline{\sigma}} = -p\underline{\underline{I}} + 2\eta \left(\underline{\underline{D}} + N_p \; Tr\left(\underline{\underline{a}} \, \underline{\underline{D}}\right) \, \underline{\underline{a}} \right) \tag{3}$$

where $\underline{\underline{D}}$ is the strain rate tensor, defined as

$$\underline{\underline{D}} = \frac{1}{2}\left(Grad\underline{v} + \left(Grad\underline{v}\right)^T\right) \tag{4}$$

η the suspension viscosity, p the pressure, $\underline{\underline{I}}$ the unit tensor, N_p a scalar parameter depending on both the fiber concentration and the fibers aspect ratio, and $\underline{\underline{a}}$ is the second order orientation tensor whose eigenvalues represent the probability of finding the fibers in the direction of the associated eigenvectors. Finally, $Tr()$ denotes the trace operator.

Due to the introduction of the fiber orientation tensor, we also need the equation governing the evolution of the fiber orientation. One possible description is given by

$$\frac{d\underline{\underline{a}}}{dt} = Grad\underline{v} \; \underline{\underline{a}} + \underline{\underline{a}} \; \left(Grad\underline{v}\right)^T - 2Tr\left(\underline{\underline{a}} \, \underline{\underline{D}}\right) \, \underline{\underline{a}} \tag{5}$$

All these equations are defined in the fluid domain Ω_f which in injection simulations is changing with time, i.e. $\Omega_f(t)$. On the boundary of $\Omega_f(t)$, which is denoted $\Gamma_f(t)$, we consider 3 parts: (i) the inflow boundary (injection nozzle) Γ^- where a non-zero velocity and an initial fiber orientation are imposed; (ii) the contact surface between the fluid and the mold walls $\Gamma_0(t)$ where the non-slip condition, due to the viscous character of the flow, prescribes a null fluid velocity, and (iii) the moving boundary (fluid flow front) $\Gamma_{mb}(t)$ where a null traction is usually applied, i.e. $\underline{\underline{\sigma}}\underline{n}\big|_{\Gamma_{mb}} = \underline{0}$, where $\underline{n}$ is the outwards unit vector, normal to the moving boundary.

1.2. Numerical modeling

The numerical model involves the coupling of an elliptic problem (anisotropic extension of the Stokes's equations) defined by equations [1], [2] and [3] (the fiber orientation is assumed known), with a hyperbolic equation [5] governing the fibers orientation evolution inside the suspension. If we consider in the solution of equation [5] the velocity field as known, then equation [5] results in a non-linear advection equation whose characteristics coincide with the flow path lines.

In the previous discussion we have assumed that the solution of the flow kinematics is carried out from a known fiber orientation field, and that the actualization of the fiber orientation is computed from the flow kinematics just obtained. This is an appropriate algorithm if one uses an explicit strategy in the mold filling simulation. Thus, if we assume the fiber orientation known in the fluid domain at time $t, \Omega(t)$, we can compute the flow kinematics solving equations [1], [2] and [3] with the corresponding boundary conditions. From the just computed velocity field $\underline{v}(t)$, we can update the fluid domain at time $t + \Delta t$, $\Omega(t + \Delta t)$, as well as solve equation [5] in order to actualize the fiber orientation $\underline{\underline{a}}(t + \Delta t)$. Obviously, implicit formulations, solving the flow kinematics and the fiber orientation simultaneously in the fluid domain, are possible, however as the fluid domain is usually explicitly updated, the implicit formulation does not introduce significant advantages.

There are two kinds of computational methods in mold filling simulations. The first one concerns moving mesh techniques which require a mesh updating at each position of the flow front, i.e. for each fluid domain actualization. The main difficulty of these techniques is related to the necessity of remeshing, which increases significantly the CPU time. The second family of methods concerns the fixed mesh techniques. In this case the mesh associated with the whole domain is fixed, and the part of the domain occupied by the fluid is identified with particular strategies: donor-acceptor, control volume, the volume of fluid technique combined with the introduction of a pseudo-behaviour in the empty region of the mold domain etc [CHI 99a].

If one uses a moving mesh, where the mesh is convected with the velocity field, the orientation equation may be accurately integrated along the nodal flow path lines by using the method of characteristics. However, it is well known that due to the high material deformations the convected mesh becomes highly distorted losing the interpolation accuracy required for the integration of the momentum equations. So other strategies have been introduced, as for example the arbitrary Lagrangian Eulerian (ALE) techniques. In this case, to avoid high mesh distortions, a new mesh is created after some time steps, and a projection of the orientation solution on the new mesh is performed. However, these projections introduce high numerical diffusion as proved in [CHI 99b], [POI 00].

On the other hand, fixed mesh strategies allow us to integrate the orientation equation by means of either fully Eulerian strategies (SUPG, SU, discontinuous finite elements, etc.) [CHI 99b] or the method of characteristics. Eulerian techniques require assumption of an initial orientation state in the elements when they start the filling process. This assumption is arbitrary and remains a delicate matter. The use of the method of characteristics implies the backward path line tracking, for each point $\underline{x}$ where we want to obtain the orientation at time t, until reaching, at the time $t - \Delta t$, the position $\underline{x}$. As the orientation solution at time $t - \Delta t$ is known, the solution at point $\underline{x}$ can be interpolated $\underline{\underline{a}}(\underline{x}$). Now, from this initial value $\underline{\underline{a}}(\underline{x}$) we can compute $\underline{\underline{a}}(\underline{x})$ at time t using the method of characteristics. The main disadvantage of this strategy is the high number of interpolations required during the filling process, which introduces a high numerical diffusion in the solution.

A good choice which keeps the advantages of moving meshes for accurate integration of the advection equation along the nodal flow path lines, avoiding the necessity of mesh updating, is the use of a meshless technique. However, standard meshless techniques (eg moving least squares) have the disadvantage of defining non nodal interpolations, which make the imposition of essential boundary conditions difficult. One possibility to avoid this situation is to consider the natural element method (NEM) to solve the equations of motion. This technique defines a nodal interpolation (that allows us to impose essential boundary conditions) without the necessity of a mesh support.

1.3. *The natural element method in the context of meshless techniques*

In the last decade considerable research efforts have been devoted to the development of a series of novel numerical tools that have been referred to as meshless or meshfree methods. These methods do not need an explicit connectivity information, as required in standard FEM. The geometrical information is generated in a process transparent to the user, thus alleviating the preprocessing stage of the method. They also present outstanding advantages when modeling complex phenomena, such as large deformation problems, forming processes, fluid flow, etc, where traditional and more experienced techniques like the FEM fail due to excessive need of remeshing.

The Natural Element Method (NEM) is one of the latest meshless methods applied in the field of linear elastostatics [BRA 95], [SUK 98], [CUE 00]. It has unique features among meshless Galerkin methods, such as interpolant character of shape functions and exact imposition of essential boundary conditions. These, and its inherent meshless structure make the NEM an appealing choice also for application in the simulation of fluid flows. Here, we apply this meshless technique in short fiber suspension flow simulations. It is shown that the NEM has an appealing potential in application to the simulation of complex industrial flows.

2. The natural element method

2.1. *Standard formulation*

The NEM is based on the natural neighbour interpolation scheme [SIB 81], [WAT 81], which in turn relies on the concepts of Voronoi diagrams [VOR 08] and Delaunay triangulations [DEL 34] (see Figure 1), to build Galerkin trial and test functions. These are defined as the Natural Neighbor coordinates (also known as Sibson's coordinates) of the point under consideration, that is, with respect to Figure 2, the value, at point x of the shape function associated with the node 1 is defined as:

$$\phi_1(x) = \frac{A_{abfe}}{A_{abcd}}$$ [6]

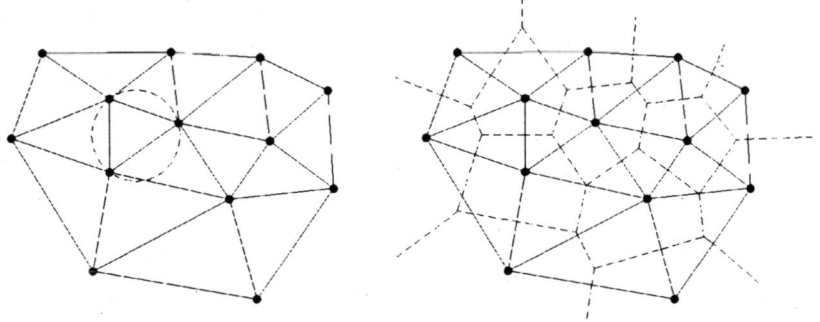

Figure 1. *Delaunay triangulation and Voronoi diagram of a set of points*

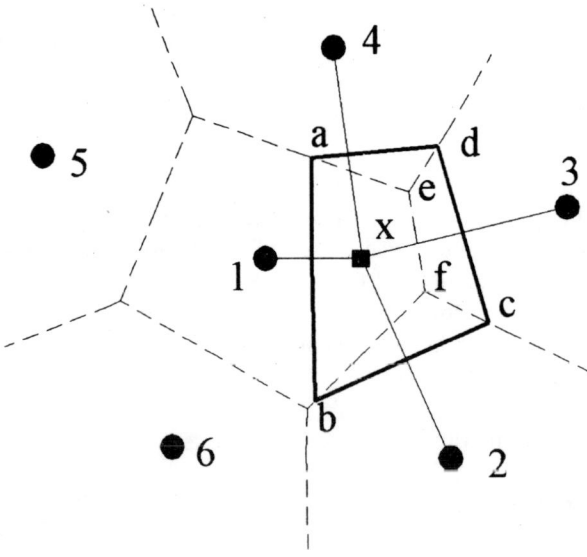

Figure 2. *Definition of natural neighbour coordinates*

These functions are used to build the discrete system of equations arising from the application of the Galerkin method in the usual way. It has been proved (see [SUK 98]) that angles of the Delaunay triangulation are not influencing the quality of the results, in opposition to the FEM.

In addition, the NEM has interesting properties such as linear consistency and smoothness of the shape functions (natural neighbour coordinates are C^∞ everywhere except of the nodes, where they are C^0). These functions are dependent on the position and density of nodes, leading to standard FE constant strain triangle shape functions, bilinear shape functions or rational quartic functions in different situations.

But perhaps the most interesting property of the Natural Element Method is the Kronecker delta property, i.e. $\phi_i(x_j) = \delta_{ij}$. In opposition to the vast majority of meshless methods, the NEM shape functions are strictly interpolant. This property permits an exact reproduction of linear (even bilinear in some 3D cases) displacement fields on the boundary of convex domains, since the influence of interior points vanishes along convex boundaries (this is not true in non-convex ones). Proof of this behaviour can be found in [SUK 98] and references therein.

2.2. The α-shape based natural element method

Recently, a modification of the way in which the Natural Neighbour interpolant is built has been proposed in order to achieve linear interpolation also over non-convex boundaries [CUE 00]. This modification has been based on the concept of α-shapes. These are a generalization of the concept of the convex hull of a cloud of points and are widely used in the field of scientific visualization and computational geometry to give a shape to a set of points. The concept was first developed by Edelsbrunner [EDE 94] and can be summarized as follows.

In essence, an α-shape is a polytope that is not necessarily convex nor connected, being triangulated by a subset of the Delaunay triangulation of the points. Thus, the empty circumcircle criterion holds. Let N be a finite set of points in $\Re^3$ and α a real number, with $0 \leq \alpha \leq \infty$. A k-simplex σ_T with $0 \leq k \leq 3$ is defined as the convex hull of a subset $T \subseteq N$ of size $|T| = k+1$. Let b be an α-ball, that is, an open ball of radius α. A k-simplex σ_T is said to be α-exposed if there exist an empty α-ball b with $T = \partial b \cap N$ where ∂ denotes the boundary of the ball. In other words, a k-simplex is said to be α-exposed if an α-ball that passes through its defining points contains no other point of the set N.

Following this, we can define the family of sets $F_{k,\alpha}$ as the sets of α-exposed k-simplices for the given set N. This allows us to define an α-shape of the set N as the politope whose boundary consists on the triangles in $F_{2,\alpha}$, the edges in $F_{1,\alpha}$ and the vertices or nodes in $F_{0,\alpha}$. As remarked before, it can be easily seen that an α-shape is a polytope that can be triangulated by a subset of the Delaunay triangulation or tetrahedrization, that is, by an α-complex. For different values of the parameter α

different shapes of the cloud of points are obtained, ranging from the convex hull of the set of points if $\alpha = \infty$ to the cloud of points itself if $\alpha = 0$. In [CUE 00] it is demonstrated that this modification makes the NEM able to exactly interpolate linear fields along convex as well as non-convex boundaries.

Consider a non-convex boundary (see Figure 3) defined by nodes A, B and C. For the node A and point x to be neighbours, thus breaking the desired linearity, a triangle of radius α' must connect nodes A and C. Since the neighbourhood is determined over an α-shape and not over the convex hull of the points – as is done inherently in the standard NEM –, we can restrict this neighbourhood to lower values (in particular, α), appropriate to reproduce the local radius of curvature in the considered portion of the boundary. In this way both the geometry of the domain and the approximation conditions are properly obtained. This leads to the shape function depicted in Figure 4. Note also that the inclusion of the α-shape concept in the NEM also allows constructing models without an explicit (CAD) definition of the boundary. The geometry of the domain is also extracted by using the α-shape concept.

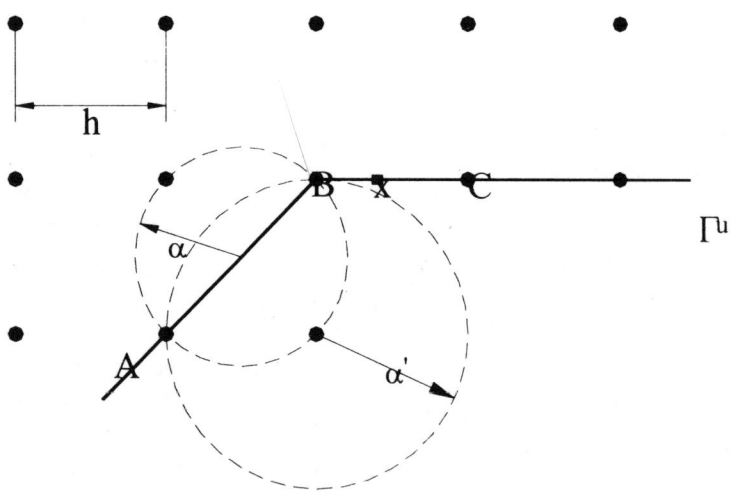

Figure 3. *Neighbourhood in a non-convex boundary*

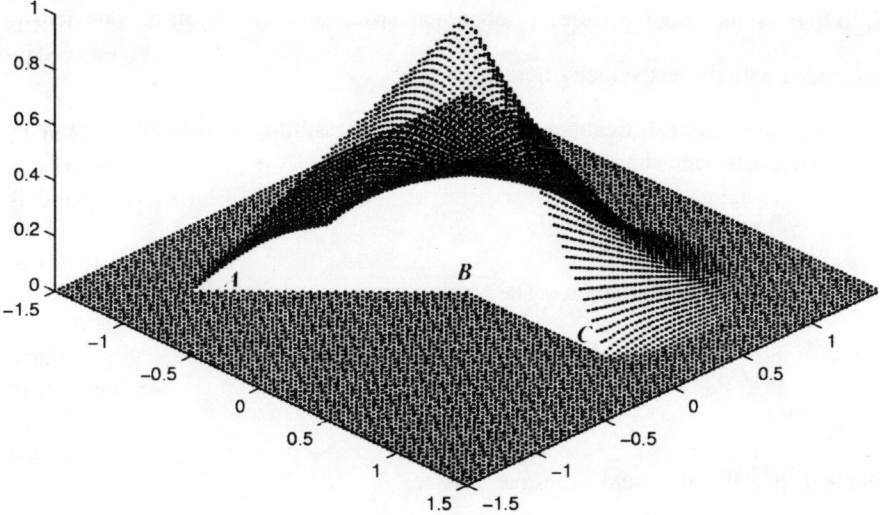

Figure 4. *Shape function associated with node B*

3. Fluid flow simulation

3.1. *Variational formulation*

The variational formulation of the equations of motion [1] is the following:

Find the kinematically admissible velocity field $\underline{v} \in U$, where U denotes the trial functional space

$$U = \left\{ \underline{v} / \underline{v} \in \left(H^1 (\Omega(t)) \right)^3, \underline{v} \big|_{\Gamma - \cup \Gamma_0(t)} = \underline{v}_g \right\} \qquad [7]$$

such that

$$\int_{\Omega(t)} \underline{\underline{\sigma}} : \underline{\underline{D}}^* \, d\Omega = \int_{\Gamma_{mh}(t)} \underline{\underline{\sigma}} n \cdot \underline{v}^* \, dS \qquad \forall v^* \in U^* \qquad [8]$$

where the stress tensor is given by equation [3] and U^* is the functional space of admissible test velocities:

$$U^* = \left\{ \underline{v} / \underline{v} \in \left(H^1 (\Omega(t)) \right)^3, \underline{v} \big|_{\Gamma - \cup \Gamma_0(t)} = \underline{0} \right\} \qquad [9]$$

$H^1(\Omega(t))$ is the usual Sobolev's functional space and $\underline{\underline{D}}^*$ the strain rate tensor associated with the test velocity field $\underline{v}^*$.

For the numerical treatment of the incompressibility constraint, a penalty formulation has been chosen:

$$-Div\underline{v} + \varepsilon\,p = 0 \qquad\qquad [10]$$

with ε a small enough parameter. The resulting variational formulation is

$$\int_{\Omega(t)} (-Div\underline{v} + \varepsilon p)p^*\,d\Omega = 0 \qquad \forall p^* \in L^2(\Omega(t)) \qquad [11]$$

where $L^2(\Omega(t))$ is the usual Lebesgue's space.

3.2. Mixed natural element formulation

A Galerkin technique is considered for the discretization of the previous mixed variational formulation (equations [8] and [11]). In this work we have chosen a C^0-C^1 interpolation for velocity and pressure fields, respectively. Natural neighbour shape functions are smooth everywhere except at the nodes, where they are C^0. A discontinuous field has been considered for the pressure approximation, given by

$$p^h(\underline{x}) = \sum_{l=1}^{n} \frac{1}{n} p_l \qquad [12]$$

where n is the number of natural neighbours of point $\underline{x}$ and p_l the nodal pressures.

No spurious pressure modes and no locking have been detected in the implementation of the method. Although it is known that this approximation does not verify the LBB condition, its behaviour has been proved to be similar to the bilinear quadrilateral (BLQ) finite element with discontinuous constant pressure interpolation for incompressible elasticity for regular nodal distributions.

3.3. Recovery of nodal variables

Recovery of the nodal stresses is made by means of a least-squares projection from the values at the integration points. If $\underline{\sigma}_{ij}^*$ is the vector of the nodal values of the $ij-$ stress component, this stress component at any point of the domain may be written in the form:

$$\sigma_{ij}^* = \underline{\Phi}^T \underline{\sigma}_{ij}^*$$ [13]

where $\underline{\Phi}$ represents the shape functions vector. Thus, the recovery problem can be formulated as a functional minimization

$$\min_{\Omega(t)} \int \left(\sigma_{ij}^* - \sigma_{ij}^h \right)^2 d\Omega$$ [14]

where σ_{ij}^h is the ij-stress component obtained by direct derivation from the velocity field $\underline{v}$, which can be computed without particular difficulties everywhere except at the mesh nodes.

Although the minimization process results in a poorly conditioned system of equations due to the usually high number of nodes of a natural element's connectivity, no numerical difficulty has been encountered using a conjugate gradient solver.

4. Numerical examples

4.1. *Newtonian fluid flow simulation*

Firstly, we present an example of a purely Newtonian fluid flow. In the following examples fibers will be introduced in the flow. The constitutive equation results from taking $N_p = 0$ in equation [3].

In the present case we refer to the filling simulation of a rectangular mold. The geometry of the domain is depicted in Figure 5, where the initial portion of the domain occupied by the fluid is also shown. The fluid is injected through the inflow boundary $x=0$, with a prescribed parabolic velocity given by:

$$v_x(x = 0, y) = \frac{4V}{h^2}(h - y)y$$ [15]

with $V = 0.1$. The initial position of the flow front is given by $b=0.01$. Dimensions of the mold are $h = 8 \cdot 10^{-3}$ and $L = 2.5 \cdot 10^{-2}$. Constant time increments of 5.10^{-3} have been chosen in an explicit time integration algorithm. The penalty parameter ε has been set to 10^{-12}. The cloud of points has initially 99 particles, shown in Figure 6, and a constant inlet of points takes place at each time increment. No-slip boundary conditions have been imposed at the mold walls.

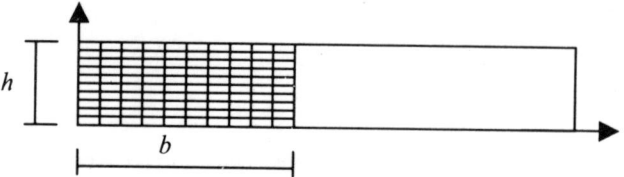

Figure 5. *Geometry of the mold in the example in Section 4.1.*

Plots of the configuration of the cloud of points at different time steps are shown in Figure 6 (a) to (d). Little arrows placed at each node represent the material velocity.

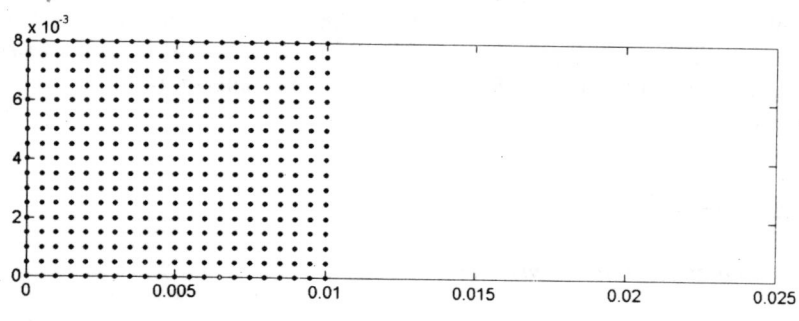

(a) Initial configuration

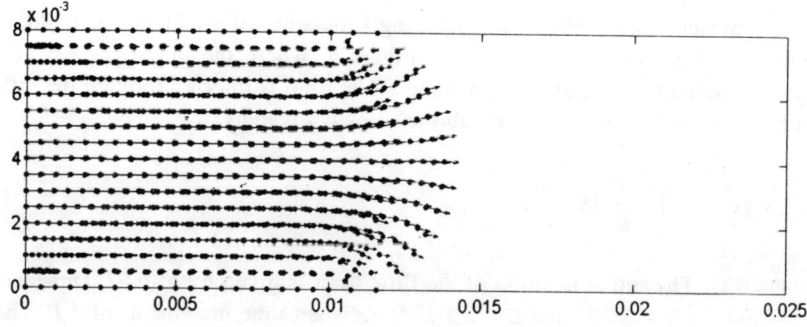

(b) Configuration at the 10th time increment

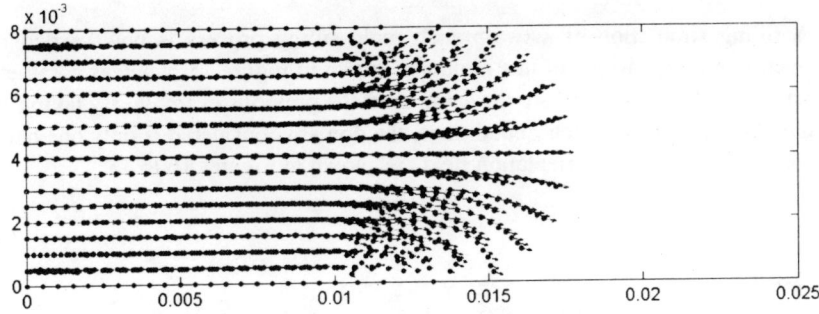

(c) Configuration at the 20th time increment

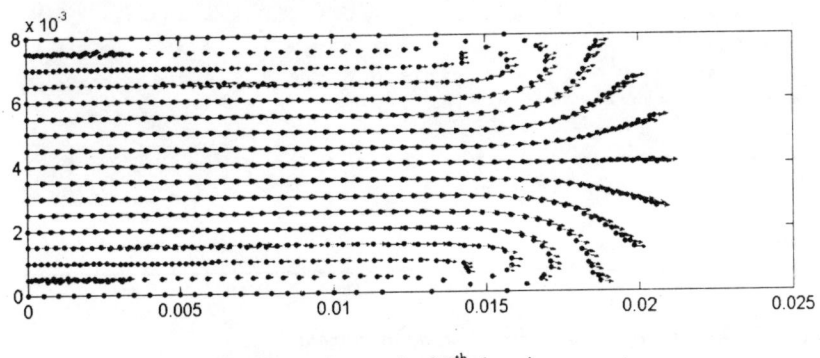

(d) Configuration at the 30th time increment

Figure 6. *Position of the nodes for several time stations*

4.2. Short fiber molten composite flow simulation

The geometry considered in the preceding example is now used to simulate a short fiber suspension flow. Nodes on the inflow boundary are assumed to have an isotropic fiber orientation and a velocity given by equation [15]. Boundary conditions and the initial cloud of points are the same as in previous example. The particle number N_p in equation [3] has been taken equal to 10.

Results of the simulation at several time stations are shown in Figures 7 to 14. The graphical representation of the fibers orientation is made from an ellipse whose axes lengths are proportional to the orientation tensor eigenvalues and their directions correspond with the associated eigenvectors. It can be noticed, as expected, that shear flow tends to align the fibers in the flow direction whereas in the central zone the orientation remains quite isotropic.

4.3. *Two-branch mold filling process*

A filling simulation of a two-branch mold filling process is now carried out. Geometry and dimensions of this mold are shown in Figure 15, whereas parameters associated to the simulation are the same as in the previous example. By taking into account appropriate symmetry conditions, the domain considered is only one half of the mold. Results for the orientation field are shown in Figures 16 to 18.

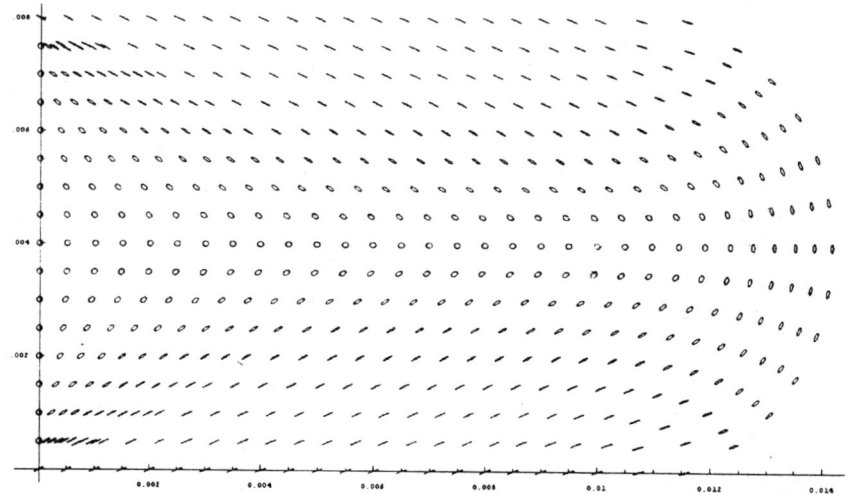

Figure 7. *Orientation field at the 10th time increment*

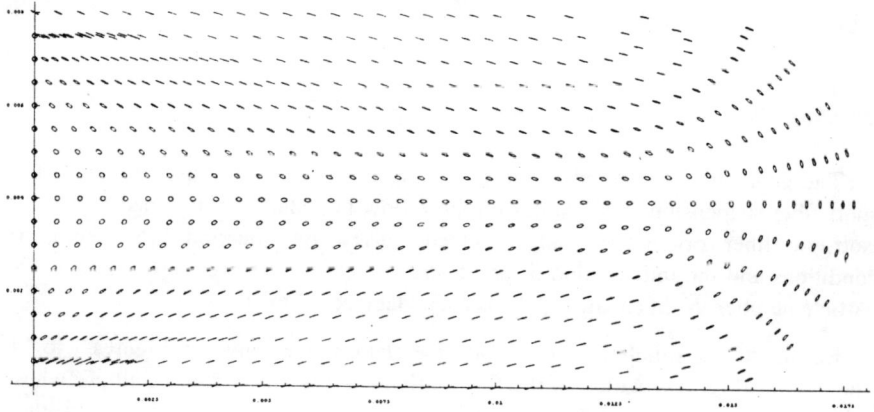

Figure 8. *Orientation field at the 20th time increment*

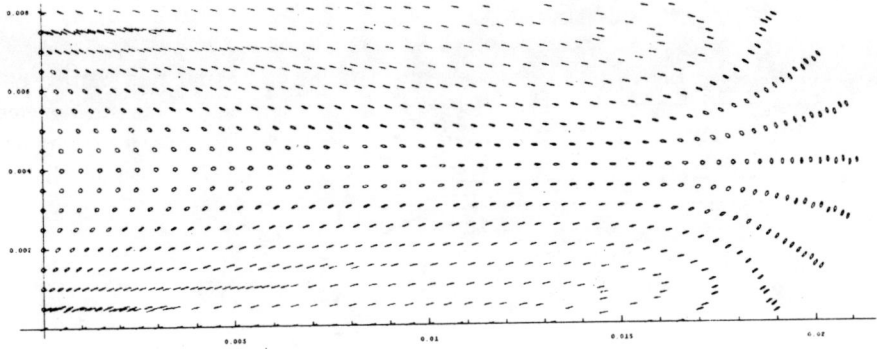

Figure 9. *Orientation field at the 30th time increment*

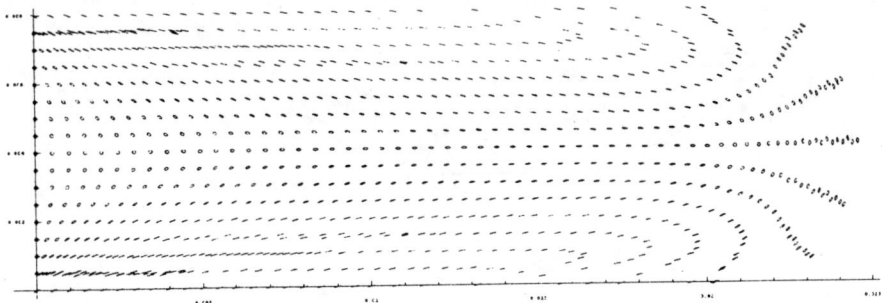

Figure 10. *Orientation field at the 40th time increment*

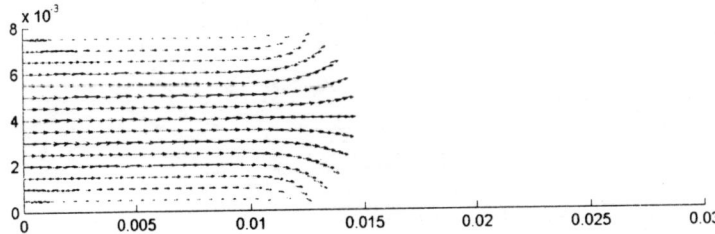

Figure 11. *Velocity field at the 10th time increment*

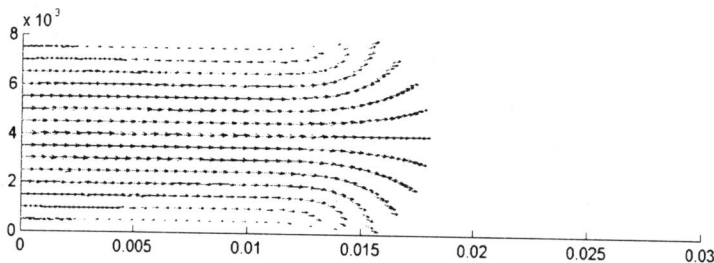

Figure 12. *Velocity field at the 20ᵗʰ time increment*

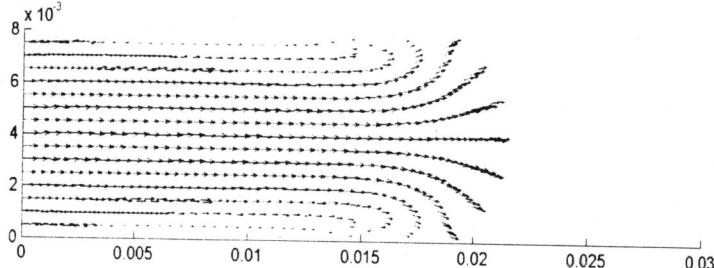

Figure 13. *Velocity field at the 30ᵗʰ time increment*

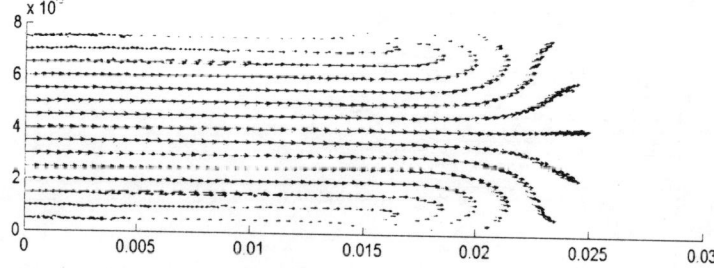

Figure 14. *Velocity field at the 40ᵗʰ time increment*

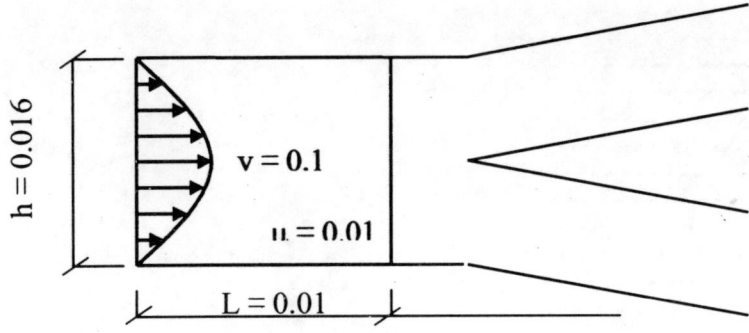

Figure 15. *Geometry of the two-branch mold considered in Section 4.3.*

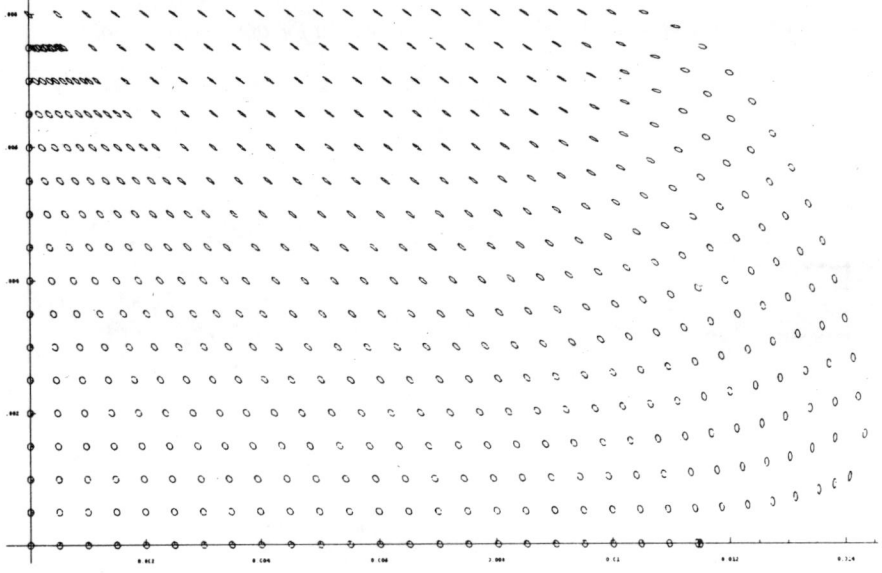

Figure 16. *Orientation field at the 10th time increment for the Y-shaped mold*

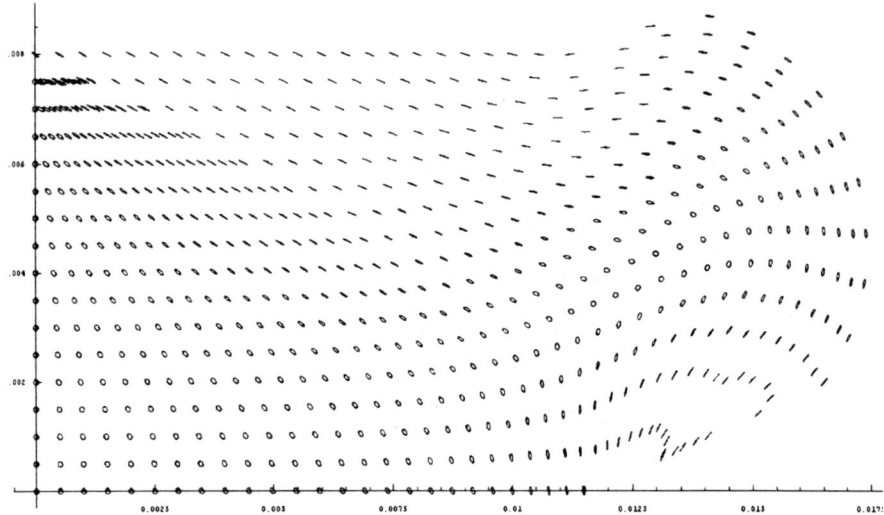

Figure 17. *Orientation field at the 20th time increment for the Y-shaped mold*

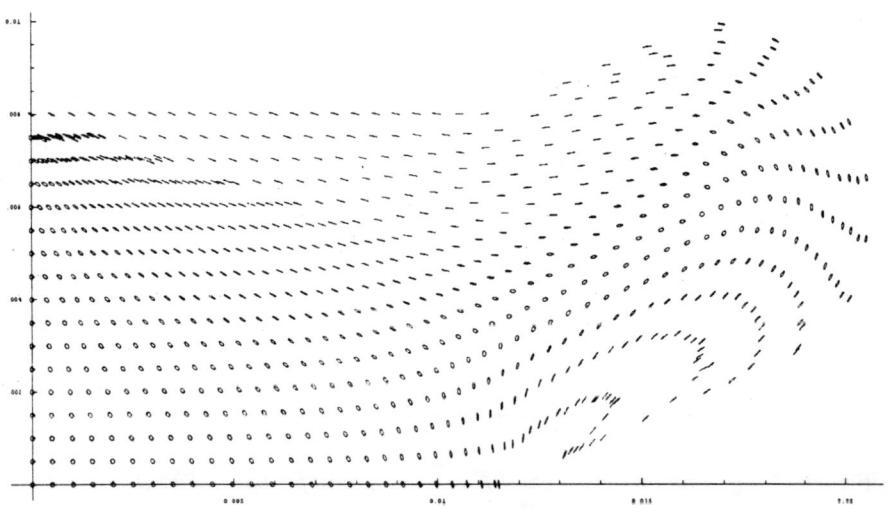

Figure 18. *Orientation field at the 30th time increment for the Y-shaped mold*

5. Conclusion

The Natural Element Method, although being a very novel method, has proven successful in a number of problems in the field of solid mechanics. Here, we presented a first result concerning mold filling process simulation with a non-

Newtonian fluid. It has been shown that the NEM seems to be equally well suited for application to fluid mechanics problems, due to its simplicity and the absence of mesh updating. The fact that the α-NEM automatically "extracts" the shape of the nodal cloud at each time step could also be used eventually to introduce surface tension in the simulation. Moreover, the fact that particles (nodes) move with the flow avoids the necessity to treat convective terms, which is a difficult matter in standard Eulerian discretizations. Thus, we can carry out accurate integrations of the constitutive equation along the nodal flow path lines. This type of meshless method seems very well adapted to treat general viscoelastic flows.

Acknowledgements

Support to Dr. Martinez from the *Programa Europa* by CAI is gratefully acknowledged. The authors also acknowledge the help of Dr. B. Calvo in programming the NEM.

References

[BAT 70] BATCHELOR, G.K., "Slender-body theory for particules of arbitrary cross-section in Stokes flow", *J. Fluid Mech.*, 44, 419–440, 1970.

[BRA 95] BRAUN J., SAMBRIDGE M.A., "Numerical method for solving partial differential equations on highly irregular evolving grids", *Nature*, 376, 655–660, 1995.

[CHI 99a] CHINESTA, F., "Modélisation numérique en mise en forme des polymères et céramiques: différents problèmes de transport", *HDR*, Université Pierre et Marie Curie, 1999.

[CHI 99b] CHINESTA F., POITOU A., TORRES R., "Numerical simulation of the steady state reinforced thermopastic flows", *Revue Européenne des Eléments Finis*, 8, 4, 355–374, 1999.

[CUE 00] E. CUETO, DOBLARE M. and GRACIA L., "Imposing essential boundary conditions in the NEM by means of density scaled alpha-shapes", *International Journal for Numerical Methods in Engineering*, 49, 4, 519–546, 2000.

[DEL 34] DELAUNAY, B., *Sur la Sphère Vide*, A la mémoire de Georges Voronoi. Izvestia Akademii Nauk SSSR, Otdelenie Matematicheskii i Estestvennyka Nauk, 7, 793–800, 1934.

[EDE 94] EDELSBRUNNER H., MÜCKE E., "Three dimensional alpha shapes", *ACM Transactions on Graphics*, 13, 43–72, 1994.

[HIN 75] HINCH E.J., LEAL L.G., "Constitutive equations in suspension mechanics. Part I", *J. Fluid. Mech.*, 71, 481–495, 1975.

[HIN 76] HINCH E.J., LEAL L.G., "Constitutive equations in suspension mechanics. Part II", *J. Fluid. Mech.*, 76, 187–208, 1976.

[MES 97] MESLIN F., "Propriétés rhéologiques des composites fibres courtes à l'état fondu", PhD Thesis, Ecole Normale Supérieure de Cachan, 1997.

[MON 00] MONTÓN I., TORRES R., CHINESTA F., POITOU A., "A 3D mold filling simulation for short fibers reinforced thermoplastics", *European Congress on Computational Methods in Applied Sciences and Engineering (ECCOMAS)*, Barcelona, 2000.

[POI 00] POITOU A., CHINESTA F., TORRES R., "Numerical simulation of the steady recirculating flows of fibers suspensions", *J. Non-Newtonian Fluid Mech.*, 90, 65–80, 2000.

[SIB 81] SIBSON R.A., "Brief description of natural neighbour interpolation. Interpreting multivariate data", V. Barnett (Ed.), 21–36, John Wiley, Chichester, 1981.

[SUK 98] SUKUMAR N., MORAN B., BELYTSCHKO T., "The natural element method in solid mechanics", *International Journal for Numerical Methods in Engineering*, 43, 5, 839–887, 1998.

[TEI 98] TEICHMAN M., CAPPS M., "Surface reconstruction with anisotropic density-scaled alpha shapes", *1998 IEEE Visualization Conference*, 1998.

[VOR 08] VORONOI G. M., "Nouvelles applications des paramètres continus à la théorie des formes quadratiques. deuxième mémoire: recherches sur les parallélloèdres primitifs", *J. Reine Angew. Math.*, 134, 198–287, 1908.

[WAT 81] WATSON D., "Computing the n-dimensional Delaunay tessellation with application to Voronoi polytopes", *The Computer Journal*, 24, 2, 162–172, 1981.

Chapter 3

Semi-solid Processing of Engineering Alloys by a Twin-screw Rheomolding Process

Z. Fan, S. Ji and M. J. Bevis
Wolfson Centre for Materials Processing, Brunel University, Middlesex, UK

1. Introduction

Semi-solid metal (SSM) processing is a new technology for near net-shape production of engineering components, in which metal alloys are processed at a temperature in the interval between solidus and liquidus (Flemming, 1991; Kirkwood, 1994). The critical characteristics of semi-solid alloys are their globular or non-dendritic grain structures formed during solidification under forced convection (Flemming, 1972). SSM slurries exhibit distinctive deformation characteristics, namely, thixotropy and pseudoplasticity (Joly, 1976). The advantages of SSM processing compared with conventional die-casting processes are lower operating temperatures, laminar flow during mold filling and reduced solidification shrinkage. Consequently, one may expect longer die life, shorter cycle time, improved mechanical properties, reduced gas entrapment and porosity and lower tendency to hot tearing (Flemming, 1991; Kirkwood, 1994).

Since the early 1970s, many alternatives to the original MIT rheocasting process (Flemming, 1976) have been developed, such as thixoforming (Kirkwood, 1994) and thixomolding (Pasternak, 1992). However, the high cost of pre-processed raw materials and the difficulties in process control are the main obstacles to their acceptance by industry. Based on the extensive experience in processing of polymeric materials offered by the polymer processing community (Rauwendaal, 1985), a twin-screw rheomolding technology for SSM processing has been developed recently (Fan,1999). In this process, the liquid alloy is converted into semi-solid slurry under high shear rate and high intensity of turbulence provided by a twin-screw extruder, and the semi-solid slurry is subsequently injected into a mold cavity to fabricate components with high integrity. In this paper we report the twin-screw rheomolding process and microstructural development during the rheomolding process.

2. Twin-screw rheomolding process

The twin-screw rheomolding process developed is schematically illustrated in Figure 1. It consists of a liquid alloy feeding device, a twin-screw extruder, an injection unit and a central control system. The liquid alloy feeding device is designed to supply liquid alloy with a predetermined temperature and a right dose for a specified component. The twin-screw extruder, the core of the rheomolding system, consists of a barrel and a pair of closely intermeshing, self-wiping and co-rotating screws. The screws have a specially designed profile to achieve high shear rate and high intensity of turbulence and to enhance the positive displacement pumping action. Once in the passageway of the twin-screw extruder, the liquid alloy is quickly cooled to a predetermined SSM processing temperature, while being mechanically sheared by the screws, to achieve the desired volume fraction of the solid phase. The semi-solid slurry is then transferred into a shot sleeve and subsequently injected at a controlled velocity into a mold cavity. The fully solidified component is finally released from the mold. All these procedures can be performed in a continuous cycle and controlled by a central control system. The typical cycle time is 20–40s, depending on the size of the components. The temperature control is achieved by balancing the heating and cooling power input by a central control unit.

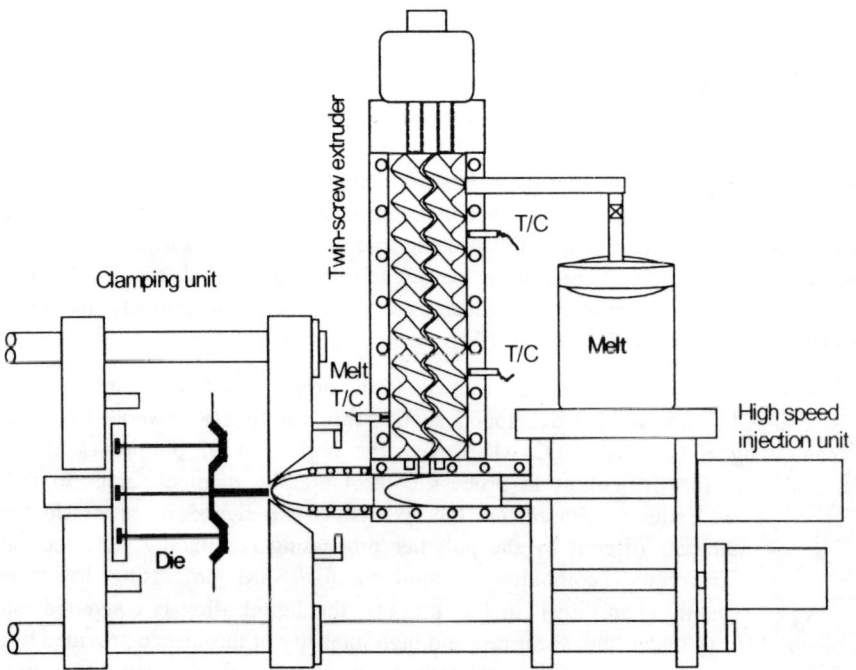

Figure 1. *The schematic illustration of the twin-screw rheomolding system*

The fluid flow characteristics in the closely intermeshing, self-wiping and co-rotating twin-screw extruder are unique. Unlike the viscous drag-induced type of materials transport in a single screw extruder, the transport behaviour in a closely intermeshing twin-screw extruder is to a large extent a positive displacement type of transport, being more or less independent of the viscosity of the molten materials. Meanwhile, the velocity profiles in the twin-screw extruder are quite complex and more difficult to describe. There are four scaling groups; each consisting of different forces acting on the materials in a twin-screw extruder. The first group relates to the scales of inertia forces and centrifugal forces; the second group concerns the scale of gravity force; the third comprises the scale of internal friction and the forth group refers to the scales of elastic and plastic deformation behaviour of the materials being processed. The principal forces acting on the liquid or semi-solid alloys during the rheomolding process between two screws and between screw and barrel are compression, rupture, shear and elasticity. Extensive study by the polymer processing community has confirmed that the fluid moves in figure "8" motions around the periphery of the screws (Erdmenger, 1964), and the figure "8" moves from one pitch to the next one, forming a figure "8" shaped helix and pushing the fluid along the axial direction of the screws, as illustrated schematically in Figure 2a. This is usually referred to as the positive displacement pumping action. In this continuous flow field, the fluid undergoes cyclic stretching, folding and reorienting processes (Andersen, 1994). This is shown schematically in Figure 2b with respect to the streamlines during the take-over of the materials from one screw to the other one. Another feature of fluid flow in the closely intermeshing twin-screw extruder is shown in Figure 2c by the circular flow pattern on the axial section. In consideration of the relatively lower viscosity of liquid or semi-solid alloys, compared with that of the polymer melt, the intensity of turbulence inside the barrel should be very high. In addition, Figure 2 indicates that the fluid is subjected to a cyclic variation of shear rate due to the continuous change of the gap between screw and the barrel. The lowest shear rate is found at the gap between the root of the screw flight and the inner surface of the barrel; the highest shear rate is offered by the intermeshing regions between two screws if there is a leakage path, although the exact shear rate can not be calculated because of the complexity of the screw geometry. However, all the materials in the twin-screw extruder undergo a shear deformation with cyclic variation of shear rate. For the viscous flow, the shear rate is proportional to the rotation speed and the outer diameter of the screw and inversely proportional to the gap between screw flight and the barrel surface. Therefore, the fluid flow in a closely intermeshing, self-wiping and co-rotating twin-screw extruder is characterized by high shear rate, high intensity of turbulence and cyclic variation of shear rate. Such fluid flow characteristics are responsible for the unique microstructural evolution during the twin-screw rheomolding process (Ji et al. , 2000).

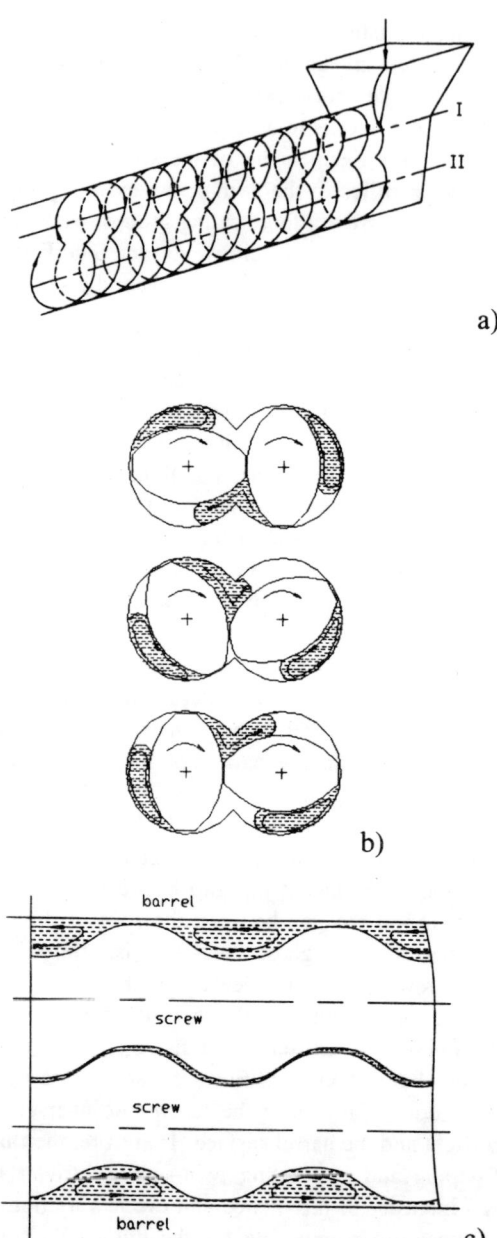

Figure 2. *Schematic diagrams showing the fluid flow characteristics in a closely intermeshing and co-rotating twin-screw extruder. (a) figure "8" flow pattern along the screw channels; (b) the flow pattern on the cross-section of the extruder; (c) the flow pattern on the axial section*

3. Microstructural characteristics of the rheomolded alloys

Aluminium alloys, magnesium alloys, zinc alloys and other low melting point alloys can be processed by the twin-screw rheomolding technology. For magnesium alloys, an Mg-30wt.%Zn alloy was tested in the system developed.

The alloy was melted in an iron crucible at a predetermined temperature with 50°C overheat under SF_6 gas protection. The liquid alloy was processed for 30s after reaching the predetermined SSM temperature. The microstructure of the rheomolded Mg-30wt.%Zn alloys obtained under different processing conditions is shown in Figure 3. The volume fractions of the primary particles in the rheomolded Mg-30wt.%Zn alloys are 0.28 and 0.5, as shown in Figures 3a and 3b, respectively. The primary particles have a mean intercept length less than 20μm. The primary particles are uniformly distributed in the Mg-MgZn eutectic matrix. The measured distribution of the intercept length of the primary particles and the corresponding equivalent particle diameter on the polished surface are presented in Figure 4. The distribution curves for the intercept length and the equivalent diameters are very close to each other, indicating again that the particle morphology is very close to spherical.

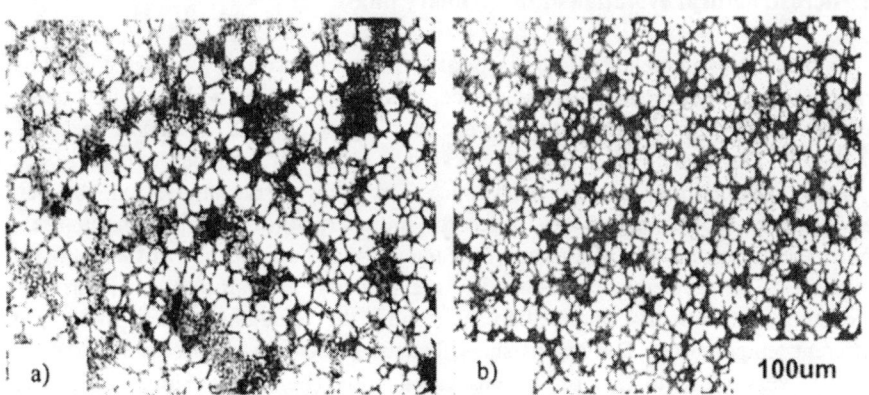

Figure 3. *Optical micrographs showing the microstructures of rheomolded Mg-30wt.%Zn alloys sheared for 30s with a shear rate of $3625s^{-1}$. (a) $f_S = 0.32$; (b) $f_S = 0.50$*

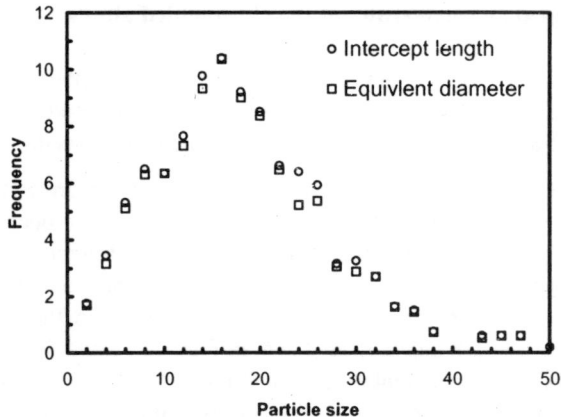

Figure 4. *The distribution of the intercept length and equivalent diameters of the primary particles of Mg-30wt.%Zn alloy sheared at 380°C for 30s and 3625s⁻¹*

4. Microstructural evolution of the primary phase

In order to examine the microstructural evolution of primary particles under high shear rate and high intensity of turbulence, liquid Sn-15wt.%Pb alloy at 220°C was fed into the twin-screw extruder with a pre-set temperature of 215°C and a shear rate of 3625s⁻¹. The melt was isothermally sheared at 215°C for 5min before cooling with a controlled cooling rate of 0.5°C/min. The alloy was sampled at different temperatures by quenching into water for microstructural examination. Figure 5 shows the microstructural evolution of the intensively sheared Sn-15wt.%Pb alloys at different SSM temperatures. Generally, the primary tin particles are distributed uniformly in the matrix obtained by quenching the remaining liquid phase from different temperatures. The results show that the nucleation temperature of sheared alloy is very close to that of the unsheared alloy (Figure 5a). Meanwhile, it is obvious that the primary phase at the nucleation temperature has a spherical morphology at high shear rate and high intensity of turbulence, which indicates that at the early stages of solidification high shear rate and high intensity of turbulence promote formation of spheroids. There are no dendrites or rosettes observed at the early stages of solidification under intensive shearing. With the decrease of the shearing temperature, as shown in Figures 5b to 5d, the volume fraction increases significantly. The measured volume fractions of the primary phase as a function of shearing temperature are very close to those predicted by the lever rule. The measured average intercept length of the primary particles is plotted in Figure 6 as a function of shearing temperature. The primary particle size is about 45 μm and does not vary with decreasing shearing temperature. The measured shape factor of the primary particles, plotted in Figure 7 as a function of the shearing temperature, shows that the shape factor has no apparent variation for different shearing temperatures and is very close to that for perfectly spherical particles.

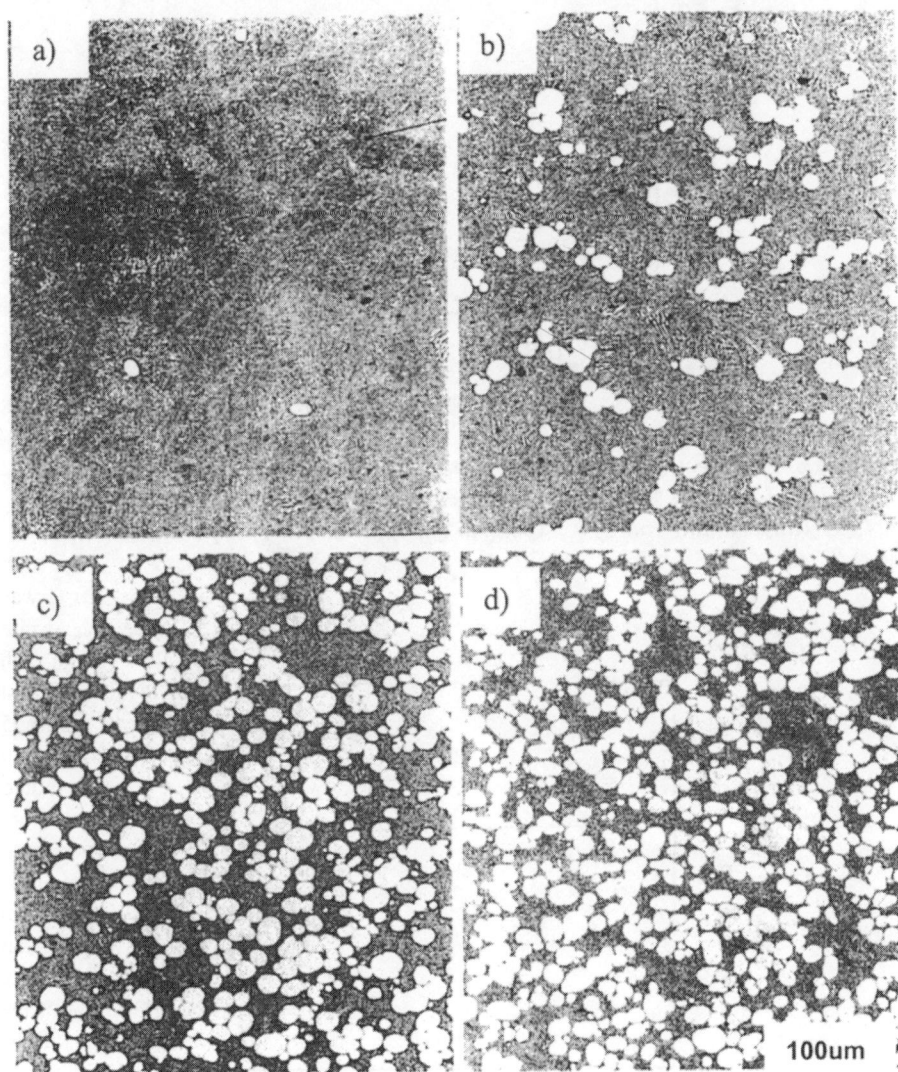

Figure 5. *Optical micrographs showing the water-quenched microstructures of Sn-15wt.%Pb alloy sheared continuously at a shear rate of $3625s^{-1}$ and cooled from $215°C$ with a cooling rate of $0.5°C/min$ to different temperatures. (a) $210°C$; (b) $208°C$; (c) $197°C$; (d) $195°C$*

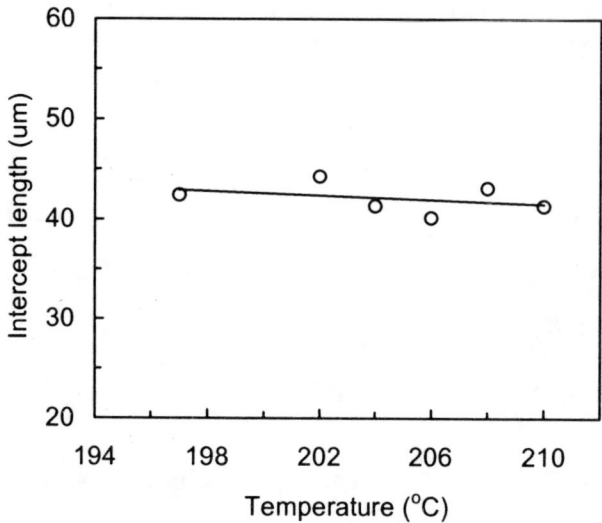

Figure 6. *The effect of shearing temperature on the intercept length of the primary particles of Sn-15wt.%Pb alloy shared continuously at 3625s^{-1} and cooled from 215 ℃ with a cooling rate of 0.5 ℃/min to different sampling temperatures*

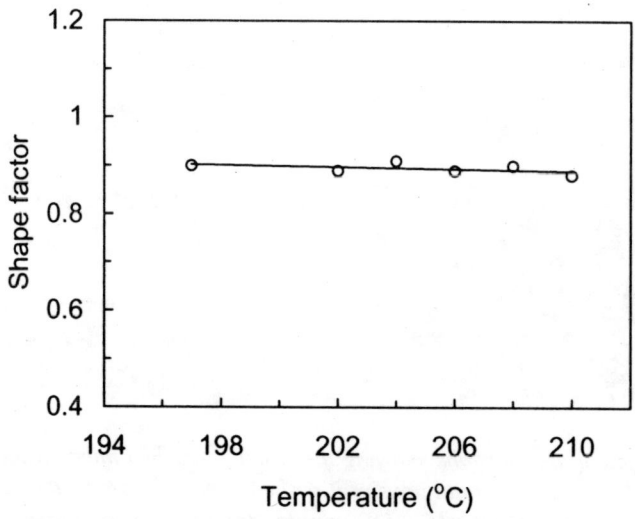

Figure 7. *The shape factor of the primary particles in Sn-15wt.%Pb alloy sheared continuously at a shear rate of 3625s-1and cooled from 215 ℃ with a cooling rate of 0.5 ℃/min to different sampling temperatures*

The theoretical analysis of the stability of solid/liquid interface during solidification under forced convection with various shear rates and intensity of turbulence indicates that pure shear flow promotes equiaxed dendritic morphology if particle rotation is ignored, and that the growth rate increases with the increase of shear rate. With the increase of both shear rate and the intensity of turbulence, the growth morphology would change from equiaxed dendrite to spheroid *via* rosette (Qin *et al.*, 2000). The results from this theoretical analysis can be used to explain the microstructural characteristics observed in the rheomolded samples. As discussed in the previous section, the fluid flow in a closely intermeshing, self-wiping and co-rotating twin-screw extruder is characterized by high shear rate and high intensity of turbulence. Such flow conditions will enhance the compositional uniformity, and therefore reduced constitutional undercooling at the solid/liquid interface during solidification, which eventually promotes the planar growth of the primary particles. Below the nucleation temperature, due to the shear enhanced particle growth rate, all the nuclei grow very quickly to a size which corresponds to the predetermined solid volume fraction dictated by the processing temperature. Because of the uniformity of both temperature and composition fields within the fluid, the growth conditions for all the particle are identical; they should consequently grow into mono-sized panicles. In addition, once the predetermined solid volume fraction is reached, according to Gibbs-Thomson effect, any possible size difference between particles will be demolished due to the enhanced mass transport through the remaining liquid phase.

To investigate the effect of shear rate on the solidification behaviour under continuous cooling and shearing conditions, 15wt.%Pb alloy was cooled from 215°C to 205°C with a cooling rate of 0.5°C/min at different shear rates. The experimental results indicated that all the samples contained the same volume fraction of the primary phase, but had different particle size and particle density. Figure 8 shows the measured average intercept length and particle density as a function of shear rate. Particle size decreases very rapidly with increasing shear rate when shear rate is below about 500/s, but levels off when the shear rate is increased beyond 1000/s. In contrast, the particle density increases very quickly with increasing shear rate in the low shear rate region, but levels off at the high shear rates. It can be therefore concluded from Figure 8 that under the current experimental conditions increasing shear rate increases particle density but decreases particle size.

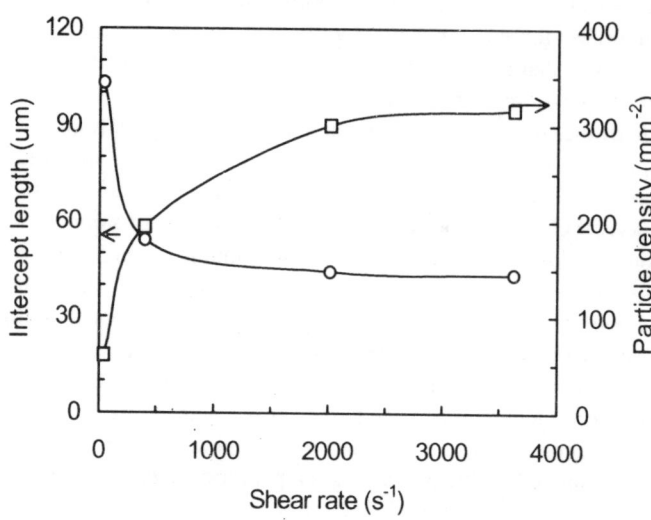

Figure 8. *The effect of shear rate on the intercept length and particle density of the primary phase in Sn-15wt.%Pb alloy shared continuously at different shear rates and cooled from 215 ℃ to 205°C with a cooling rate of 0.5 ℃/min*

The experimentally observed grain refinement (Figure 8) under high shear rate and high intensity of turbulence can be explained by a copious nucleation mechanism. In a conventional casting process, after pouring, solidification starts by heterogeneous nucleation at the mold wall through the so-called "big bang" mechanism. Some of the nuclei formed at this stage contributed to the formation of the chilled zone, but the majority of the nuclei are transferred into the hotter bulk liquid and remelted. The final solidified microstructure depends largely on the number of nuclei survived after the "big bang" nucleation. Under the intensive mixing action offered by the twin-screw rheomolding machine, both temperature and composition fields of the liquid alloy are extremely uniform. During the continuous cooling under forced convection, heterogeneous nucleation takes place instantaneously throughout the whole liquid alloy. Compared with conventional solidification, the actual nucleation rate in the rheomolding process may not be increased, but all the nuclei formed will survive due to the uniform temperature field, resulting in an increased effective nucleation rate. In addition, the intensive mixing action may also disperse the clusters of potential nucleation agents, giving rise to an increased number of potential nucleation sites. In fact, this situation may not only occur under intensive turbulent flow in the rheomolding machine, but also applies to nucleation under forced convection in general. However, it seems that laminar flow is much less effective for homogenizing the temperature and composition and for dispersing the potential nucleating agents. Consequently, laminar flow is less powerful for structural refinement, which is in good agreement with the experimental observations in the literature.

5. Secondary solidification of intensively sheared semisolid slurry

The formation of the primary phase during semisolid processing can be basically divided into two stages. In the initial stage, the primary phase is formed under forced convection during cooling and shearing from the liquidus temperature to the semisolid processing temperature, as discussed in previous section. This is called *primary solidification*. In the second stage, the retained liquid phase solidifies without shearing to produce more primary phase grains and the final eutectic matrix, which can be referenced as *secondary solidification*.

The behaviour of the secondary solidification is also affected by shear rate, shearing time and shearing temperature. Figure 9 shows the microstructures of Sn-15wt.%Pb alloy sheared at different processing conditions, which illustrates the effects of shear rate, shearing time and shearing temperature. Table 1 summarizes the experimental conditions and microstructural information revealed in Figure 9. The effects of shear rate and shearing time are complex and strongly depend on the shearing temperature. The effect of shear rate can be seen by a comparison between Figures 9a and 9b. High shear rate promotes the formation of spherical primary particles during both initial solidification and secondary solidification, as shown in Figure 9a. The effect of shearing time can be found by comparing Figure 9b with Figure 9c. A short shearing time favours the formation of rosettes during the primary solidification and dendrites during the secondary solidification, as shown in Figure 9c. The particle size formed under low shear rate in Figure 9c is apparently larger than that formed under high shear rate in Figure 9b. Comparing Figure 9b with Figure 9c, it is found that neither high shear rate with short shearing time nor low shear rate with prolonged shearing time can produce spherical primary particles. In these cases, the initially solidified primary phase is rosettes or particles and the primary phase formed by the secondary solidification is usually small dendrites or rosettes. The effect of shearing temperature can be seen by comparing Figure 9c with Figure 9d. When the shearing temperature is close to solidus, low shear rate and short shearing time can produce spherical particles from the initial solidification, as shown in Figure 9d. However, the primary phase formed from the secondary solidification is not visible, because the composition of retained liquid is very close to that of the eutectic composition of the Sn-Pb system, and the retained liquid in this case undergoes a eutectic solidification. To sum up, it seems that fine and spherical particles formed during the secondary solidification are promoted by high shear rate, high degree of turbulence, longer shearing time and higher solid fraction during the primary solidification under forced convection.

Table 1. *Summary of the processing conditions and the solidification microstructures of Sn-15wt.%Pb alloy used in Figure 9*

Sample		Figure 9a	Figure 9b	Figure 9c	Figure 9d
Pouring temperature (°C)		220	220	220	220
Shearing temperature (°C)		204	204	204	185
Shear rate (s^{-1})		5200	80	80	80
Shearing time (s)		180	180	20	20
Initially solidified primary phase	Volume fraction	0.24	0.22	0.24	0.73
	Grain size (μm)	43	65	123	102
	Morphology	particle	Rosette	Rosette	Particle
Secondly solidified primary phase	Volume fraction	0.47	0.50	0.46	0
	Grain size (μm)	27	57	78	–
	Morphology	Particle	dendrite	dendrites	–
Total Volume fraction		0.71	0.72	0.70	0.73

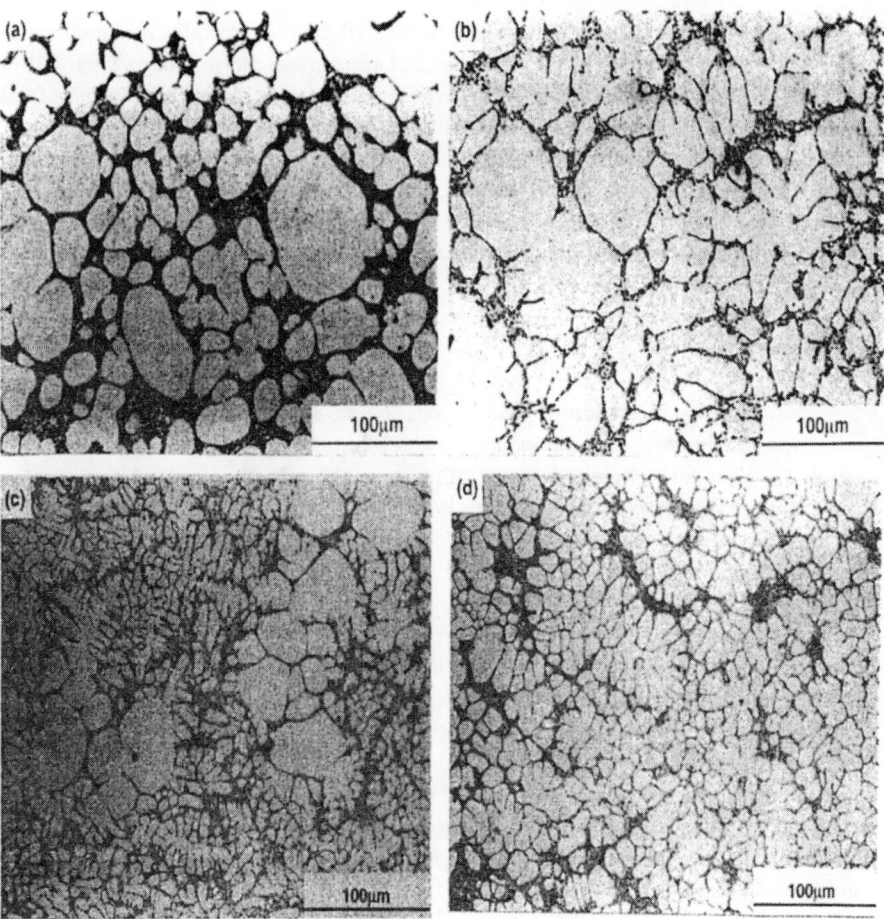

Figure 9. *Micrographs showing the microstructures of Sn-15wt.%Pb alloy processed at different conditions, as summarized in Table 1, to illustrate the effects of shear rate, shearing time and shearing temperature on the morphology of the primary phase*

6. Advantages and potential application for twin-screw rheomolding process

As a near net-shape production technology, the advantages of the twin-screw rheomolding process over other existing SSM technologies are mainly derived from its unique fluid flow characteristics and its high flexibility in terms of processing conditions and requirements for feedstock materials. Firstly, the process is one-step and relatively simple. SSM slurry production and component shaping are integrated into one single operation. It can be fully automated for high volume production, just as in the injection molding processes in the plastics industry. Secondly, the high shear rate and high intensity of turbulence offered by the specially designed twin-screw extruder ensures a fine particle size and a spherical morphology. This will in turn result in a finer microstructure and improved mechanical properties of the final component. Finally, twin-screw rheomolding offers a lower overall component cost. There are no special requirements for the feedstock material except the chemical composition and temperature. Thus, the raw materials cost should be low. In addition, the new process has a shorter cycle time, resulting in a higher production rate. All these factors contribute to the lower component cost.

7. Summary

A twin-screw rheomolding process has been developed for near net-shape production of engineering components, in which liquid alloy is melted and fed into a twin-screw extruder, where the liquid alloy is intensively sheared and cooled to a temperature in the interval of liquids and solidus, and the sheared semi-solid slurry is then transferred to a shot sleeve and subsequently injected into a mold cavity for component shaping. The fluid flow in the twin-screw rheomolding process is characterized by high shear rate, high intensity of turbulence and cyclic variation of shear rate. The experimental results of rheomolded alloys have demonstrated that the developed rheomolding process is capable of producing fine and spherical primary particles uniformly distributed in the matrix. Solidification under high shear rate and high intensity of turbulence can result in substantial grain refinement due to the increased effective nucleation rate caused by the increased temperature and composition uniformity. It is also found that fine and spherical particles formed during the secondary solidification are promoted by high shear rate, high degree of turbulence, longer shearing time and higher solid fraction during the the primary solidification under forced convection.

Acknowledgements

Financial support from HEFCE, EPSRC, Ford Motor Co. and Dessett Process Engineering Ltd is gratefully acknowledged.

References

Andersen P.G., *Mixing and Compounding of Polymers*, ed. by I. Manas-Zloczower and Z. Tadmor, Hanser Publishers, New York, 1994, pp. 679–705.

Erdmenger R., *Chem. Ing. Tech.*, 36 (1964), 175.

Fan Z., Bevis M. J. and Ji S., *International Patent*, WO 01/21343 A1, 1999.

Flemings M.C., *Metall. Trans.*, 22A, 1991, 269.

Flemings M.C., Rick R.G. and Young K.P., *Mat. Sci. Eng.*, 125, 1976, 103.

Ji S. and Fan Z., *Proc. the 6th Intl Conf. on Semi-Solid Processing of Alloys and Composites*, G. L. Chiarmetta and M. Rorro (Ed.), Turin, Italy, 2000, pp. 723–728.

Joly P.A., Mehrabian R., Mater J., *Sci.*, 11, 1976, 1393.

Kirkwood D.H., *Inter. Mater. Rev.*, 39, 1994, 173.

Pasternak L., Carnaha R.D., Decker R. and Kilbert R., *Proc. the 2nd Intl Conf. on Semi-Solid Processing of Alloys and Composites*, ed. by S. B. Brown and M. C. Flemings, MIT, Cambridge, 1992, pp. 159–169.

Qin R., and Fan Z., *Proc. the 6th Intl Conf. on Semi-Solid Processing of Alloys and Composites*, G. L. Chiarmetta and M. Rorro (Ed.), Turin, Italy, 2000, pp. 819–824.

Rauwendaal C., *Polymer Extrusion*, 3rd Edition, Hanser Publisher, New York, 1985.

Spencer D.B., Mehrabian R., Flemings M.C., *Metall Trans.*, 3, 1972, 453.

Young K.P., *US Patent*, No. 4,687,042, 1987.

Chapter 4

Micro Injection Molding

Christian G. Kukla

Industrial Liaison Department, University of Leoben, Austria

1. Introduction

In microelectronics we are aware of the fact that everything is becoming smaller and smaller. This trend influences more and more mechanical components and complete machines and systems. In the down scaling of systems and their components the limits of conventional manufacturing techniques are reached. This initiates the improvement of conventional methods and further development of new techniques. For the shaping of micro parts the main area for developments lies within the process step of injection molding.

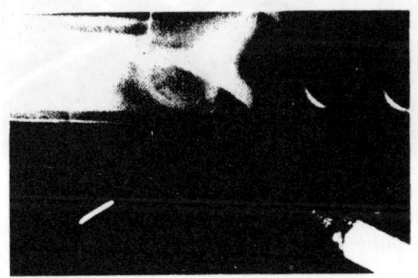 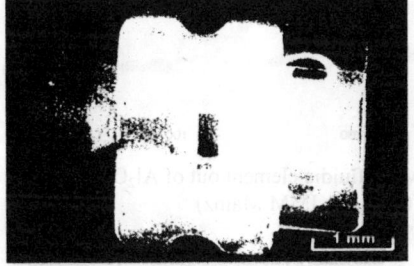

Micro gears for watches – Out of POM – 0,8 [mg]
(Courtesy: Fa. Rolla, CH; Fa. Battenfeld, A)

Orthodontic brackets – Out of Al_2O_3
(Courtesy: SPT Roth Ltd)

Figure 1. *Examples of micro-injection molded parts*

2. Definition of micro parts and examples

With regard to micro injection molding different concepts of a micro part exist. It is very important to distinguish between these concepts because different production equipment and processing are connected with each of them. The following types of parts related to the concept of "micro" exist [KUK 98, REI 98, KUK 00]:

  – micro-injection molded parts,

  – injection molded parts with micro-structured areas ("Micro on Macro").

A micro-injection molded part is defined by a volume of less than around 0,3 cm^3. This limit is set by injection molding machines. It defines the lower limit of shot volumes realizable by standard machines. Such parts may, but need not, have dimensions in the range of micrometers. Examples are shown in Figure 1.

The second type are injection molded parts with micro-structured areas (short name: "Micro on Macro"). These cover injection molded parts of conventional dimensions with areas having dimensions in the range of micrometers. One example is a CD or DVD. Further examples are shown in Figures 2 and 3.

Here mainly micro-injection molded parts are discussed, especially the necessary machinery. Certain problems related to tooling and the molding process are similar for the two types of micro part.

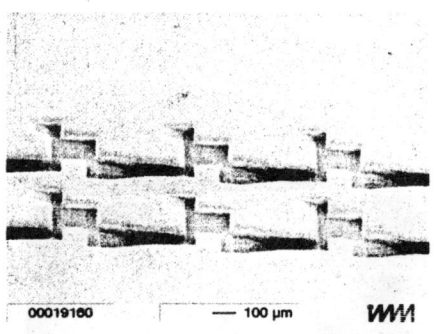
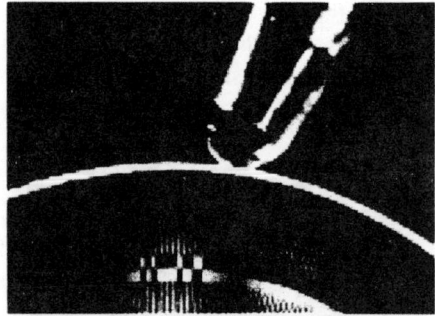

Microfluidic element out of Al_2O_3
(Courtesy: IMM Mainz)

Sensor disc with microstructures mold inserts by LIGA
(Courtesy: Fa. Scholz, D)

Figure 2. *Examples of injection molded parts with micro-structured areas*

3. State of the art

Conventional injection molding machines utilize screws with diameters down to 14 mm (Figure 5). The limit is the mechanical strength of the screw material. The screw should be able to process resins with standard pellet sizes of the resin. Thus the depth of the screw channels should have at least the dimensions of a single pellet of resin – 2 to 3 millimetres. The remaining inner diameter of the screw is thus limited.

As a rule of thumb 5% of the maximum shot weight can be taken as the minimum shot weight of an injection molding machine. Both facts determine the minimum shot weight of conventional injection molding machines which lie in the range of tenths of cm^3 for stable production. For smaller shot weights, machines like Babyplast or Butler are available in the market. Usually they utilize a plunger type injection unit. They are limited in precision due to a pneumatic drive system (Babyplast) or the limited capability of the machine control system. Parts which can be defined as micro injection molded ones are already produced. To achieve such production big runner systems are used. These runner systems are bigger than the parts by a factor of 100 and more. Production is thus possible but connected with certain disadvantages for cavity filling and recycling. For injection molding an oversized runner system means that the parts themselves are small appendages at the end of the flow path. In this case the holding pressure compensates for the shrinkage of the runner system but not for that of the parts. A highly sophisticated control system is required for a perfectly repeatable injection molding process and only in this case the process leads to high quality runner systems with high quality parts associated with them.

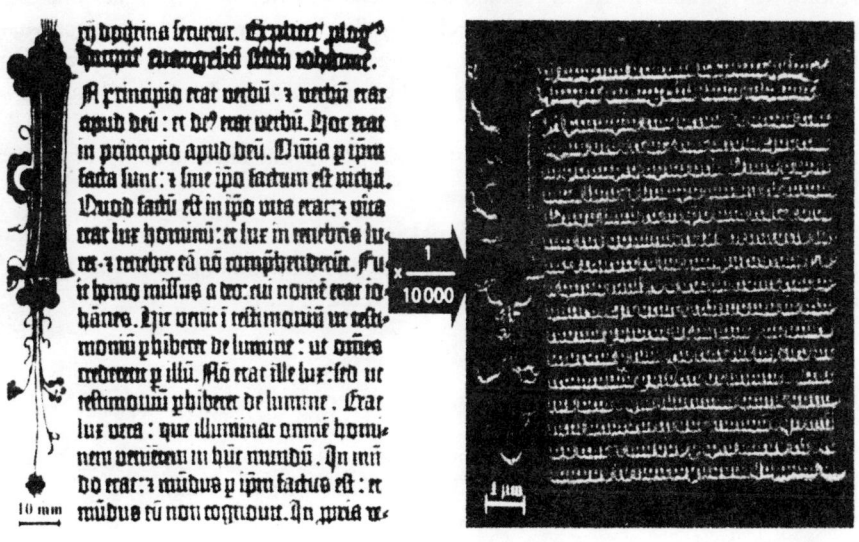

Figure 3. *Bible of Gutenberg. Left: original bible. Right: REM-picture of a 10,000-times minimized copy, produced by injection molding of plastics*
(Courtesy: Paul Scherrer Institut Villigen / Fachhochschule Aargau Windisch)

A further disadvantage is that the recycling of the runner system becomes more difficult. A runner system being a hundred times bigger than the parts means that the resin has to be injection molded a hundred times when being recycled directly. To process a resin thus often implies degradation and therefore is impossible. For recycling it is necessary to use the recycled resin for other, bigger products. Regarding the small volumes of micro-injection molded parts conventional injection molding machines are oversized for this type of part. But if nothing better is available for high precision micro parts one has to work with conventional machines.

4. Micro injection molding

Starting with the topic of "micro" for the production of parts one has to be aware of the fact of entering a different world. It is not enough *e.g.* to buy a proper machine, set it up and to start a production of micro parts. Doing this would lead to failure. The reason for this lies in a completely different way of thinking which is necessary especially when dealing with very small parts. The operator cannot see if a part is completely molded or if some material is sticking in the cavity after ejection. In some cases he must not handle the part manually because it would break. Additionally, micro injection molded parts behave differently to bigger ones, *e.g.* they do not fall downwards when ejected by the ejector pins. Due to electrostatic forces they can adhere to the vertical mold platens. All these facts illustrate some details of the "micro"-world where the workers, operators and engineers have to act differently than in the standard sized world. Therefore one key factor of molding micro parts successfully is the right staff with the right understanding of a micro production.

To point out the main differences between the normal world and the micro world the following detail the existing technology for injection molding and what has to be changed for the production of micro-injection molded parts.

4.1. *Injection molding machine*

As mentioned before the minimum shot volume of conventional injection molding machines is around 0,3 cm^3. This exceeds the part volume of *e.g.* gears for watch industry or orthodontic brackets by far (Figures 2 and 4). To produce such gears, relatively huge runner systems are used to increase the shot volume and thus compensate for this disadvantage. In addition, when producing parts at the lower limit of the machinery, stable and reproducible production is harder to achieve.

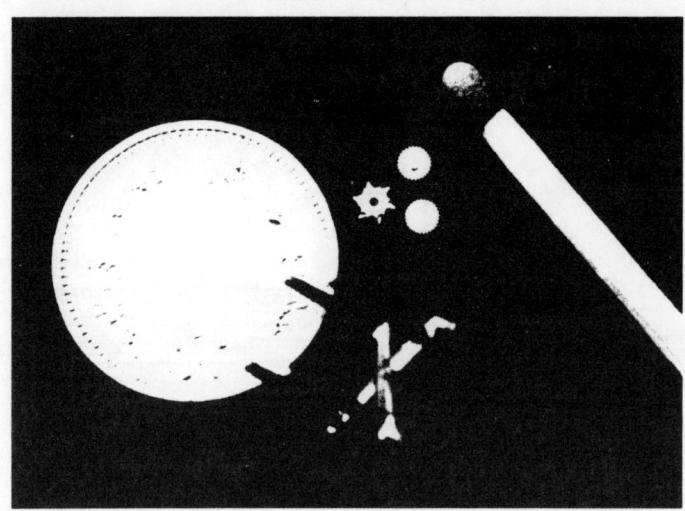

Figure 4. *Micro gear wheels for the watch industry (Photo courtesy: Ronda SA,CH)*

Therefore the first idea is a further reduction of the screw diameter. But, as mentioned above, this line can not be followed because of the mechanical strength needed for plasticising the material. To overcome the aforementioned problems the design of the injection unit has to be changed. The concept of a screw for melting and injecting polymers combines four functions in one unit (Figure 5):

– plasticising,
– homogenising,
– metering and,
– injection.

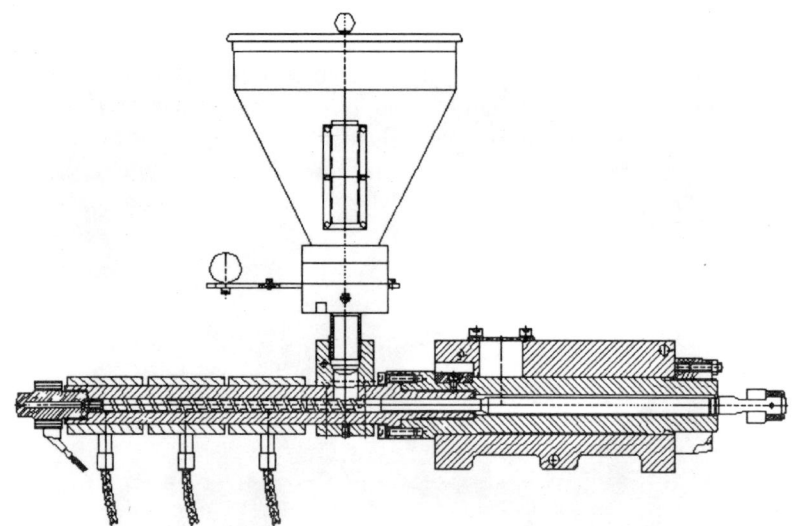

Figure 5. *Micromelt'-injection unit with a screw diameter of 14 mm (Photo courtesy: Battenfeld IMT)*

For further down scaling the diameter of the injection piston these functions have to be fulfilled by at least two different units: a screw for plasticising and homogenising and a piston for metering and injection. This provides the advantage of reducing the diameter of the injection piston down to *e.g.* 5 mm and thus reducing the minimum shot volume of the injection molding machine. In Figure 6 the design of a dedicated micro-injection molding machine is shown. It consists of a screw for plasticising and homogenising the feedstock, one piston for metering and one piston for injection. This design reduces cycle times by parallel movements. Metering can start directly after the injection piston is in the front position. The injection speed of the piston is up to 0,8 m/s.

In designing a micro-injection molding machine there are further points to be taken into account. For a high and repeatable product quality the drive system is important. Therefore servo-electrical motors are the best choice. Usually micro-injection molded parts are thin walled, so the cycle times could be very short regarding the required cooling time. For state-of-the-art micro-injection molded parts cycle times are 10 seconds and longer. This time is necessary for cooling the relatively big runner system. By making the right choice of the injection molding machine the cycle times can be reduced to around 3 seconds thus greatly increasing the productivity.

Additionally, the control unit of the machine has to be adapted for small shot weights and a high injection speed which is required to achieve the same volume rates as with larger piston/screw diameters. In micro injection molding the injection itself takes place within some milliseconds. Therefore a closed-loop-controlled injection profile is hard to achieve. Table 1 shows a comparison of the screw stroke needed for 1 mm^3 shot volume with the resolution of the A/D-converter used in injection molding machines. Here one can find that the piston diameter has to be smaller than 7 mm to provide the needed resolution for a closed-loop-control.

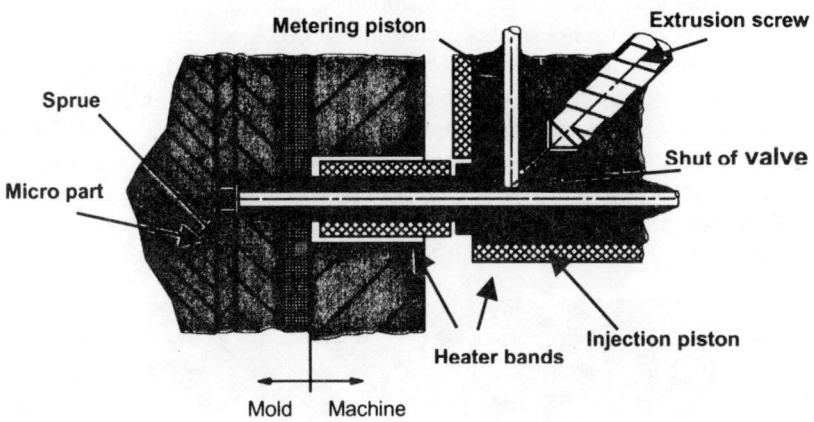

Figure 6. *Screw/piston injection unit for micro-injection molding*

Table 1. *Accuracy of the control system for injection molding*

Screw/Piston diameter mm	Stroke 4D mm	Stroke volume 4D cm^3	Screw stroke for 1 mm^3 mm	Resolution limit of a A/D converter – 4D stroke mm
22	88	33.45	0.0024	0.0430
18	72	18.32	0.0034	0.0352
14	56	8.62	0.0056	0.0273
10	40	3.14	0.0112	0.0195
7	28	1.08	0.0228	0.0137
5	20	0.39	0.0446	0.0098
3	12	0.08	0.1239	0.0058

4.2. Molds

For the production of molds for small and micro parts by injection molding conventional manufacturing techniques have reached their limits in realising part dimensions in the µm-range. Therefore they have to be extended to smaller dimensions (micro machining, micro EDM) or new techniques have to be developed (LIGA). The choice of which technology to use is dependent on the shape and dimensions of the part and the related costs. Roughly one can say that mold inserts with a 2½ D geometry, sharp edges, very smooth walls, a height of less than 1 mm and tolerances under 1 µm can be produced by LIGA technologies. These are standard technologies for the production of computer chips and can be used for micro parts and mold inserts too. Depending on the radiation used for the lithography several LIGA-techniques are available starting with UV-LIGA as the cheapest one. Here the height of the structures is limited to 0,4 mm. By using other sources of radiation, e.g. synchrotron, higher and more accurate structures can be achieved. But the price is also increasing. Mold inserts produced by LIGA-technique are showing a roughness of 30 to 40 nm. Such a mold insert is shown in Figure 7.

Figure 7. *Mold insert for micro parts produced by LIGA (Source: IMM Mainz, IFWT, Vienna)*

Conventional technologies like machining or EDM can not compete with LIGA regarding the dimensions and tolerances. On the other hand they are cheaper. Therefore a vast amount of research is undertaken to expand the capabilities of machining or EDM into the micro dimensions. Micro-machining is mainly used for parts with micro-structured areas on them with a high surface quality (*e.g.* mirrors for laser applications). Beside conventional techniques like turning or milling in micro-machining others like fly-cutting are used. Figure 8 shows a mold insert produced by micro-machining.

In micro-machining, tools made of mono-crystalline diamond are mainly used because of the high forces occurring. The big disadvantage of this technology is the restriction to certain materials like brass, Ni, Cu, Al or plastics. For other harder materials, *e.g.* steel, micro-EDM has to be used. Here the electrodes used can be produced by LIGA too. Figure 9 shows such an electrode with the according structure produced by micro-EDM [WOL 97]. Similar opportunities are shown by micro-ECM.

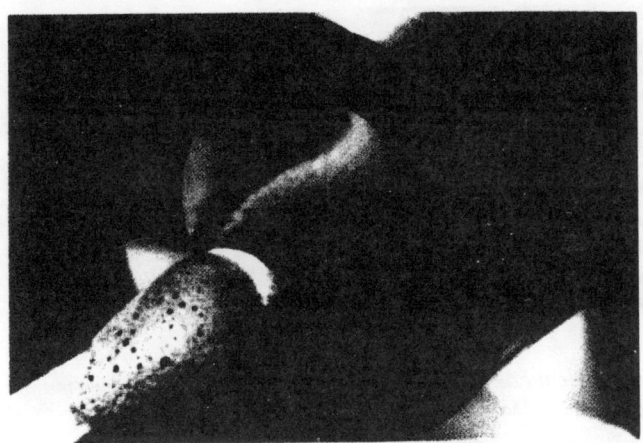

Figure 8. *Micro-machined mold insert for the production of contact lenses (Source: FWT – University of Kaiserslautern)*

4.3. *Injection molding process*

For the molding of parts with dimensions of less than 0,2 mm filling problems arise because the melt is freezing before the cavity is completely filled. One solution is the so called "variotherm"-process. Before injection the mold itself is heated up to a temperature above the melting point of the resin. Then the melt is injected and the mold is cooled down to the ejection temperature where the part can be ejected without distortions. The main disadvantage of this process is the comparatively long cycle time in the range of one minute when using oil for temperature control. To shorten the cycle time a new mold concept for the "variotherm"-process has been developed and tested [REI 98]. Figure 10 shows the open test mold. The insert for the runner system for the 4-cavity-mold is situated in the middle. The cavity itself can be heated electrically and cooled by air. To utilize electrical resistance for heating the mold insert, the mold insert itself has to be insulated against the rest of the mold. Thus a fast "variotherm"-process can be achieved. With this mold design cycle times down to around 15 s are possible. This is nearly the same time as with conventional injection molding.

Usually the air can leave the cavity through channels formed by the roughness and tolerances of the mold. For micro-injection molding the dimensions are in the same range as the channels mentioned. Therefore the melt would fill these channels especially in the "variotherm"-process. To avoid flashing the mold tolerances have to be very tight. Additionally, the mold design must allow the evacuation of the cavity prior to injection.

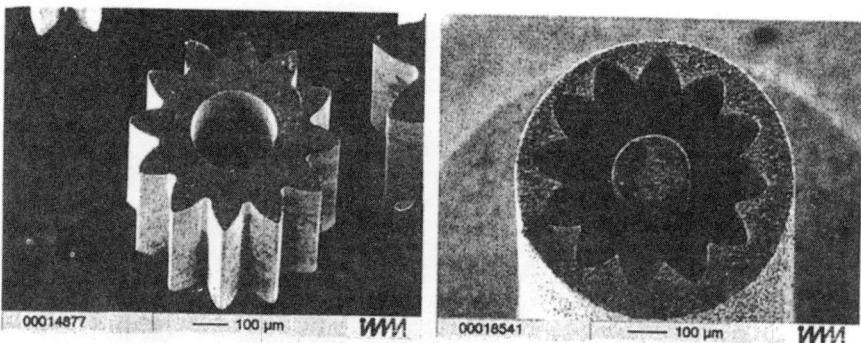

Figure 9. *Electrode made by LIGA (left) and hard metal structure made by micro-EDM (right) (Source: IMM Mainz)*

4.4. Simulation

Simulation of the injection molding process is highly developed and accurate enough for normal sized parts. For the simulation of micro injection molding there is a lack of knowledge about material behaviour in small channel heights and high shear rates. A channel of 50 μm height and a volume rate of some cm^3 per second easily gives shear rates of 10^7 s^{-1} and higher. There is only a small amount of literature available on how a polymer melt is behaving under very high shear rates [HAD 00]. It seems that ewtonian flow behaviour apperas at shear rates above 10^7 s^{-1}. Rheology is not the only open question in very thin channels for injection molding. Another question is e.g. thermal conductivity. Thermal conductivity is highly depending on the orientation of the molecules. A high orientation can be assumed to take place in thin channels. Therefore it is necessary to take this fact into account for simulation of micro injection molding.

4.5. *Quality control*

Micro injection molded parts have very small dimensions in the range of some tenth of millimetres and less. This makes it impossible to check if a part is completely filled with the naked eye. A further problem is to check if specified dimensions are within the tolerances especially if they are within or below the micrometer range.

In-line quality control is normally done by scales. For very small parts with a weight below 1 mg this is impossible in an industrial production environment. Every truck going by the factory would cause errors in the measurement.

According to the problems mentioned and the demand to realize a 100% quality control of the molded parts, a computer based video system is the most efficient way to check the quality of the molded parts. Designing such a system requires a suitable handling system to show the part to the camera in the right position. Afterwards a good/bad part separation can be done. Thus 100% quality checked parts are delivered.

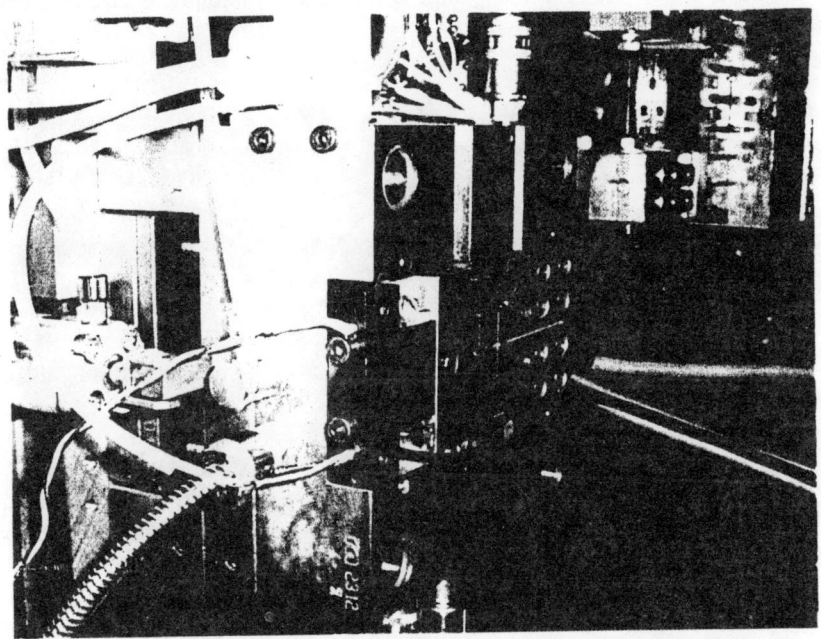

Figure 10. *Mold for the "variotherm"-process (Source: FoTec GmbH)*

4.6. Ejection and handling of parts

A further problem is the ejection of the parts since ejector pins must be especially designed to be part of the cavity because of the "variotherm"-process. The normal tolerances would allow the melt to flow in between mold platen and ejector pin.

The dimensions of micro injection molded parts ejection and handling of the micro parts are a special problem. Because of the small volume of such parts the electrostatic forces can let them adhere to vertical walls or elsewhere but would not let them just fall down. Additionally, handling them manually or by a gripper must be done very carefully so as not to damage the parts. Therefore an especially designed handling system is necessary to take the parts out of the cavity. Figure 11 shows a powder injection molded part damaged during ejection.

Additional problems arise especially for PIM parts when injection molding such small structures. The green strength of feedstocks is relatively low. If a green part break during ejection and some material remains in the cavity it is hard to clean the cavity. Ultrasonics with the right solvents seem to be the best cleaning technology at the moment. Here further investigation is necessary to improve cleaning under production conditions.

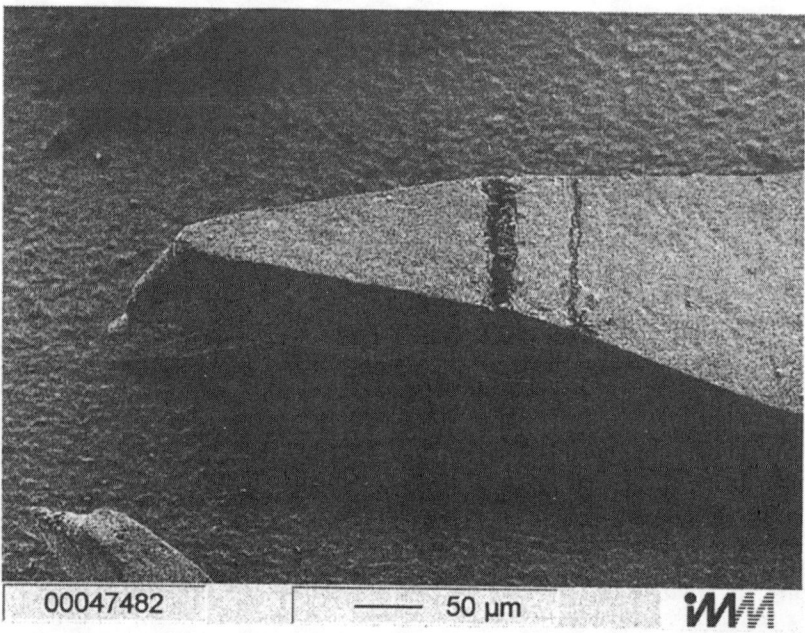

Figure 11. *Part damaged during ejection (Source: IFWT, Vienna, A; IMM Mainz, D)*

4.7. *Packaging*

As with electronic components it is hard to have micro injection molded parts separated and oriented. This advantage is available directly after molding and should not be lost in further processing of the parts. Otherwise it is necessary to separate them automatically. Utilising bull feeders for this separation is not practical. Due to electrostatic forces micro injection molded parts can leap everywhere and not stay within the feeder. One way to have such small parts separated and oriented for further processing steps is to package them directly after molding into a blister such as is used for electronic components.

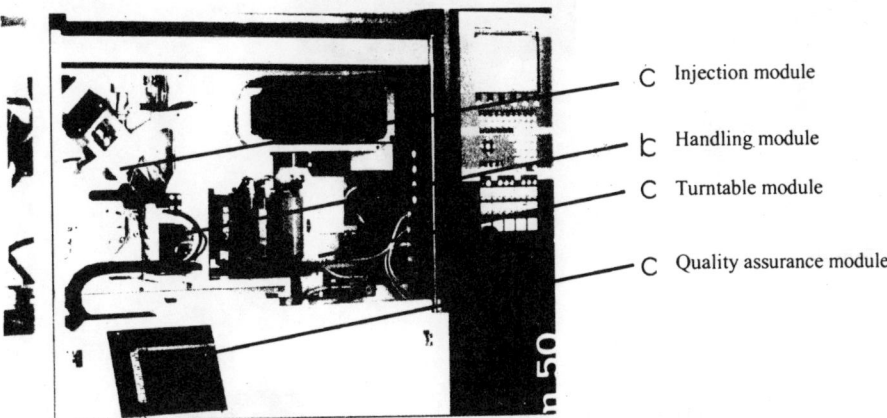

C Injection module

c Handling module

C Turntable module

C Quality assurance module

Figure 12. *Screw/piston injection unit for micro-injection molding*

5. Summary

For the production of micro parts not only has the injection molding machine to be improved but also other problems related to the small dimensions of the parts have to be solved. These are *e.g.* how to eject a micro part from the mold, how to handle it or how to check the quality.

As Figure 14 shows, a proper system for micro injection molding is not a mere injection molding machine any more, but a complete manufacturing cell for micro parts which includes injection molding, handling, quality control and packaging. If required, this production cell can be equipped with a clean room module. For highest accuracy and repeatability this manufacturing cell is operated fully electrically. Figure 12 gives some details of this manufacturing cell. One can see the injection unit (a) which consists mainly of three parts: the plasticising unit with an optimized screw (diameter: 14 mm), the metering unit and the injection unit. The maximum injection speed is 1,3 m/s. The turntable module (c) allows for a minimum opening stroke so that just half the mold can be rotated by 180°. The ejection into the handling unit (b) takes place in the rotated position while at the same time the next parts are molded. The handling unit presents the molded parts to a camera for quality assurance. The monitor is shown in Figure 12 (d).

Figure 13. *Packaging of micro parts*

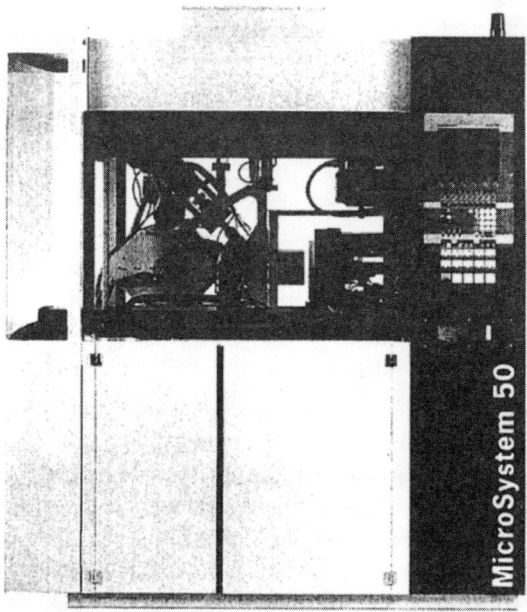

Figure 14. *Micro manufacturing cell, type Microsystem 50 (Photo: Battenfeld IMT)*

This manufacturing cell is the beginning of a new type and understanding of production which is dedicated to the "micro-world". A vast area for research and development is open.

References

[HAD 00] Haddout A., Villoutreix G., "Polymer Melt Rheology at High Shear Rates", *International Polymer Processing* , XV, 2000, 3, pp. 291–296.

[KUK 98] Kukla C., Loibl H., Detter H., Hannenheim W., "Micro-Injection Moulding – The Aims of a Project Partnership", *Kunststoffe Plast Europe,* Vol. 88, No. 9, 1998, pp. 6–7.

[KUK 00] C. Kukla, PIM of Micro Parts, *Proceedings of PM^{2}TEC*, New York, 2000.

[REI 98] Kukla C.G., Hannenheim W., Loibl H., "Manufacturing of Micro Parts by Micro Injection Moulding", *Micro System Technologies 98*, compiled by H. Reichl and E. Obermeier, VDE-Verlag GmbH Berlin Offenbach, 1998, pp. 337–342.

[WOL 97] Wolf A., Ehrfeld W., Lehr H., Michel F., Nienhaus M., Gruber H.-P., "Combining LIGA and Electro Discharge Machining for the Generation of Complex Micro Structures in Hard Materials", *Proceedings 9-IPES/UME4 Conference*, Braunschweig, 1997.

Chapter 5

Materials Characterization Methods and Material Models for Stamping of Plain Woven Composites

J. Chen and D. S. Lussier
Advanced Composite Materials and Textile Research Laboratory, Dept of Mechanical Engineering, University of Massachusetts Lowell, USA

J. Cao and X. Q. Peng
Advanced Materials Processing Laboratory, Dept of Mechanical Engineering, Northwestern University, Evanston, Illinois, USA

1. Introduction

Stamping of continuous fiber, woven fabric offers the promising combination of short cycle times and structural properties. The glass fibers are commingled with thermoplastic (polypropylene or PP) fibers to promote ease of handling and thorough and rapid wetting of the glass fibers. The commingled glass/PP sheets can be heated above their melting temperature, formed to a shape, and cooled back to a rigid solid without drastic changes in their original properties. Figure 1 shows an example of a hemispherical stamping tool.

Numerical simulations of forming processes can be powerful tools for materials selection, tool design, and process optimization. All simulations require the input of material parameters and correct material models. Adding complexity to the problem, the properties of textile reinforcements are constantly changing during forming. Shear frame tests, bias extension tests, and biaxial tests [BOI 97, FRI 98, LUS 00a, LUS 00b, MCG 98, MOH 00, SOU 00, WIL 99] are just some of the methods currently being utilized to quantify these evolving material properties. However, test method variations reported in the literature and proper interpretation of test results add to the uncertainty. Clearly, improvements in the predictive capabilities of stamping simulations require significant advances in areas such as standard test methods for materials characterization, micro-mechanical models for predicting effects of fabric architecture, and proper friction models for binder segments. Unfortunately, there are currently no standard tests to measure these key material responses for stamping.

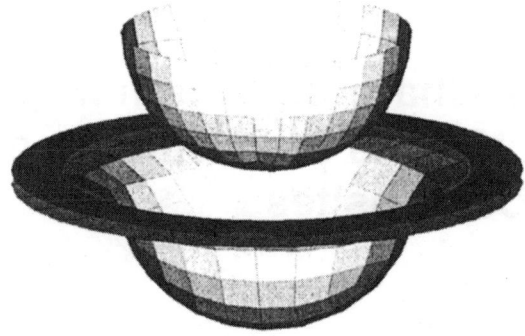

Figure 1. *Finite element model of matched die hemispherical stamping tool and binder ring (fabric removed for clarity)*

It is also recognized that "simple" standard tests can be used to inform numerical models of the material, so that more extensive material properties can be obtained. The numerical models can be interrogated more thoroughly than the experimental setup. In this paper, we present a comparison of both experimental and numerical results of the fabric shear behavior, contributing both to the development of some standard test practices and improved material models. It should be noted that the experimental and numerical approaches discussed here are applicable not only to the stamping of commingled thermoplastic sheets, but are also relevant to other forming processes (e.g., pre-consolidated sheets, prepreg thermoset fabrics, preforms for resin transfer molding).

2.   Shear frame and shear response

An example of a typical shear (or trellis) fixture is shown in Figure 2. Material is clamped onto the frame by means of toggle clamps and a grooved surface to prevent slip. The multiplier links increase testing rates roughly 4 times the crosshead speed. Tension along each of the yarn directions can be adjusted by adding springs between the bolt and the outer edges of the shear frame. In the experiments described in this paper, tensioning springs were not used. The fabric was placed in the frame such that there was no slack. Slight tension variations due to this procedure did not significantly affect the results.

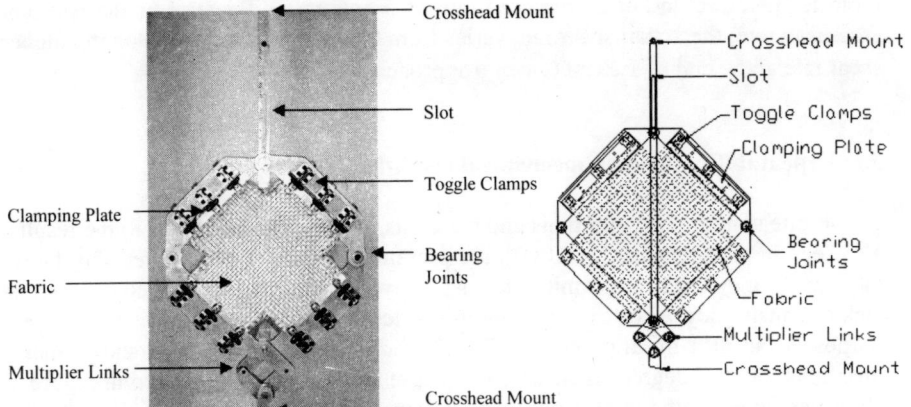

Figure 2. *Shear frame with rate multiplier (total fabric: 255 x 255mm, minus corner cutouts; pin-to-pin: 216 x 216mm; inner fabric test area: 135 x 135mm)*

Figure 3 shows a plot of typical data obtained from a plain weave fabric in terms of crosshead load vs. shear angle. The shear angle is defined as the angle through which the fabric has sheared. Thus, θ starts at $0°$ and increases as the trellis frame deforms. The typical shear response shows an initial region with little increase in load. In this region, the yarns are rotating with little resistance. The curve then reaches a point where the load increases more rapidly. The rapid increase in load occurs when the yarns begin to compress each other laterally at the crossover points. Further rotation is limited by the available yarn compaction.

The material used for the experiments presented in the rest of this paper was a commingled glass/PP plain weave fabric with a glass volume fraction of 35% and mass fraction of 60%. The yarn spacing is 2.02 yarns/cm, and the fabric has an areal weight of 745 g/m^2. Unless otherwise specified, the shear tests were run at

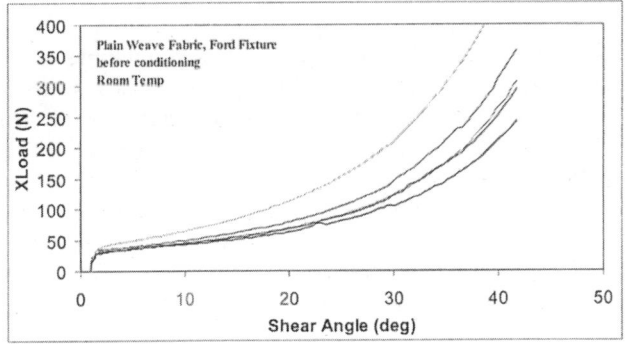

Figure 3. *Non-repeatability of shear response for plain weave fabrics before mechanical conditioning*

room temperature and at a crosshead rate of 50mm/min. Because of the constant crosshead rate, the actual shear rate varies from 0.028 – 0.05 rad/sec, with the higher shear rate at the end of the test (when θ approaches 45°).

3.   Repeatability through mechanical conditioning

Despite similar test conditions and materials, significant variability in the results, also seen in the literature [SOU 00], are present in Figure 3. Fabric tension, fabric alignment, and fabric uniformity are just a few reasons that may account for this lack of consistency. For example, additional tests were run on specimens that were purposely cut at a misalignment of 0, 2.5, and 5 degrees. Figure 4 shows that a misalignment of 5 degrees is sufficient to cause the variability displayed in Figure 3. However, even with great care in placement of the fabric in the shear frame, significant non-repeatability was observed.

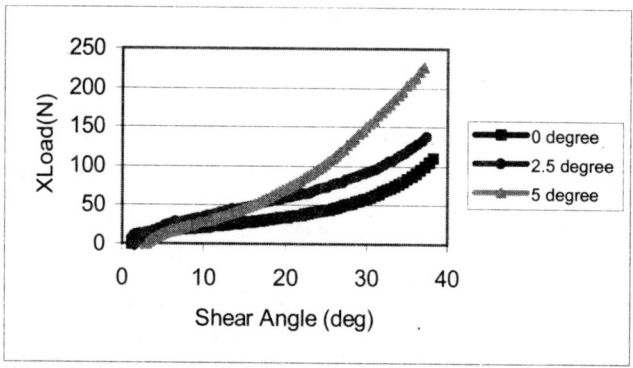

Figure 4. *Effect of misalignment on shear response*

During testing, it was noticed that a fabric sheared multiple times, without removal from the frame, would exhibit a steady decrease in loads the first two or three times and would then generate almost the identical curve for all the remaining number of runs. Based on this observation, it was believed that a form of "mechanical conditioning", or shearing of the specimen in the frame several times prior to collecting data, could address the repeatability problem. Thus, for the specimens shown in Figure 5 and all subsequent testing, the fabric was sheared in the frame to slightly less than 45° of shear angle deformation. This "conditioning" was done at room temperature between 6–8 times before running the final recorded test.

Figure 5 shows that the mechanical conditioning significantly improved repeatability, but that the loads for the conditioned fabrics were only half that of the unconditioned fabric. The conditioning step led to realignment of tows that were not uniformly spaced and straight in the material taken directly off the roll.

4.   Area normalization

Previous work [e.g. MCG 98, WIL 99] was reviewed to determine if test results from different laboratories were of similar magnitude and trend. Many different shear frame sizes and designs have been used in the past. To establish a standard test, it is important that results from the same material tested at different laboratories are consistent. Figure 6 depicts a visual interpretation, showing a distinction between two fabric test areas. A larger area is constructed of many more yarns, causing more crossovers within the fabric specimen. More crossovers would increase the friction and compaction loads needed for yarn rotation. Thus, it was expected that large differences in fabric test area would show significant effects on the shear load results.

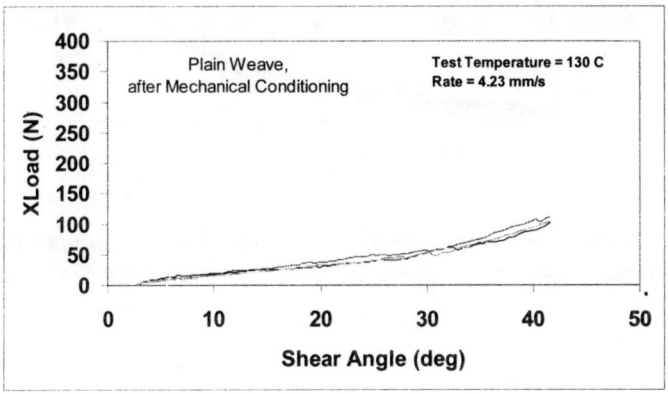

Figure 5. *Repeatability of shear response after mechanical conditioning*

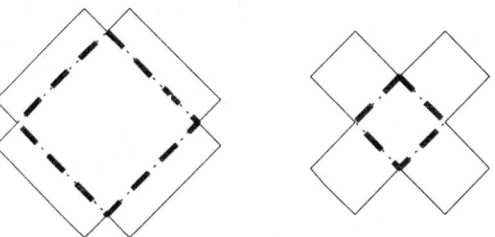

Figure 6. *Examples of two different fabric test areas used for area normalization comparison*

Tests were performed at similar temperatures and loading rates on the same shear frame with three different size fabric test areas – 0.012 m², 0.025 m², and 0.044 m². If plotted as simply crosshead load versus shear angle, the loads are much higher for the larger test areas, as expected. However, after normalizing the loads by the inner fabric test area (the area outlined by the dashed red line in Figure 6), all three sets of data coincide, as shown in Figure 7. Results from a different shear frame at the University of Nottingham, using similar materials and test conditions, also correlated well [LUS 00b, WIL 99]. These results suggest that (1) by normalizing to area or number of crossovers, data from different shear frames can be compared; and (2) the number of crossovers, as represented by the area, has a direct effect on the shear test results.

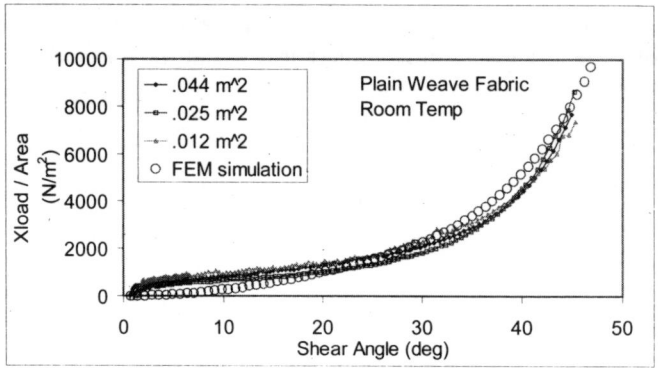

Figure 7. *Normalized shear response of plain weave fabrics on same shear frame and FEM simulation results (crosshead rate: 50 mm/min, shear rate: ranges from 0.028–0.05 rad/sec)*

5.   Comparison with bias extension test

A second test commonly used to characterize the shear behavior of fabrics is the bias extension test. This test is essentially a tensile test with the yarns initially oriented at ± 45-degrees to the loading direction. A typical setup is shown in Figure 8. As the crosshead moves, the yarns rotate towards the loading direction, resulting in a combination of shear, slip, and tension. Typical load-displacement results can depend heavily on the length-to-width ratio of the specimen, as anything less than 1:1 results in direct tensioning of yarns that are held in both top and bottom grips. Very high ratios lead to extensive slip or pull-out of the yarns, while ratios on the order of 2:1 generate reasonable amounts of shear without excessive pullout in the initial stages of the test.

Interpretation of the results can be difficult because of the combined effects of tension, shear, and slip, as well as the nonuniform response in the specimen. For example, the yarns directly gripped are not allowed to rotate, resulting in a triangle region extending from the grip that does not deform (Figure 9). The middle region is most susceptible to pull-out, but in the absence of slip, this region will experience pure shear. The four remaining regions display some shear, but to a lesser degree than the middle section. Because of this variation, it is difficult to determine an appropriate area by which to normalize the load data.

Figure 8.   Experimental setup for bias extension test

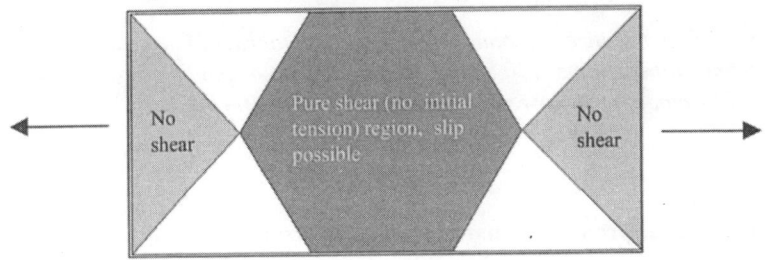

Figure 9. Schematic of deformation zones in bias extension specimen

In Figure 10, the crosshead load is normalized by the surface area of the pure shear region, in an attempt to compare with the normalized loads from the shear frame test. Unfortunately, good repeatability of the bias extension test is difficult to obtain. From observations of the shear frame and bias extension tests, this variability is most likely due to initial local misalignment of yarns in the test piece. Because a great deal of slip can occur in the bias extension tests, it is not practical to mechanically condition the specimens to obtain better repeatability, as was done with the shear frame specimens. In addition, it is difficult to compare the results in Figures 7 and 10 directly, because it is unclear how the load values should be normalized.   However, if the bias extension loads are normalized by the pure shear area, the resulting values are an order of magnitude higher than those for the shear frame test on the same material.   Even if the regions experiencing partial shear are included, approximately doubling the area, the bias extension loads will still be significantly higher at shear angles above 20°.   This is because of the contribution of tensile loading of high modulus yarns in the bias extension test.   The numerical modeling of the material captures this difference well, as will be discussed in the following section.

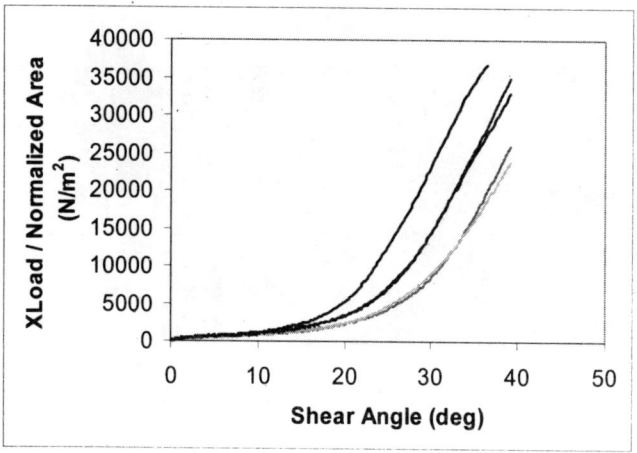

Figure 10. *Normalized response of plain weave fabric in bias extension test (specimen dimensions: 190.5mm x 95.3mm (LxW), room temperature, crosshead rate: 254 mm/min, shear rate: ranges from 0.05–0.10 rad/sec)*

6.   Numerical approach for material characterization

A multi-scale approach shown in Figure 11 is proposed to characterize the material behavior of the plain weave composite. The homogenization method is first employed to obtain the effective elastic constants of a single fiber composite based on the properties of the constituent phases [PEN 00, PEN 01]. The composite is E-glass fiber and polypropylene (PP) resin and their properties under room temperature are listed in Table 1. The homogenized elastic constants for the fiber yarns are shown below the table.

Table 1. *Material Properties of E-glass/PP Composite*

Property	Unit	E-Glass	Polypropylene
Tensile Modulus	GPa	73.1	1–1.4
Poisson's Ratio	–	0.22	0.3
Axial Shear Modulus	GPa	30.19	
Density	Kg/m³	2540	900

$$E_l = 51.92GPa, \quad E_t = 21.97GPa, \quad v_{lt} = 0.2489$$
$$v_{tt} = 0.2143, \quad G_{lt} = 8.856GPa, \quad G_{tt} = 6.250GPa$$

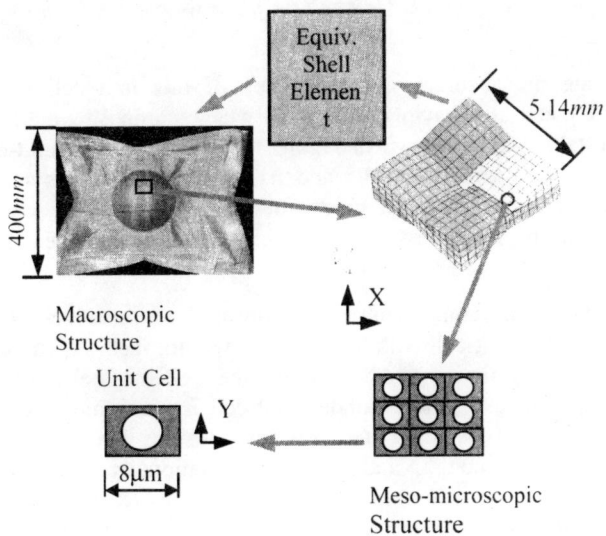

Figure 11. *Multi-scale approach for material characterization*

A unit cell modeled by 3D 8-node continuum elements, as shown in Figure 12, is then designed to represent the periodic patterns in the plain weave composite. The unit cell is defined by four sinusoidal curves in terms of yarn width w, yarn spacing s, and fabric thickness h. The characteristic values of w, s and h are 3.72mm, 5.14mm and 0.39mm, respectively. The effective elastic constants obtained from the homogenization method are imposed on each element. The pin-jointed net idealization is assumed along the four corner lines of the unit cell, which means each fiber yarn can rotate freely along the corner line with no relative slippage over the contact point of the fiber yarns. Contact conditions are prescribed between the possible interlacing surfaces of the fiber yarns under loading. A Coulomb friction model is used and the friction coefficient on the contacting surface is assumed to be 0.05 as measured from experiment.

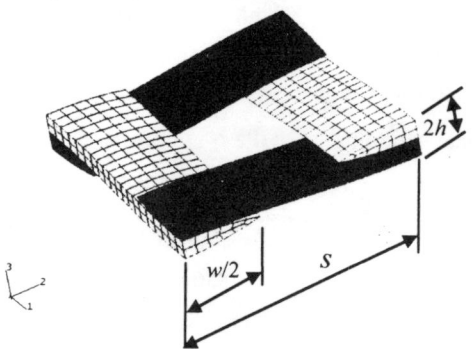

Figure 12. *Unit cell for plain weave composite*

To investigate the accuracy and efficiency of this unit cell (Figure 12) in representing the material behavior of the plain weave composite, another expanded unit-cell was established as shown in Figure 13. This expanded unit-cell encloses four of the previous small unit cells and has strict geometric symmetry. Since trellising is the main deformation mode for the plain weave composite during stamping, only the trellising test was conducted on this expanded unit-cell. The deformed shape is shown in Figure 14.

Figure 15 shows the comparison of the normalized force versus shear angle curve of the expanded unit-cell with that of the small unit-cell. As can be seen, both unit-cells provide very similar predictions for the trellising behavior of the plain weave composite. However, the expanded unit-cell is much more computationally costly, about 20 times of the small one. This demonstrates that the small unit-cell is accurate and efficient enough, and more computationally economical, for the process of determining the material behavior of the plain weave composites. Hence, in the following analysis, only the small unit cell was used to obtain the effective material constants of the plain weave composite.

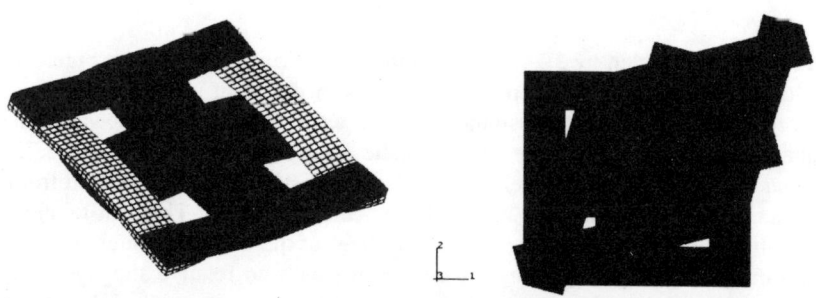

Figure 13. *Expanded unit-cell*   **Figure 14.** *Deformed unit-cell in trellising*

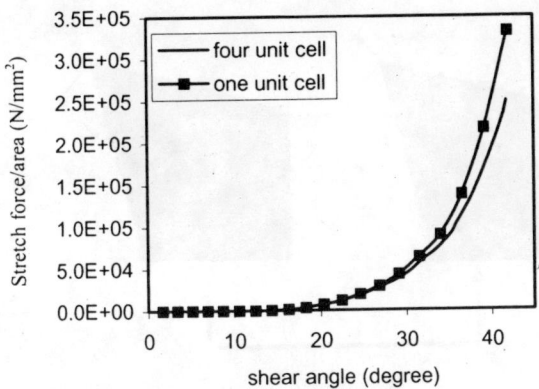

Figure 15. *Comparison of trellising results*

Numerical tests, such as uni-axial extension and bi-axial extension and trellising, are imposed on the unit cell. Large deformation, large rotation and geometric non-linearity are considered. Boundary conditions are modified to reflect the periodic boundary conditions of the macrostructure under different loading conditions. The corresponding load-displacement curves are obtained from each test. A four-node shell element with the same outer size as the unit cell is built to obtain the effective mechanical stiffness tensor of the plain woven composite. The composite is assumed to be and will remain[1] orthotropic during evolution. The corresponding constitutive equation is given as:

$$\begin{Bmatrix} \sigma_1 \\ \sigma_2 \\ \sigma_{12} \end{Bmatrix} = \begin{bmatrix} \dfrac{E_1}{1-\nu_{12}\nu_{21}} & \dfrac{\nu_{12}E_2}{1-\nu_{12}\nu_{21}} & 0 \\ \dfrac{\nu_{12}E_2}{1-\nu_{12}\nu_{21}} & \dfrac{E_2}{1-\nu_{12}\nu_{21}} & 0 \\ 0 & 0 & G_{12} \end{bmatrix} \begin{Bmatrix} \varepsilon_1 \\ \varepsilon_2 \\ \gamma_{12} \end{Bmatrix} \tag{1}$$

where

$$\frac{E_1}{\nu_{12}} = \frac{E_2}{\nu_{21}} \tag{2}$$

[1] A non-orthogonal material model has been recently developed and was submitted to 'Composites Part A' for review [XUE 01].

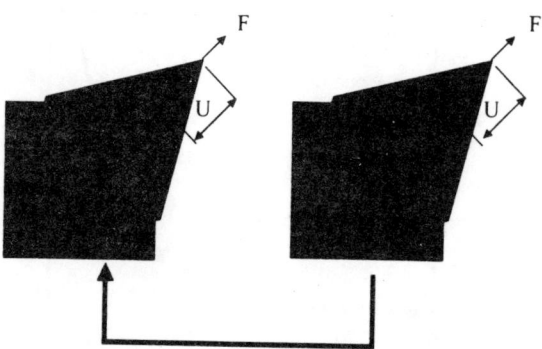

Figure 16. *Matching force and displacement results to obtain effective shear modulus*

The mechanical stiffness tensor is numerically determined by matching the force-displacement curves of the unit cell to those of the corresponding single shell element under trellising (Figure 16) and simple extension tests. The details can be found in [PEN 01]. Therefore, the effective elastic shear modulus G_{12}, tensile modulus E_i and Poisson's ratio v_{12} are obtained as:

$$G_{12} = \begin{cases} 1.7444\gamma_{12} + 4.96 \times 10^{-2} & \gamma_{12} \leq 0.207 \\ C_3\gamma_{12}^3 + C_2\gamma_{12}^2 + C_1\gamma_{12} + C_0 & \gamma_{12} > 0.207 \end{cases}$$

where

$$C_3 = 11.3103, \quad C_2 = 22.303,$$
$$C_1 = -8.9429, \quad C_0 = 1.2059.$$

$$(3)$$

$$E_i = \begin{cases} 10\gamma_{12}^2 + 0.1 & \varepsilon_i < 0 \\ A_3\varepsilon_i^3 + A_2\varepsilon_i^2 + A_1\varepsilon_i + A_0 + \\ |\varepsilon_1 - \varepsilon_2| \exp(B_1\varepsilon_i + B_0) & 0 \leq \varepsilon_i < 0.04 \end{cases}$$

$$(4)$$

where

$$A_3 = 7.60033 \times 10^7; \quad A_2 = -9.81521 \times 10^6;$$
$$A_1 = 4.35522 \times 10^5; \quad A_0 = 4.64523 \times 10^3;$$
$$B_1 = -87.34587; \quad B_0 = 13.02737$$

and

$$v_{12} = \begin{cases} 0.02 & \varepsilon_1 < 0 \\ 346.63\varepsilon_1^2 - 25.96\varepsilon_1 + 0.544 & 0 \le \varepsilon_1 < 0.04 \end{cases}$$

$$(5)$$

The obtained material models are validated on two scales. First, the material models are integrated into the single shell element used in the material characterization. A combined extension and shear test is conducted on both the unit cell and the shell element. Figure 17 shows the deformed shape of the unit cell under mixed extension and shear deformation. The obtained total reaction force at point A in Figure 17 versus displacement curve is shown in Figure 18 with circles. The solid line in Figure 18 represents the results from the shell element. It can be seen that good agreement is obtained between the plain weave unit cell model and the equivalent shell element.

A

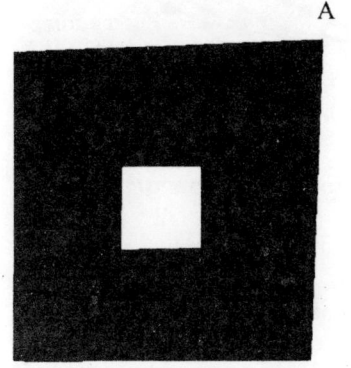

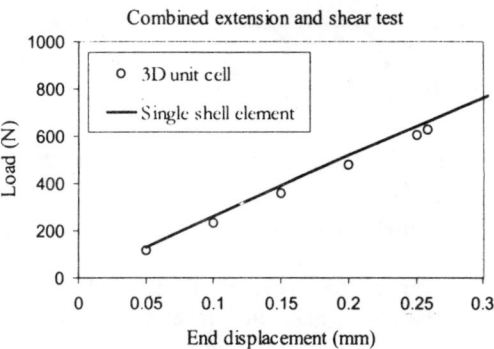

Figure 17. *Deformed unit cell under combined extension and shear*

Figure 18. *Comparison of reaction force in combined extension and shear test*

Secondly, a multi-element FEM model is used to simulate the experimental bias extension tests conducted on a composite patch with size of 190.5*mm* × 95.8*mm* as illustrated in Figure 19. The comparison of force versus displacement curves is shown in Figure 20. The FEM simulation result is shown in Figure 20 as a solid line with square symbols. As can be seen, the experimental results are quite scattered, as discussed previously. Nevertheless, the simulation result with shell elements provided a reasonable and, as expected because of slip in the experiment, slightly higher load prediction.

Another validation is conducted by simulating the experimental trellising tests shown in Figure 2 with shell elements based on the above material model. A FEM simulation is implemented on a composite patch with a size of 240*mm* × 240*mm*. The composite patch is discretized by 10 × 10 identical shell elements. Boundary conditions are prescribed to reflect the behavior of the experimental trellising shear frame. The obtained reaction force is normalized by the patch area and is shown in Figure 7 with circles to compare with experimental shear frame data. As can be seen, a good agreement is obtained by using the present material model.

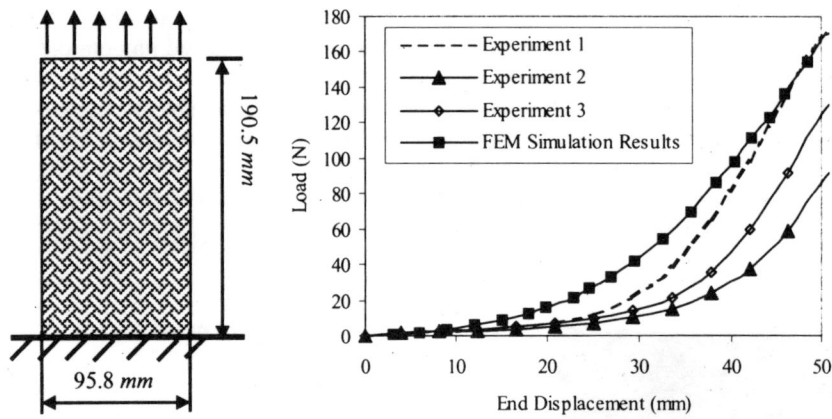

Figure 19. *Bias extension*   **Figure 20.** *Comparison of bias extension tests*

7.   Conclusion

Over the past several years, a substantial body of experimental and modeling data has been generated from studies of composite sheet forming. The research efforts are at a critical point where benchmarking will lead to major advances in our understanding of the strengths and limitations of existing experimental and modeling approaches. The results discussed in this paper suggest that methods such as mechanical conditioning, area normalization, and multi-scale modeling should be considered in order to generate valid, instructive comparisons between various researchers, as well as between theory and experiment.

Acknowledgements

The support of the National Science Foundation, Division of Design, Manufacture, and Industrial Innovation (DMI-9900185, DMI-9800483) and the Ford Motor Company is gratefully acknowledged. The authors would also like to thank the University of Nottingham and the University of Orleans (ESEM/ENSAM) for productive discussions and collaboration.

References

[BOI 97]   BOISSE, P., BORR, M., BUET, K., and CHEROUAT, A., "Finite Element Simulations of Textile Composite Forming Including the Biaxial Fabric Behaviour", *Composites Part B: Engineering* v 28B n 4, p 453–464, 1997.

[FRI 98]   FRIEDRICH, K. and HOU, M., "On Stamp Forming of Curved and Flexible Geometry Components from Continuous Glass Fiber/Polypropylene Composites", *Composites Part A*, 29A:217–226, 1998.

[LUS 00a]   LUSSIER, D.S. and CHEN, J., "The Investigation of Thermal Effects in Thermoforming of Composites", *15th Ann Tech Conf of the American Society for Composites*, College Station, TX, Sept 2000.

[LUS 00b]   LUSSIER, D.S. and CHEN, J., "Shear Frame Standardization for Stamping of Thermoplastic Woven Fabric Composites", *32nd Ann Intl SAMPE Tech Conf*, Boston, MA, November 2000.

[MCG 98]   MCGUINNESS, G. B. and Ó'BRÁDAIGH, C.M., "Characterization of Thermoplastic Composite Melts in Rhombus-Shear: The Picture Frame Experiment," *Composites Part A*, 29A:115–132, 1998.

[MOH 00]   MOHAMMED, U., LEKAKOU, C., DONG, L., and BADER, M.G., "Shear Deformation and Micromechanics of Woven Fabrics," *Composites Part A*, 31:299–308, 2000.

[PEN 00]   PENG, X.Q. and CAO, J., "Numerical Determination of Mechanical Elastic Constants of Textile Composites", *15th Ann Tech Conf of the Amer Soc for Composites*, College Station, TX, Sept 2000.

[PEN 01]   PENG, X.Q. and CAO, J., "A Dual Homogenization and Finite Element Method for Material Characterization of Textile Composites", accepted for publication by *Composites Part B*, 2001.

[SOU 00]   SOUTER, B.J., LONG, A.C., ROBITAILLE, F., and RUDD, C.D., "Modelling the Influence of Reinforcement Architecture on Formability," *9th Eur Conf Comp Matl*, Brighton, UK, June 2000.

[WIL 99]   WILKS, C.E., RUDD, C.D., LONG, A.C., and JOHNSON, C.F., "Rate Dependency During Processing of Glass/Thermoplastic Composites", *Proc. 12th Intl Conf Comp Matls (ICCM-12)*, Paris, July 1999.

[XUE 01]   XUE, P., PENG, X.Q., and CAO, J. "A Non-orthogonal Constitutive Model for Characterizing Woven Composite", submitted to *Composites Part A*.

Chapter 6

Characterization and Modeling of Fabric Deformation During Forming of Textile Composites

Andrew Long
School of Mechanical, Materials, Manufacturing Engineering and Management,
University of Nottingham, UK

1. Introduction

Textile composites, consisting of a textile reinforcement within a thermoplastic or thermoset polymer matrix, can be processed via a number of techniques. Components can be produced directly by forming of thermoset or thermoplastic plies [BAT 97]. For thermosets, the fibres are combined with a partially cured polymer to form a pre-impregnated sheet (prepreg), with an appropriate temperature and pressure cycle used to fully cure the component after forming. Thermoplastic composites can be supplied in either fully impregnated forms or as partially impregnated materials such as commingled or powder impregnated fabrics. These materials are processed by heating to above the polymer melt temperature, and then stamping/consolidating by application of pressure. Cooling usually takes place during the consolidation phase, as molding tools are held at a temperature significantly below the polymer melting point. Alternatively, liquid composite molding (LCM) processes such as resin transfer molding can be used, in which a dry reinforcement is impregnated with a thermosetting resin, which then cures to form a rigid composite [RUD 97]. The reinforcement is usually formed to the component shape in a separate operation to produce a textile preform.

Whatever the material and process of choice, a forming operation is required to convert the two-dimensional layers (plies) into the required three-dimensional geometry. For fabrics based on orthogonal yarns (or tows), a number of deformation mechanisms may occur during this operation (as described for example by Potter [POT 79]. Within individual layers intra-ply (in-plane) shear, corresponding to rotation of tows about their crossovers, is considered to be the dominant deformation mechanism as very high strains can be achieved in the bias direction (ie. 45° to the fibres) at relatively low applied force. Inter-tow slip can also occur, and may be significant when forming to tight radii where the intra-ply shear angle varies significantly [WAN 98]. For multi-layer components, inter-ply slip is required to accommodate curved surfaces. Tensile forces along the fibres may lead to fibre straightening for woven fabrics, whilst compressive forces can cause buckling leading to wrinkling of the fabric. Experimental measurements suggest that wrinkling occurs at a limiting degree of intra-ply shear known as the locking angle, typically ranging from 20° to 65° [BRE 96, PRO 97, WAN 98, MOH 00]. This value is considered important when modeling fabric forming, as it is used to identify areas of wrinkling which would lead to a defective component.

This paper will describe the intra-ply shear behaviour of dry textiles during the manufacture of preforms for LCM. In section 2, the shear compliance (shear force versus shear angle) and locking angle will be measured for a range of reinforcements including woven and warp knit, non-crimp fabrics. In section 3 this data will be used to develop an iterative model for forming/draping of textile composites, which minimizes fabric shear energy to account for resistance to intra-ply shear. This will be shown to provide more accurate results than obtained using the traditional modeling approach based on a geometric mapping, whilst the associated computation times are significantly lower than those required for non-linear finite element analysis. Results from a number of forming experiments will be presented to validate the iterative model.

2. Characterization of intra-ply shear

In previous studies, resistance to intra-ply shear has been characterized using two approaches. Uniaxial extension of relatively wide samples in the bias direction is favoured by a number of researchers [SKE 76, POT 79, WAN 98], as the testing procedure is relatively simple. However, the deformation field within the sample is non-uniform, with maximum shear observed in the central region and a combination of shear and inter-tow slip observed adjacent to the clamped edges. In addition, the shear angle cannot be obtained directly from the crosshead displacement, so that the test must be monitored visually to measure deformation. An alternative is the picture-frame shear test [CAN 95, BRE 96, LON 97, PRO 97, MOH 00], in which the fabric is clamped within a frame hinged at each corner, with the two diagonally opposite corners displaced using a mechanical testing machine. Although this test may be sensitive to small variations in material alignment, it is used in this study as it produces uniform shear deformation (if performed with care).

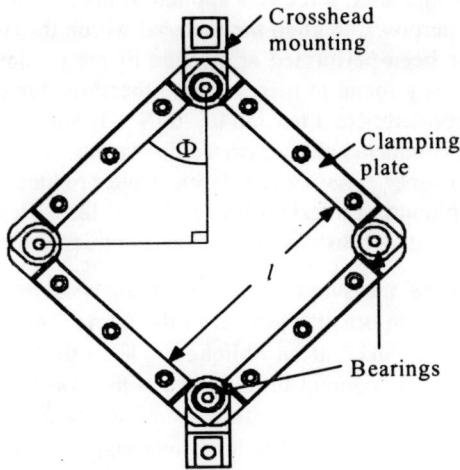

Figure 1. *Schematic of the picture-frame shear rig. The distance between the bearings (l) was 145 mm, whilst the clamping length was 115 mm so that the initial area of the sample was 13225 mm²*

The picture-frame shearing equipment used is illustrated in Figure 1. Crimped clamps were used to ensure that the fabric did not slip from the grips during testing. The apparatus was operated using a Hounsfield mechanical testing machine, which monitored both load and displacement during the experiment. The results were converted into shear force versus shear angle using the following relationships. The shear force (F_s) can be obtained from the measured force in the direction of extension (F_x) using:

$$F_s = \frac{F_x}{2\cos\Phi} \tag{1}$$

where the frame angle Φ can be determined from the crosshead displacement D_x and dimensions of the shear frame using:

$$\Phi = \cos^{-1}\left[\frac{1}{\sqrt{2}} + \frac{D_x}{2l}\right] \tag{2}$$

The shear angle is given by:

$$\theta = \frac{\pi}{2} - 2\Phi \tag{3}$$

A pre-tensioning rig was used to position dry fabrics within the picture-frame. This consisted of a frame within which fabric samples were clamped prior to mounting within the picture frame. Two adjacent sides of the pre-tensioning frame were hinged, to which a measured force was applied to impart tension to the fabric. This device served two purposes, to align the material within the rig and to enhance repeatability. Tests have been performed at a range of pre-tension levels, with an increase in tension generally found to increase both the shear force and the locking angle. For the results reported here a tension of 200N was applied to fibres in each direction. All tests were conducted at a crosshead displacement rate of 100 mm/min, although experiments at other rates for dry fabric have produced almost identical results [SOU 01]. A minimum of 6 tests were conducted for each fabric, with error bars produced using the student t-distribution at 90% confidence limit.

Initial tow widths were measured using image analysis from digital images obtained using a video camera oriented normal to the fabric. The locking angle was measured with the camera placed at an oblique angle to the plane of the fabric. Samples were marked with horizontal lines, which were observed to buckle when wrinkling occurred. A range of fabric reinforcements was tested, including woven fabrics and non-crimp fabrics (NCFs). The latter materials consist of aligned fibres held together using a polyester stitching thread, produced using a warp-knitting process (see Figure 2). All fabrics consisted of glass fibre tows. Descriptions of the fabrics tested are given in Table 1, which also includes locking angles averaged for at least four samples.

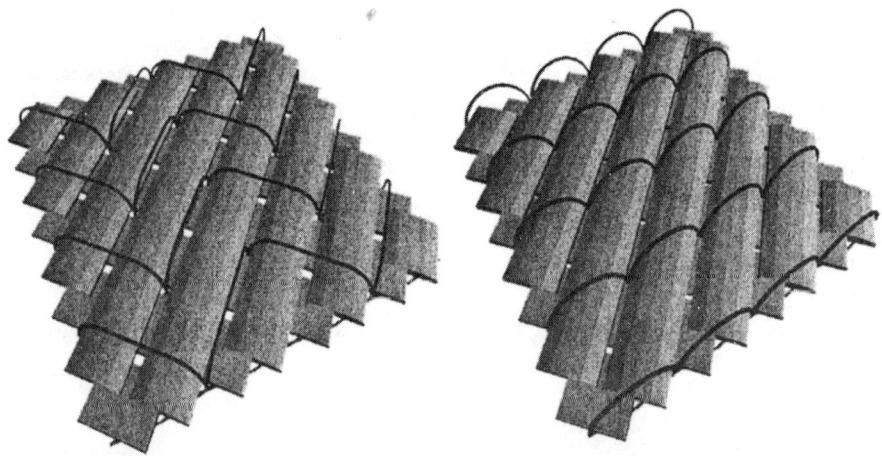

Figure 2. *Schematic representations of ±45° non-crimp fabrics. Left – Tricot 1&1 warp-knit (Ebx936); right – Pillar warp-knit (Ebx318)*

Table 1. *Woven and non-crimp fabric reinforcements characterized in shear*

Fabric ID	Style	Tow Angles	Surface Density (g/m²)	Tow Pitch (mm)	Tow Width (mm)	Tow Width/ Pitch	Locking Angle (s.d.)
P150	Plain weave	0/90	147	2.04	1.22	0.60	68° (-)
P800	Plain weave	0/90	800	4.01	3.09	0.77	60° (1.4)
S800-01	4 harness Satin	0/90	800	0.68	0.62	0.91	55° (1.7)
S800-02	4 Harness Satin	0/90	788	3.70	2.80	0.76	61° (1.6)
T800	2:2 Twill weave	0/90	790	6.30	4.40	0.70	62° (1.7)
Ebx936	Tricot 1&1 NCF	±45	936	0.82	0.70	0.86	62° (-)
Ebx318 *	Pillar NCF	±45	318	0.62	0.47	0.77	37° / 64° (-)

* Locking angle for this material depends on direction of testing. Value for fabric sheared parallel to stitching is significantly lower than value when sheared perpendicular to stitch.

As shown in the table, for woven fabrics shear angles of at least 55° were achieved before wrinkling was observed. Studies relating to prediction of fabric locking would appear to be relatively limited. Skelton [SKE 76] proposed that a lower limit may be determined when adjacent tows come into contact (ie. tow width is equal to tow pitch). Application of this model to the fabrics described in Table 1 is of limited use, as the predicted locking angles are typically 20 ° lower than those measured. For these materials fabric locking occurred some time after adjacent tows came into contact. For the light plain weave (P 150), the relatively large spacing between the tows permitted large shear angles, and in fact only three of the six samples wrinkled before the end of the shear test. The locking angle was lower for the heavier plain weave (P 800). Woven fabrics can be ranked for locking angle from the ratio between initial tow width and pitch, with a low ratio indicating a high locking angle. The same procedure appears to apply to non-crimp fabrics, although ratios here are generally higher than those for woven fabrics with similar locking angles. In addition for these materials the locking angle can depend on the direction of testing, with very different behaviour observed when the fabric is sheared parallel or perpendicular to the stitching thread (as discussed below).

Typical shear compliance curves for woven fabrics are shown in Figure 3. This Figure compares the behaviour of fabrics with similar surface densities but different fibre architectures. The plain weave (P 800) requires the highest force to achieve a particular shear angle, whilst the twill weave (T 800) is the most compliant. This is related to the fact that the ratio between tow width and pitch is greatest for this material. However, for all fabrics tested two distinct regions may be identified. The initial resistance to shear is relatively low, and is likely to be caused by friction at the tow crossovers. Once adjacent tows come into contact, the resistance increases significantly as the tows are compressed together. This is the region where wrinkling is usually observed. If the test were continued, the curve would tend towards an asymptote corresponding to maximum tow compaction (*i.e.* close packing of the filaments within each tow).

Also included in Figure 3 are predictions obtained using a mechanical model (described in detail elsewhere [LON 00]). This is based on a generalized geometric model for textile reinforcements, which describes yarn paths for both woven and non-crimp fabrics. Deformation of the fibre architecture during in-plane shear is also represented, allowing tow cross-sections to be modified when lateral contact occurs. Tow contact areas are calculated, over which Coulomb friction is assumed to determine the local torque contribution. Contact force is calculated by integrating the tow pressure over the contact area, with a semi-empirical model used to relate pressure to tow volume fraction and fibre modulus. The total torque over the specimen is used to calculate the intra-ply shear force. Comparison of predictions with experimental results illustrates that the model is capable of representing the effects of fibre architecture on compliance.

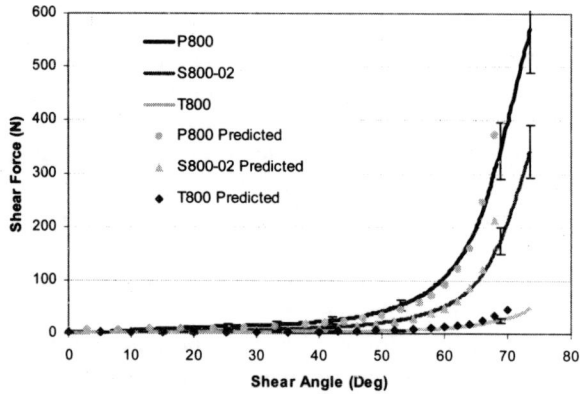

Figure 3. *Experimental and predicted shear compliance curves for woven fabrics with different architectures: Plain weave (P800), 4-harness satin weave (S800-02), and 2:2 Twill weave (T800)*

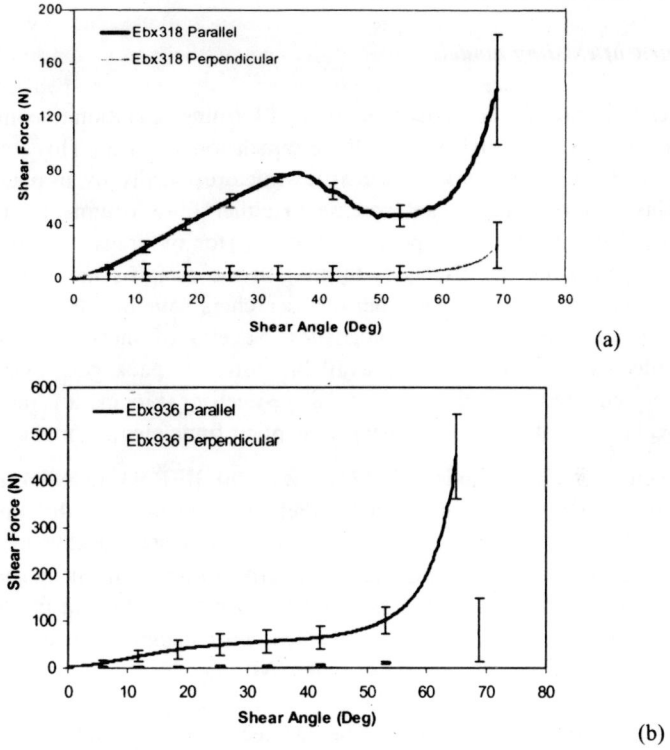

Figure 4. *Shear compliance curves for +/–45° non-crimp fabrics tested parallel and perpendicular to the stitch. (a) Tricot 1&1 warp-knit (Ebx936). (b) Pillar warp-knit (Ebx318)*

Figure 4 shows typical shear compliance curves obtained for non-crimp fabrics with both a tricot and pillar warp-knitted threads. The tricot warp-knit results in a "zig-zag" stitching thread pattern, whereas the pillar warp-knit is similar to a chain stitch (as illustrated in Figure 2). In both cases it is apparent that the compliance is lower when the fabric is sheared parallel to the stitching direction. Testing in this direction results in a tensile strain within the stitch, which causes an increase in shear force. The effect is more pronounced for the pillar warp-knit, where testing parallel to the stitch results in a linear increase in force until the stitching thread snaps. After this point the force is reduced until inter-tow compaction occurs. A predictive model for the shear compliance of non-crimp fabrics is currently being developed. The directionality exhibited by non-crimp fabrics during shear can result in non-symmetric fibre patterns during draping, as described in the following section.

3. Fabric forming simulation

3.1. *Review of existing models*

Processing of textile composites involves a forming operation to convert the two-dimensional base material into a three-dimensional form. This results in re-orientation of the fibres within the textile structure usually by in-plane (intra-ply) shear. This in turn leads to an increase in either fibre volume fraction (for dry textiles used in LCM) or component thickness (for prepregs). Consequently it is important to consider the effects of forming at the design stage. To anticipate the deformed fibre architecture, a number of researchers have developed simulations of draping or forming for textile composites. Several of these models have been implemented within commercially available software packages. Two approaches have been adopted, based on either a geometric (kinematic) mapping, or a mechanical representation solved using an explicit finite element method.

Geometric/kinematic models [ROB 81, WES 90, BER 93, LON 94] represent the fabric structure as a pin-jointed net, which is mapped onto the surface of the component/forming tool by assuming that tow segments are able to shear at the joints (tow crossovers). A unique draped pattern can be obtained by specifying two intersecting tow paths, referred to as generators, on the surface of the forming tool. The remaining tows are positioned using a geometric mapping. Several strategies are available for specifying the generator paths, with geodesic or projected paths used typically. Correct specification of the generators is critical, as these will determine the positions of all remaining fibres. The intersection point should correspond to the starting point for draping, with the highest point of the tool or the initial point of contact between the tool and ply typically used. The kinematic approach provides a very fast solution, with run times typically less than 10 seconds. However, this approach is unable to differentiate between materials other than in the specification of the locking angle, which is used to indicate possible areas of wrinkling. Consequently an identical fibre pattern is obtained, regardless of variations in material forming characteristics or processing technique.

The mechanical approach involves simulation of the entire forming process over a number of time steps. At each stage equilibrium equations are solved, usually using an explicit finite element technique. This approach has been applied to both dry fabrics [BLA 97, BOI 97, DON 00, BIL 01] and thermoplastic or thermoset prepregs [OBR 91, LUC 98, CHE 99]. Provided that accurate processing property data are specified, it is possible to represent material specific behaviour. In addition, ply/tool and ply/ply friction can be modelled. Ply wrinkling is anticipated by the occurrence of in-plane compressive forces, rather than the specification of a locking angle. However, this approach is time consuming, both in terms of CPU time and in the collection of the large set of materials data required.

3.2. Iterative draping simulation

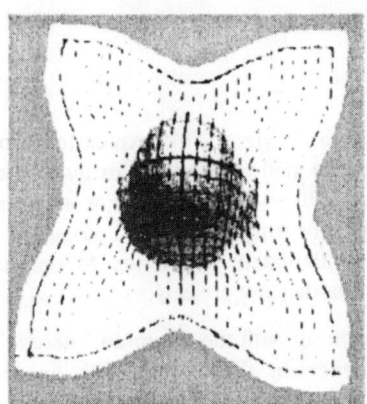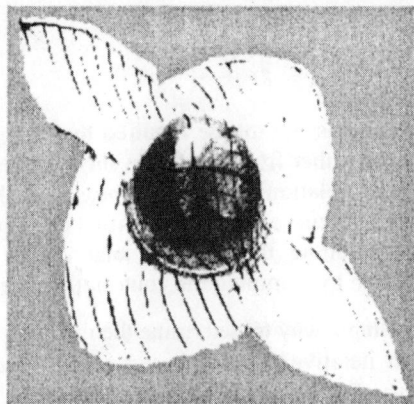

Figure 5. *Hemispherical preforms produced using two glass fabrics. Left: 4 harness satin weave (S800-02). Right: ±45° pillar warp-knit (Ebx318)*

Generally, the geometric/kinematic approach to drape simulation works well for symmetric shapes draped with balanced materials, but for other cases the approach is less satisfactory. This is demonstrated in Figure 5, which shows the fibre patterns for hemispherical preforms produced using a woven reinforcement (S800-02) and a non-crimp fabric (Ebx 318). The latter material contains glass tows oriented at ±45° to a polyester stitching thread. Samples were produced using matched tooling mounted in a mechanical testing machine, with the fabric supported around the periphery by a simple blank-holder (a ring applying a constant pressure). Deformation was observed by drawing lines parallel to one group of fibres on each surface. These were visible as a grid after the components were impregnated with a thermosetting resin. Despite the component symmetry, two very different fibre patterns were obtained. The warp-knitted fabric was restricted in quadrants sheared parallel to the stitching thread.

The use of a mechanical forming simulation may allow consideration of the directionality exhibited by non-crimp fabrics. However, this would be achieved at the expense of increased computation time. Whilst this may be considered appropriate for thermoplastic prepregs, where material properties are both rate and temperature dependent [WIL 99], an alternative approach may be sufficient for dry fabrics. The approach presented here is based on the use of a geometric mapping algorithm within an iterative scheme. The energy required to produce each mapping is calculated, with the mapping resulting in the lowest energy assumed to represent the actual behaviour of the fabric. For dry fabrics the deformation energy is the sum of several components related to the mechanisms identified in Section 1. At present the model considers only intra-ply shear, as this is thought to be most indicative of the effect of fabric construction on deformation. The fabric shear energy (work done during shearing, U_s) can be calculated simply from the area under the torque-shear angle curve:

$$U_s(\theta) = \int_0^\theta T(\gamma)d\gamma \qquad\qquad [4]$$

where $T(\gamma)$ is the torque required to reach a shear angle γ. This expression can be evaluated either from the fabric shear model mentioned in Section 2 or by fitting an empirical relationship to the measured shear compliance curve. For non-crimp fabrics, two curves may be specified to represent shearing parallel and perpendicular to the stitching thread. The total energy is calculated within the fabric drape simulation by summing the contribution at each node (tow crossover).

A simple way to determine the mapping resulting with the minimum energy is to use an iterative scheme based on the two generator fibre paths. This approach involves finding the two intersecting paths that result in the lowest total energy. A Hooke and Jeeves minimization method is used [BUN 84], where the generator path is defined one step at a time from a user-defined starting point. Each successive set of nodes is optimized by iterating the generator path angle to achieve the minimum increase in shear energy. For reasons of computational efficiency, this technique is preferred to a global minimization algorithm in which the entire generator path is modified at each stage of the process. However, as nodes in contact with the tool are subject to additional constraints due to friction, the present approach is likely to be reasonably accurate for automated forming operations.

The results of this minimization algorithm are shown in Figure 6, which shows predicted fibre patterns for a hemisphere using shear data for a 4 harness satin weave (S800-02, Figure 3) and a ±45 pillar warp-knit (Ebx 318, Figure 4b). The predictions show a good correlation with the experimental results (Figure 5). These results are particularly encouraging, as a conventional geometric draping simulation would predict identical fibre patterns for these materials.

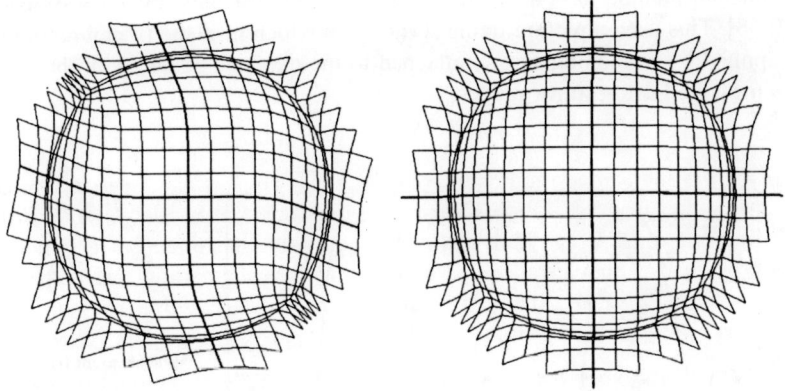

Figure 6. *Results from drape simulation over a hemispherical tool using the shear data for S800-002 (left) and Ebx318 (right)*

3.3. Validation for nose cone

To validate the model for a more challenging component, a jet engine nose cone geometry was used. This component required significant depth of draw, preventing the majority of textiles from being formed without defects. A ±45° tricot 1&1 warp-knit (Ebx 936) was used, as this could be formed to the geometry without significant wrinkling. Preforms were produced by forming of reinforcement layers over a male tool, with the material supported in an adjustable blank-holder, operated using a mechanical testing machine. This allowed preforms to be produced in a repeatable manner. The pressure supporting the fabric was varied using eight spring-loaded petals around the blank-holder. Clamping pressure was varied by adjusting the spring compression for each petal. Results presented here were obtained with either a uniform *low pressure* (corresponding to 4 N force for each spring) or a uniform *high pressure* (126 N). Fabric layers were marked with an orthogonal grid, and shear angles were determined by measuring the relative position of grid points using digital vernier callipers. The results represent the shear angle measured around the cone at a fixed distance from the tip.

Figure 7 compares predicted and measured fabric shear angles. For a uniform low clamping pressure, Figure 7(a), the correlation is reasonably good. In particular, the observed preferential shear behaviour in quadrants sheared perpendicular to the stitching thread is clearly illustrated. However, under high clamping pressure, Figure 7(b), the correlation is poor. In this case a more uniform fibre pattern was observed, so that maximum shear angles in each quadrant varied by less than 5 °. To represent changes in boundary conditions such as these, the effects of tensile forces should be included in calculation of the deformation energy. One simple approach may be to measure the shear compliance as a function of in-plane tension. A suitable experimental method to conduct such measurements was suggested by Breuer et al [BRE 96]. This used a picture frame shear rig in which in-plane (membrane) tension was applied by tensioning screws attached to the clamped edges, with the resulting strain measured using strain gauges.

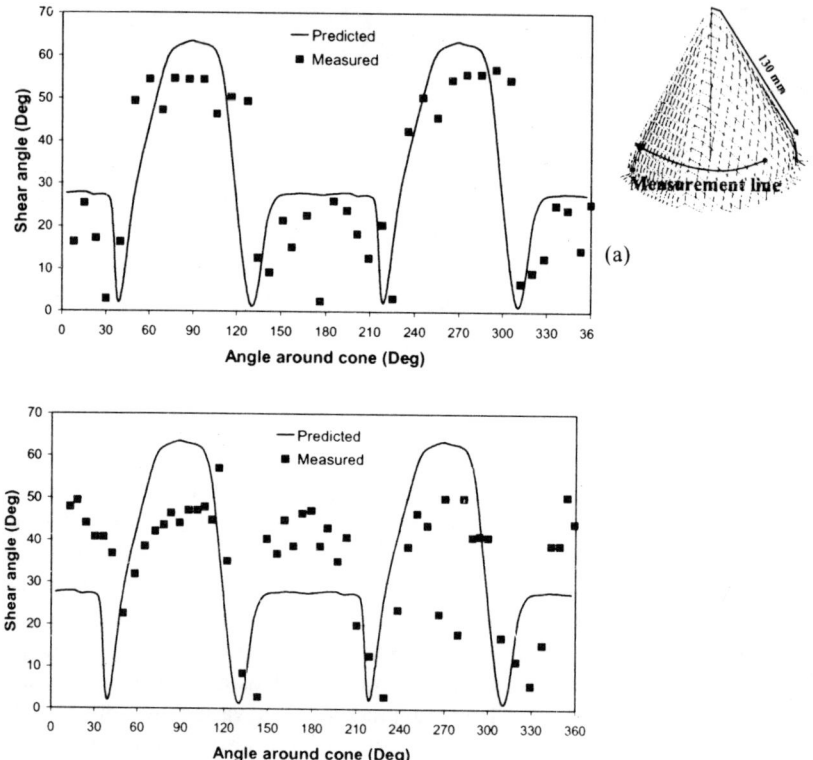

Figure 7. *Comparison of predicted and measured shear angles for Ebx936 around the nose-cone at 130mm from the tip. (a) Uniform low clamping pressure. (b) Uniform high clamping pressure*

3.4. *Validation for automotive transmission tunnel*

An automotive component was used to evaluate the iterative draping simulation for a component with no axial symmetry. This was based on a transmission tunnel for a high performance vehicle. Preforms were produced by hand lay-up over a male former. An arrangement of several discrete forming pads was developed to hold the fabric onto the component surface during lay-up. These assisted in the lay-up process and resulted in improved repeatability during preform manufacture.

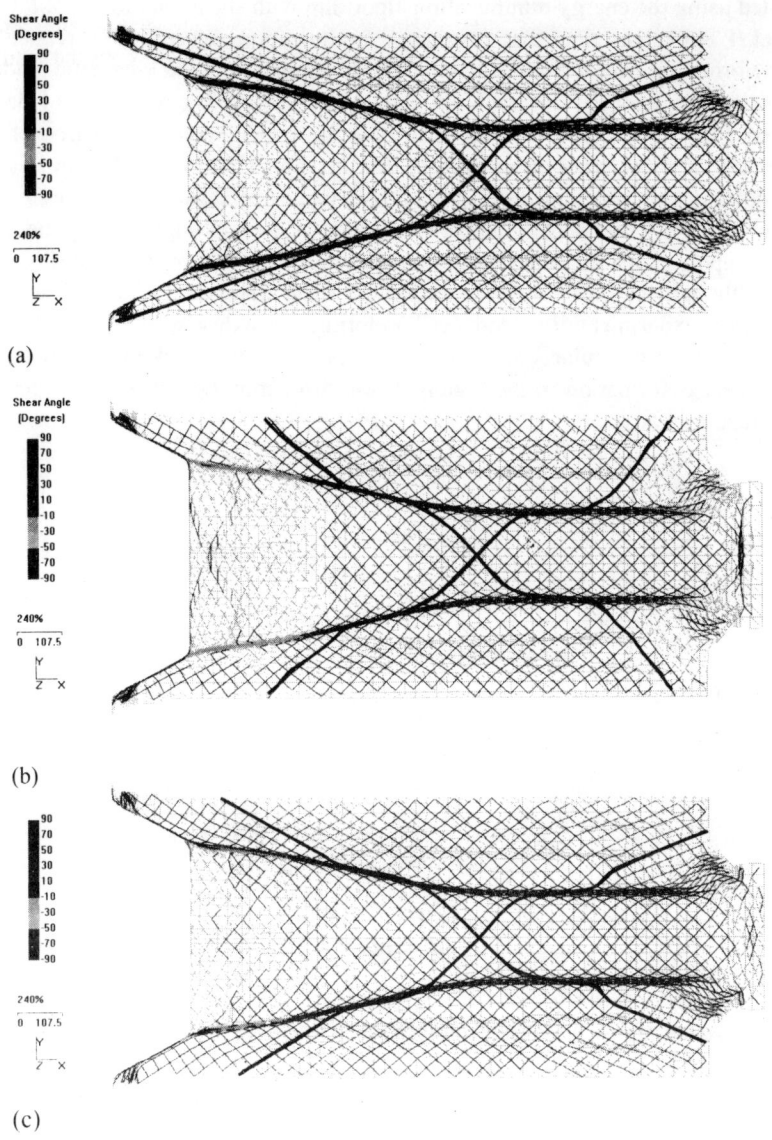

Figure 8. *Deformed fibre pattern for a prototype automotive component using the energy minimization algorithm. a) Geodesic generator paths. b) Shear data for a plain weave fabric. c) Shear data for a ±45° tricot stitched fabric*

Figure 8 compares predicted fibre patterns for this geometry obtained using a variety of techniques. Figure 8 a) was obtained using the purely kinematic algorithm, where generator paths were defined as geodesics. Figure 8 b) was generated using the energy minimization algorithm with shear data for a plain weave material (P 800). It is clear that the energy minimization algorithm has reduced the shear deformation over the surface significantly compared to the kinematic model; if the shear strain energies for each simulation are analysed a reduction of 35% is recorded. Figure 8 c) was based on shear data for a ±45 ° non-crimp carbon fabric, having similar shear compliance data to that for Ebx 936 (Figure 4a). In this case fabric deformation was increased in quadrants that were sheared in the preferential direction. Figure 9 compares predicted and measured shear angles along the length of the component for the latter material. The results agree over the majority of the length, although the model over-estimates the shear deformation at the rear of the tunnel. For experimentally produced preforms, wrinkles were present in this location. Darts (triangular cuts) were used to alleviate wrinkling, reducing the overall shear deformation in the region. These discontinuities were not represented in the drape analysis.

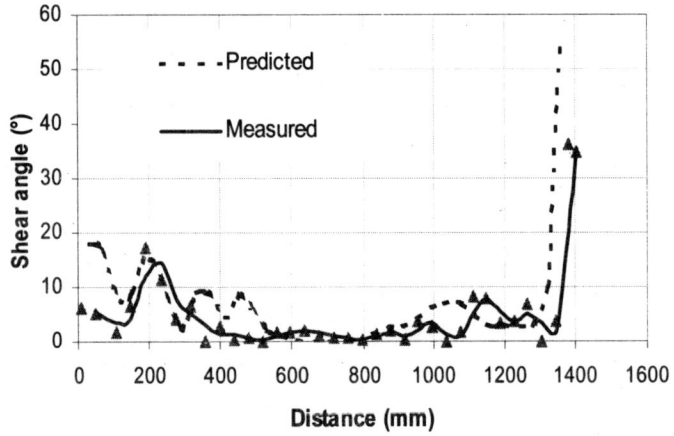

Figure 9. *Comparison of predicted and measured shear angles along the length of the transmission tunnel, measured on the centre line from left to right with reference to Figure 8 c)*

4. Conclusion

This paper has analysed the deformation mechanisms exhibited by dry textiles during the manufacture of polymer composites. Although a number of mechanisms exist, this study has focused on intra-ply (in-plane) shear, as this is generally accepted as the most important mechanism. Experimental characterization was performed using a picture-frame shear rig, with results obtained in terms of fabric locking angle and shear compliance. Both measures were shown to be a function of fibre architecture, with the ratio between tow width and spacing (pitch) shown to provide a convenient value for ranking of materials. For woven fabrics, the shear compliance curve consists of two regions, corresponding to inter-tow friction at low shear angles and lateral tow compaction at higher angles as the fabric approaches locking. Non-crimp fabrics exhibit similar compliance curves, although here a preferential shearing direction can be identified, with a far higher compliance (lower shear force) for deformation perpendicular to the warp-knit stitching thread.

Experimental evidence was presented to show that the deformation behaviour exhibited by non-crimp fabrics resulted in a non-uniform fibre pattern, which could not be predicted using a standard kinematic draping simulation. To account for this behaviour, shear compliance data was incorporated within an enhanced draping simulation, which used a geometric (pin-jointed) mapping within an iterative algorithm to minimize shear energy. This approach also eliminates the need for the user to estimate the paths of two generator fibres on the component surface, potentially improving the accuracy for symmetric components draped with balanced fabrics. Validation experiments for a range of materials and geometries showed the model to be reasonably accurate. In particular, the model is able to predict accurately the difference in deformed fibre patterns for alternative fabric styles. Some discrepency was observed when a large blank-holder pressure was used, suggesting that further developments are required to incorporate boundary conditions accurately within the model.

This modeling approach is more efficient than non-linear finite element analysis. Despite the relatively crude optimization algorithm employed, run times are typically several orders of magnitude lower than equivalent FE simulations. However, this model is unlikely to be as accurate as an FE simulation. For example the effects of tool/ply and ply/ply friction cannot be represented accurately without modeling the forming process sequentially. For textile composite component and process design, a pragmatic approach would be to use the iterative model developed here for initial component design and materials selection, with an FE simulation used to optimize the manufacturing process.

Acknowledgements

The author would like to acknowledge the work of a number of colleagues, in particular Professor Chris Rudd, Dr Francois Robitaille and Ben Souter. The following organizations are also thanked for their support: The Engineering & Physical Sciences Research Council, Brookhouse Patterns, Dowty Aerospace Propellers, Flemings Industrial Fabrics, Ford Motor Company, Rolls Royce.

References

[BAT 97] BHATTACHARYYA D., *Composite sheet forming*, Elsevier Science B.V., 1997.

[BER 93] BERGSMA O.K., "Computer simulation of 3D forming processes of fabric reinforced plastics", *Proc 9th Int Conf on Composite Materials*, Madrid, July 1993, IV: 560–567.

[BIL 01] BILLOËT J.L., CHEROUAT A.H., "A specific finite element model for composite fabric forming processes", *Proc 4th ESAFORM Conference on Material Forming*, Liège, April 2001, pp. 139–142.

[BLA 97] BLANLOT R., "Simulation of the shaping of woven fabrics", *Proc 11th Int Conf on Composite Materials*, Gold Coast, Australia, July 1997, Vol. 160–170.

[BOI 97] BOISSE P., BORR M., BUET K., CHEROUAT A., "Finite element simulations of textile composite forming including the biaxial fabric behaviour", *Composites Part B*, Vol. 28, 1997, pp. 453–464.

[BRE 96] BREUER U., NEITZEL M., KETZER V., REINICKE R., "Deep drawing of fabric-reinforced thermoplastics: wrinkle formation and their reduction", *Polymer Composites*, Vol. 17, 1996, pp. 643–647.

[BUN 84] BUNDAY B.D., *Basic optimisation methods*, Edward Arnold Ltd, London, 1984.

[CAN 95] CANAVAN R.A., MCGUINNESS G.B., O'BRAIDAIGH C.M., "Experimental intraply shear testing of glass-fabric reinforced thermoplastic melts", *Proc 4th Int Conf on Automated Composites*, Nottingham, 1995, pp. 127–138.

[CHE 99] CHEN, J., SHERWOOD J.A., "Stamping of co-mingled glass/thermoplastic fabrics", *Proc SAMPE_ACCE-DOE Advanced Composites Conference*, Detroit, Sept 1999, pp. 251–258.

[DON 00] DONG L., LEKAKOU C., BADER M.G., "Solid-mechanics finite element simulations of the draping of fabrics: a sensitivity analysis", *Composites Part A*, Vol. 31, 2000, pp. 639–652.

[LON 94] LONG A.C., RUDD C.D., "A simulation of reinforcement manufacture during the production of preforms for liquid moulding processes", *ImechE J Engineering Manufacture*, Vol. 208, 1994, pp. 269–278.

[LON 97] LONG A.C., RUDD C.D., BLAGDON M., JOHNSON M.S., "Experimental analysis of fabric deformation mechanisms during preform manufacture", *Proc 11th Int Conf on Composite Materials*, Gold Coast, Australia, July 1997, Vol. 238–248.

[LON 00] LONG A.C., SOUTER B.J., ROBITAILLE F., "A fabric mechanics approach to draping of woven and non-crimp reinforcements", *Proc 15th Tech Conf, American Society for Composites*, College Station, Texas, September 2000.

[LUC 98] DE LUCA P., LEFÉBURE P., PICKETT A.K., "Numerical and experimental investigation of some press forming parameters of two fibre reinforced thermoplastics: APC2-AS4 and PEI-CETEX", *Composites Part A*, Vol. 29, 1998, 101–110.

[MOH 00] MOHAMMAD, U., LEKAKOU, C., DONG, L., BADER, M.G., "Shear deformation and micromechanics of woven fabrics", *Composites Part A*, Vol. 31, 2000, pp. 299–308.

[OBR 91] O Brádaigh C.M., Pipes R.B., "Finite element analysis of composite sheet-forming process", *Composites Manufacturing*, Vol. 2, 1991, pp. 161–170.

[POT 79] Potter K.D., "The influence of accurate stretch data for reinforcements on the production of complex structural mouldings", *Composites*, July 1979, pp. 161–167.

[PRO 97] Prodromou A.G., Chen J., "On the relationship between shear angle and wrinkling of textile composite preforms", *Composites Part A*, Vol. 28, 1997, 491–503.

[ROB 81] Robertson R.E., Hsiue E.S., Sickafus E.N., Yeh G.S.Y., "Fiber rearrangements during the moulding of continuous fiber composites", *Polymer Composites*, Vol. 2, 1981, pp. 126–131.

[RUD 97] Rudd C.D., Long A.C., Kendall, K.N., Mangin C.G.E., *Liquid moulding technologies*, Woodhead Publishing Ltd, Cambridge, 1997.

[SKE 76] Skelton J., "Fundamentals of fabric shear", *Textile Research Journal*, December 1976, pp. 862–869.

[SOU 01] Souter B.J., "Effects of fibre architecture on formability of textile preforms", PhD Thesis, University of Nottingham, 2001.

[WAN 98] Wang J., Page J.R., Paton R., "Experimental investigation of the draping properties of reinforcement fabrics", *Composites Science & Technology*, Vol. 58, 1998, pp. 229–237.

[WES 90] Van West B.P., Pipes R.B., Keefe M., "A simulation of the draping of bidirectional fabrics over arbitrary surfaces", *J Text Inst*, Vol. 81, 1990, pp. 448–460.

[WIL 99] Wilks C.E., Rudd C.D., Long A.C., Johnson C.F.,"Rate dependency during processing of glass/thermoplastic composites", *Proc. 12th International Conference on Composite Materials*, Paris, July 1999.

Chapter 7

Comparison of Ductile Damage Models

Cyril Bordreuil, Emmanuelle Vidal-Sallé and Jean-Claude Boyer

INSA de Lyon – Laboratoire de Mécanique des Solides, Lyon, France

1. Introduction

In previous published work, dissipation potentials for damage prediction deduced from local void growth models of Gurson [GUR 77] or Rice and Tracey [RIC 81] have b presented by several authors. Most of these discussions consider only the behaviour of em voids under hydrostatic loading. In this paper, four yield functions are applied to the v known collar test and to an upsetting of shaped workpiece between conical dies un isothermal conditions. The first two yield functions are improvements of the original work Gurson [GUR 77] formulated by Tvergaard [TVE 81] on the one hand and by Oudin *et* [OUD 95] on the other hand. The third and the fourth yield functions are identified with local Rice and Tracey model but the last is also formulated for voids initially filled with inclusion and is sensitive to each component of the stress tensor. The influence of the differ plastic dilatant stress-strain laws on the change on the void volume fraction used as dam parameter are analysed at different stages of both different forming processes.

2. Dissipation potentials

2.1. *The Gurson-Tvergaard yield function*

The classical Berg-Gurson potential, Φ_1, [GUR 77] predicts a dilatational plastic strain rate D_{kk} only sensitive to the mean stress σ_m or to the triaxiality ratio $3\sigma_m/2\sigma_0$ with σ_0 the yield stress of the sound material. The damage parameter of this model is f, the void volume fraction. In order to take into account interaction between voids, Tvergaard [TVE 81] analysed the behaviour of a network of unit cells with spherical voids and improved the original Berg-Gurson proposition with the q_1, q_2, and q_3 coefficients. The expression of the Gurson-Tvergaard yield function is written as:

$$\Phi_1 = \frac{\sigma_{vm}^2}{\sigma_0^2} + 2q_1 f \cosh\left(\frac{3q_2\sigma_m}{2\sigma_0}\right) - \left(1 + q_3 f^2\right) = 0 \qquad [1]$$

The corresponding dilatational part of the plastic strain rate is found to be:

$$D_{kk} = 3\lambda q_1 q_2 \frac{f}{\sigma_0} \sinh\left(\frac{3q_2\sigma_m}{2\sigma_0}\right)$$ [2]

with: λ : the plastic factor

and: σ_{vm} : the classical von Mises equivalent stress: $\sigma_{vm} = \sqrt{\frac{3}{2} s_{ij} s_{ij}}$

In Equation 2, as in the sequel to this work, the elastic strain rate is neglected. D_{ij} stands for the plastic strain rate.

The Tvergaard coefficient values numerically obtained are: $q_1 = 1.5$, $q_2 = 1$, $q_3 = 2.25$.

The plastic factor can be expressed as:

$$\lambda = \frac{D_{eq}\sigma_0^3}{\left(2.\sigma_{vm}^2 + 3q_1 q_2 \sigma_m \sigma_0 \sinh\left(\frac{3q_2\sigma_m}{2\sigma_0}\right)\right)}$$ [3]

where D_{eq} is the energetic equivalent plastic strain rate: $\sigma_{ij} D_{ij} = \sigma_0 D_{eq}$.

As this damage model is reversible for compressive stress state, Oudin *et al.* [OUD 95] suggested limiting its application to tensile mean stress and to consider classical non dilatant plasticity for compressive mean stress with a modified von Mises yield function:

$$\text{if } \sigma_m > 0 \quad \Phi_2 = \frac{\sigma_{vm}^2}{\sigma_0^2} + 2q_1 f \cosh\left(\frac{3q_2\sigma_m}{2\sigma_0}\right) - \left(1 + q_3 f^2\right) = 0$$

$$\text{if } \sigma_m \leq 0 \quad \Phi_2 = \frac{\sigma_{vm}^2}{\sigma_0^2} + 2q_1 f - \left(1 + q_3 f^2\right) = 0$$ [4]

For negative mean stress, the dilatational plastic strain rate is set to zero.

2.2. The Gelin yield function

For applications involving high triaxiality ratio σ_m/σ_0, Gelin [GEL 95] identified the plastic dilatant part to the corresponding Rice and Tracey model [RIC 81] developed for the growth of a spherical void in a remote plastic strain rate field. The damage parameter of this model is also f, the void volume fraction. With the same notations used in the previous section, the expression of Φ_3, the Gelin yield function is written as:

$$\Phi_3 = \sigma_{vm} - \sigma_0\left(1 - \frac{3}{2}(f - f_0)\exp\left(\frac{3\sigma_m}{2\sigma_0}\right)\right) = 0 \qquad [5]$$

where f_0 is a minimal void volume fraction. In this case, the dilatational plastic strain rate is equal to:

$$D_{kk} = \frac{9}{4}D_{eq}(f - f_0)\exp\left(\frac{3\sigma_m}{2\sigma_0}\right) \qquad [6]$$

This yield function Φ_3 is only sensitive to the mean stress σ_m or to the triaxiality ratio $3\sigma_m/2\sigma_0$ but gives a higher void volume fraction rate than the Tvergaard yield function as its volumetric plastic strain rate D_{kk} is an exponential function of the triaxiality ratio that is to be compared to a hyperbolic sine function. Furthermore, this damage model is irreversible as the corresponding plastic strain rate D_{kk} is positive for tensile as compressive mean stress states.

2.3. A specific yield function

Any dilatant plasticity constitutive laws have to fulfil the mass conservation. This principle implies a relation between the rate of the density ρ, the dilatational plastic strain rate D_{kk}, and the rate of the void volume fraction f:

$$\dot{f} = (1 - f)D_{kk} = -(1 - f)\frac{\dot{\rho}}{\rho} \qquad [7]$$

where $\dot{f}$ is the rate of the void volume fraction and, $\dot{\rho}$, the rate of the density.

As proposed in previous works of the authors [BOY 97], the mass conservation can be introduced in the von Mises yield function if the shear energy is considered at the mass element level for any stress state as for the tensile reference state. Equating the shear energy in any stress state to the tensile shear energy leads to:

$$\Phi_4 = \frac{\sigma_{vm}^2}{\rho} - \frac{\sigma_T^2}{\rho_T} = 0 \qquad [8]$$

where ρ is the actual density considered as stress dependent,

σ_T, the tensile yield stress of the voided material,

and ρ_T, the density dependency of the material under tensile loading.

The density changes can be measured for various plastic strain on samples extracted from tensile specimen or deduced from the Young modulus variations under loading-unloading cycles. Introducing, σ_0, the yield stress of the sound material, f, the actual void volume fraction and f_T, the void volume fraction dependency under tensile stress state, the yield function Φ_4 becomes:

$$\Phi_4 = \frac{1}{\rho_0}\left(\frac{\sigma_{vm}^2}{1-f} - \left(1 - f_T\right)\sigma_0^2\right) = 0 \qquad [9]$$

where ρ_0 is the density of the sound material.

In the plastic flow rule, the change of the density is considered as a function of each component of the stress tensor and not only mean stress dependent. In order to identify this variation, a link between the macroscopic behaviour of the voided material and the local behaviour of a unit cell stated by Rice and Tracey is proposed.

3. A modified Rice and Tracey model

3.1. Void growth model

Experimental evidence shows that damage is not only mean stress dependent. The deviatoric part of the stress state induces volume changes as material density variation is observed during torsion test. Finite element modeling of the behaviour of a unit cell with an initial spherical void under cylindrical shear or plane shear leads to prediction of the rate of the void volume fraction that has been compared with a modified Rice and Tracey model, [BOY 99]. Following Thomason [THO 90], the original formulation is first extended to ellipsoidal voids and then modified for a cavity initially filled by an inclusion.

In this case, the rate of the eigen radii R_i of the void can be obtained with some kinematic assumptions for unilateral boundary conditions between the void and the inclusion. The radius rate has to be zero when contact exists between the void and the inclusion and when the plastic strain rate closes the void.

$$if\ D_i \leq 0 \text{ and } R_i = R_I \quad \dot{R}_i = 0$$

$$\qquad\qquad\qquad\qquad\qquad\qquad\qquad\qquad\qquad\qquad [10]$$

$$if\ D_i > 0 \text{ or } R_i > R_I \quad \dot{R}_i = \left[\frac{5}{3}D_i + \frac{3}{4}\frac{\sigma_m}{\sigma_0}D_{eq}\right]R_i^{1/3}\left(R_1 R_2 R_3\right)^{2/9}$$

where D_i: is the plastic strain rate in the eigen direction i of the strain rate tensor

R_i: the eigen radius of the void in the i direction

and R_I: the radius of the spherical inclusion.

The power coefficients are identified from a fitting procedure with finite element results up to a void volume fraction of 1%.

With such a modification, the initial rate of void volume fraction for a spherical void can be evaluated for different loadings.

Table 1. *Rate of void volume fraction*

Stress state	Tensile	Compressive	Cyl. Shear
Empty void	$9/12D_{kk}$	$-9/12\ D_{kk}$	0
Void with inclusion	$23/12\ D_{kk}$	$14/12\ D_{kk}$	$20/12\ D_{kk}$

The results presented in Table 1 are classical for an empty void but show positive values for a void with inclusion in any stress state including cylindrical shear. This is the property of an irreversible damage model and it has to be quoted that this one is sensitive to pure deviatoric stress state. As a tensile stress state can be split into a cylindrical shear state and a pure hydrostatic one, it is important to note the main contribution of the deviatoric tensor of the tensile state on the corresponding rate of the void volume fraction.

3.2. Dilatational plastic strain rate

From Equation 10 and the specific yield function, Equation 9, a flow rule can be deduced with the following dilatational plastic strain rate:

$$D_{kk} = \frac{1}{4}\frac{D_{eq}}{\sigma_0}\sum_{i=1}^{3}\left(10s_i + 3\sigma_m\right)\alpha_i \qquad [11]$$

where s_i are the eigen values of the deviatoric stress tensor and α_i geometric coefficients for ellipsoidal cavities controlling the unilateral boundary conditions for void with inclusion:

– for a void without contact with an inclusion

$$\alpha_i = \left(\left(R_1R_2R_3\right)^{1/3}\Big/R_i\right)^{\frac{2}{3}} \qquad [12]$$

– for a void contacting an inclusion on the i direction

$$\alpha_i = \left(\left(R_1 R_2 R_3 \right)^{1/3} \Big/ R_i \right)^{\frac{2}{3}} \quad \text{if } D_i > 0$$

$$\alpha_i = 0 \qquad\qquad\qquad \text{if } D_i \leq 0 \tag{13}$$

Equation 9 is the general expression of the dilatation plastic strain rate including deviatoric stress effects. For an isotropic specific damage model only mean stress dependent, the corresponding dilatational plastic strain rate reduces to:

$$D_{kk} = \frac{9}{4} D_{eq} f \frac{\sigma_m}{\sigma_0} \tag{14}$$

4. Comparison of the damage models

4.1. Comparison of the dilatational plastic strain rates

In Figure 1, the different normalized dilatational plastic strain rate D_{kk}/D_{eq} are plotted versus the triaxiality ratio $3\sigma_m/2\sigma_0$ for a void volume fraction, f, equal to 1%. The "Gurson" key refers to Equation 2, the "Oudin" key to Equation 2 with a zero dilatational plastic strain rate for negative mean stress, the "Gelin $f_0 = 0$" and "Gelin $f_0 = 0.005$" keys to Equation 4 with the corresponding minimal void volume fraction, the "Isotropic specific" key to Equation 14 for comparison of a specific damage model only mean stress dependent. The black dot "Specific" key refers to Equation 11 for $f_T = 0.01$ and the three following particular stress states: pure compressive stress ($3\sigma_m/2\sigma_0 = -0.5$), pure cylindrical shear stress ($3\sigma_m/2\sigma_0 = 0.$), and pure tensile stress ($3\sigma_m/2\sigma_0 = 0.5$).

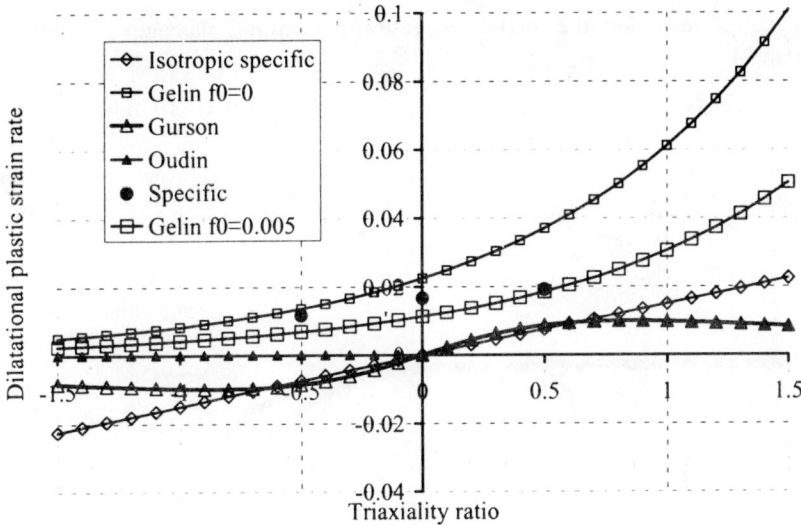

Figure 1. *Normalized dilatational plastic strain rate for f = 0.01*

Both reversible models, Gurson and isotropic specific ones, predict close values of the dilatational plastic strain rate in the range of the triaxiality ratio between pure compressive and pure tensile stress states. For higher triaxiality ratio, the dilatational plastic strain rate increases linearly with the isotropic specific damage model when it decreases slowly according to the Gurson one. These models allow the closing of voids and are suited for porosity or voids larger than their internal inclusions. Following Gelin, the dilatational plastic strain rate rises exponentially with the rate controlled by the minimal void volume fraction f_0, this irreversible damage model leads to positive dilatational plastic strain rate for positive as negative mean stress and is appropriate for void still in contact with an inclusion. The specific damage model provides the same property with higher dilatational plastic strain rate than the isotropic specific model on account of the effects of the deviatoric part of the stress tensor. For this comparison, the specific damage models consider only spherical voids but ellipsoidal voids can be considered and the dilatational plastic strain rate increases with their eccentricity ratio.

4.2. *Comparison of the yield surfaces*

The yield surfaces of the different plastic potentials, Φ_1 to Φ_3, are represented in Figure 2 in a meridian plane of the normalized eigen stress space. The abscissa, parallel to the hydrostatic axis, is the triaxiality ratio $3\sigma_m/2\sigma_0$, the ordinate, parallel to any axis of the deviatoric plane, is the normalized equivalent stress σ_{vm}/σ_0. The keys are the same as those used in the previous section. The specific yield surface Φ_4 is represented for the particular case of isotropic damage governed by Equation 14.

The Gelin potential is plotted for two different minimal void volume fractions, $f_0 = 0$ and $f_0 = 0.005$ respectively.

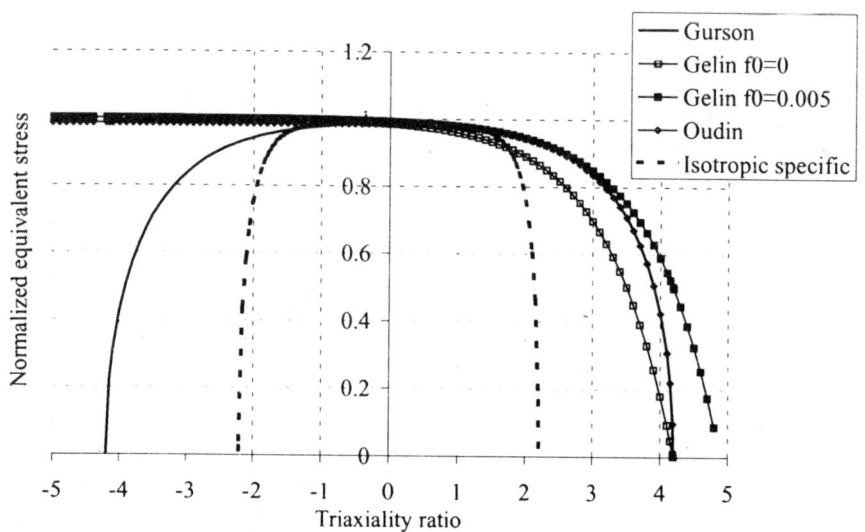

Figure 2. *Yield surfaces for a void volume fraction f = 0.01*

The y axis is a symmetric one for the reversible damage models, "Gurson" and "isotropic specific" as they are based on even functions of the triaxiality ratio, negative as positive mean stress bounds the elastic domain of the voided material when only positive mean stress closes the elastic domain for the irreversible model "Oudin" and "Gelin". It has to be noticed that the reduction of the volume of the elastic domain predicted with the specific damage model is the most sensitive to the triaxiality ratio. The four yield loci are very close in the range of low triaxiality ratio.

5. The collar test

5.1. *Geometry and material data*

For this first test as for the workability test, the behaviour of the material is considered as elastic plastic with piece-wise linear hardening. The two workpieces are manufactured in a 42CD4 grade steel. The corresponding strain-stress curves given in Figure 3 are the results of torsion tests at 950°C for three different plastic strain rates. These experimental data are fitted with piecewise linear functions and linear extrapolation are considered for intermediate plastic strain rates.

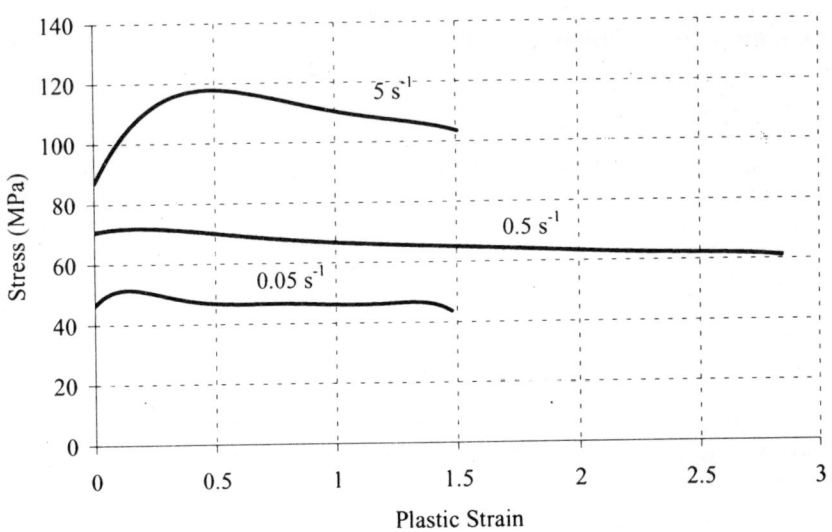

Figure 3. *Evolution of the yield stress of 42CD4*

The void volume fraction changes under pure tensile stress state are approximated by a linear function:

$$f_T = 0.056\varepsilon_p + 0.001 \qquad [15]$$

where ε_p is the equivalent plastic strain. For sake of numerical comparison, both forming processes are considered isothermal at 950 °C with stick friction between the workpieces and the tools. The finite elements used are six node axisymmetric triangles with a three point Gauss quadrature rule. The plasticity is implemented with the radial return algorithm for the four yield functions.

The initial height of the workpiece is 14 mm. Only a half of the initial and final mesh are represented.

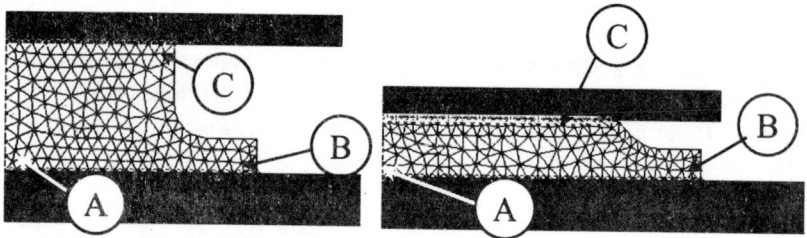

Figure 4. *Initial and deformed mesh of the collar test*

The void volume fraction is observed at three different locations: point A at the centre of the workpiece, point B near the collar surface, and point C at the initial external radius in contact with the punch. The initial void volume fraction is set to 0.1%.

5.2. Results

The void volume fraction changes versus the punch stroke are presented for the three points where the following keys: "Gurson"," Oudin" and "Gelin" stand for the Φ_1 to Φ_3 damage models respectively. The fourth model Φ_4 is presented for three different assumptions, as for the dilatational plastic strain rate, the specific isotropic model is only mean stress dependent while the two others take into account the shear stress effects in the cases of voids without inclusion or voids initially filled with an inclusion.

Table 2. *Keys*

♦ Gurson	✕ Specific isotropic model
■ Gelin	□ Specific model for empty voids
▲ Oudin	● Specific model for voids with inclusions

At points A and C, the plastic strain rises to 1.2 during the upsetting while the triaxiality ratio reaches –6. and –3. respectively. For a 4 mm stroke, at point B, the plastic strain is close to 0.2 and the triaxiality ratio keeps a constant value of 0.3.

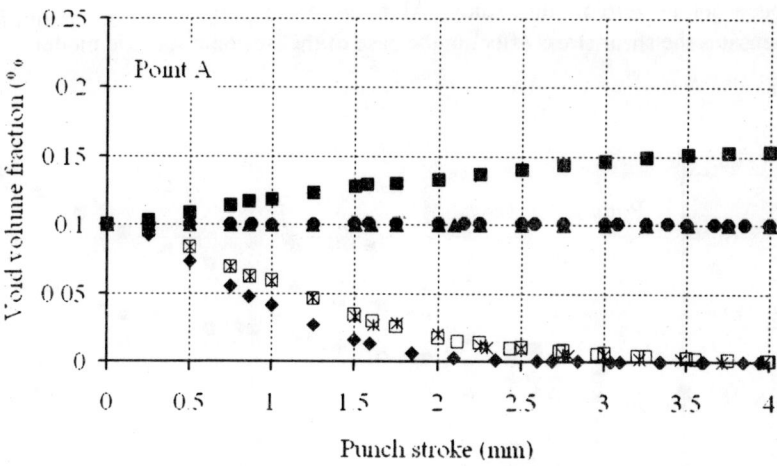

Figure 5. *Void volume fraction changes at point A during the collar test*

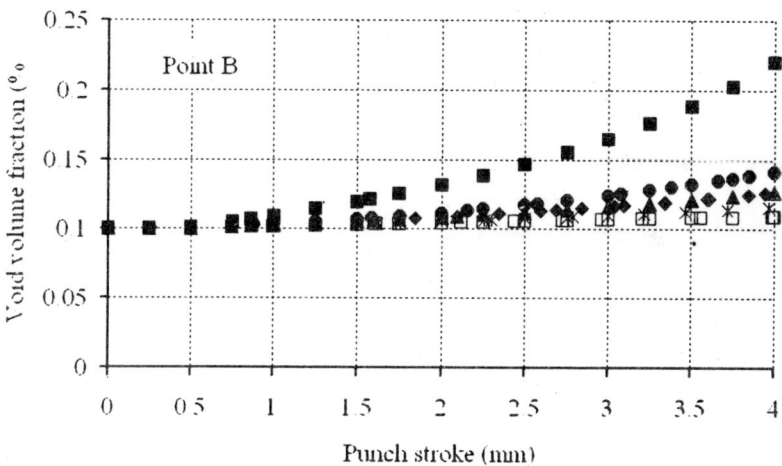

Figure 6. *Void volume fraction changes at Point B during the collar test*

The results predicted by the three reversible damage models are very similar. For positive mean stress, see Figure 6, their void volume fraction rates are smaller than the rates of irreversible models. For negative mean stress, the irreversible damage models predict very different values of the void volume fraction at points A and B but the tendencies are similar. The Gelin model is the more pessimistic but its dilatation plastic strain rate is controlled by the minimal void volume fraction f_0 that has been set to zero in this study. At point A, the high negative mean stress compensates the shear stress effect in the case of the isotropic specific model.

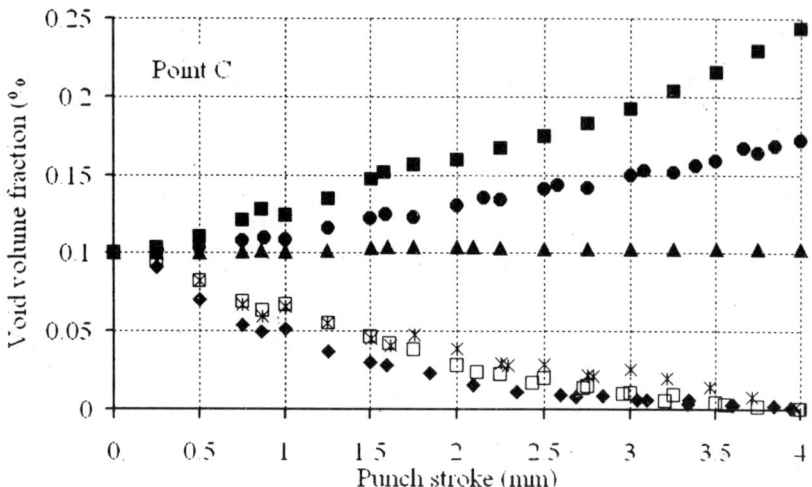

Figure 7. *Void volume fractions at point C during the collar test*

6. The workability test

6.1. *Geometry and material data*

This workability test was developed for analysing the material behaviour under more severe conditions than the collar test. Two points of the shaped workpiece are chosen for observation, one close to the equatorial diameter of the workpiece (A) and the other near the conical part of the die (B).

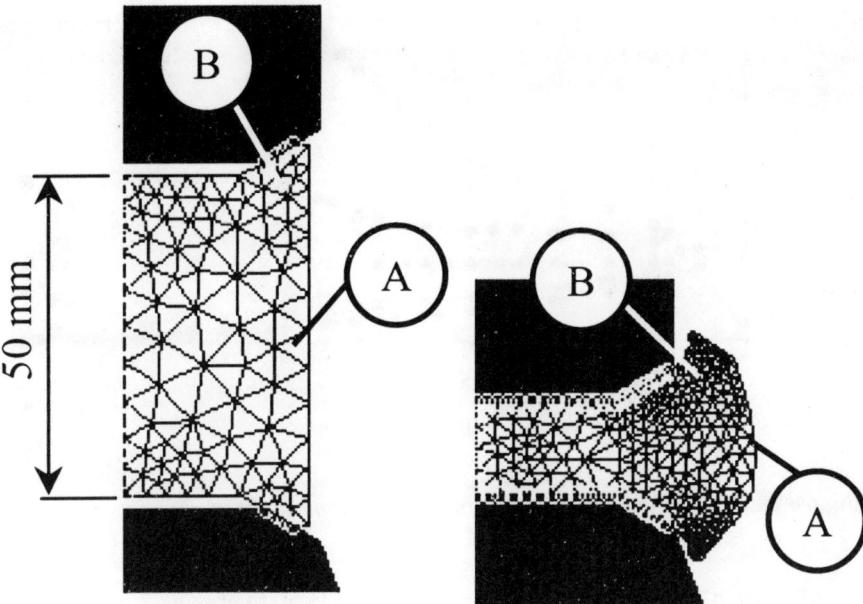

Figure 8. *Initial and deformed mesh of the workability test*

6.2. Results

The keys used for the two following figures are the same as those used in the previous section. The void volume fraction changes are also plotted versus the punch stroke.

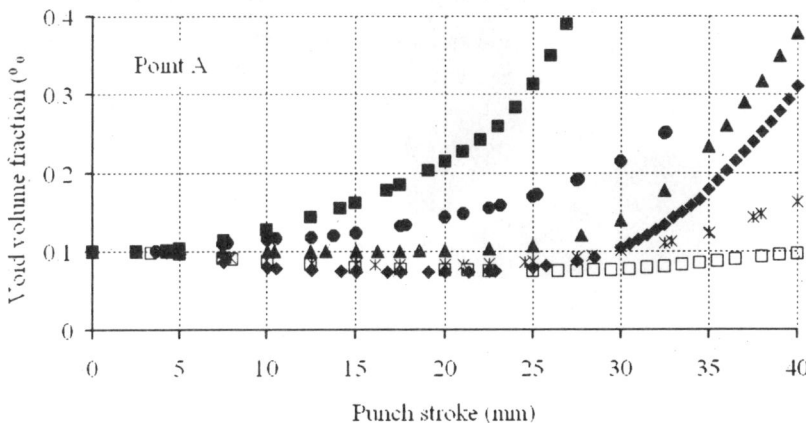

Figure 9. *Void volume fractions during the upsetting of the workability test*

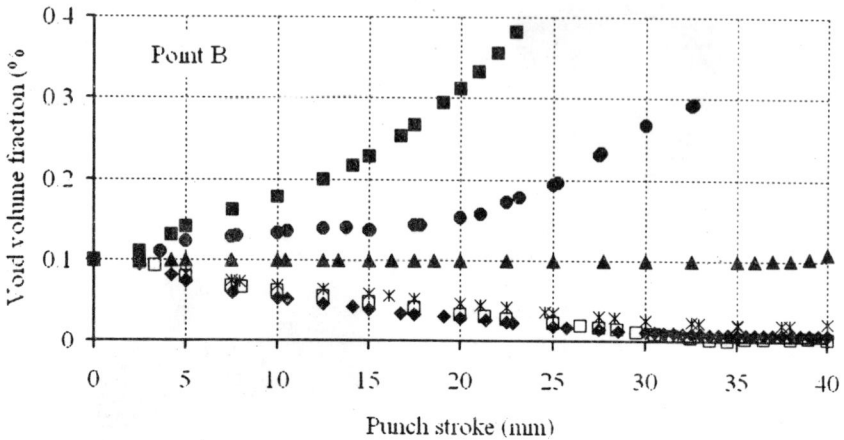

Figure 10. *Void volume fractions during the upsetting of the workability test*

After the 40 mm stroke, the plastic strain is up to 50 % at point A and 100 % at point B. The triaxiality ratio changes from –0.5 to 0.6 at point A and from –0.5 to 0.1 at point B. The Gurson-Tvergaard model, the specific isotropic model and the specific with empty voids model predict the maximum damage in the equatorial area of the workpiece as they are only sensitive to the hydrostatic stress.

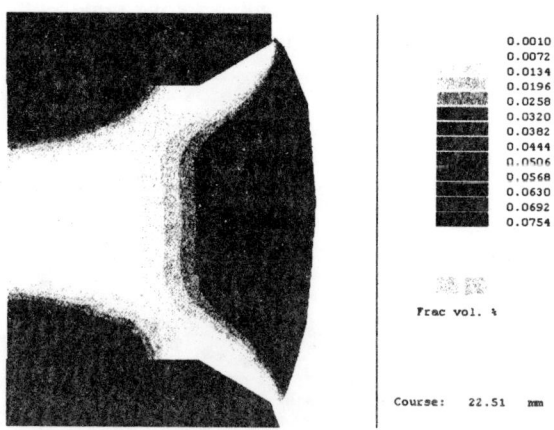

Figure 11. *Void volume fraction distribution for the Gurson model*

When reversible models predict closure of the voids at point B, for naturally irreversible models (Gelin and specific with inclusion models) it is the more damaged area although the triaxiality ratio is lower than at point A.

Actually, cracks appear first in this area. At point A, small discrepancies between the Gurson model, the isotropic specific model and the specific model with empty voids can be noted. They are induced by the sensitiveness of the growth rate to the deviatoric stresses that close systematically the empty voids.

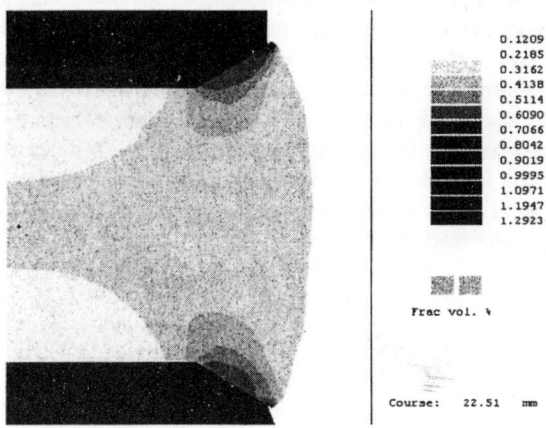

Figure 12. *Void volume fraction distribution for the Gelin model*

The specific model for voids with inclusions has a slower rate than the Gelin model used with a zero minimal void volume fraction but predict the same dangerous area where large plastic strain occurs.

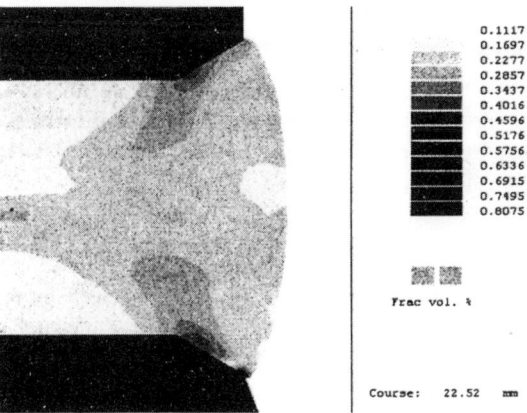

Figure 13. *Void volume fraction distribution for the specific model with inclusion*

7. Conclusion

For both cases, the highest void volume fraction is predicted by the Gelin model with a zero minimal void volume fraction. As the rate of the damage parameter is a function of the minimal void volume fraction f_0, the values obtained in this study show only tendency without more accurate identification. The Gurson-Tvergaard model is close to the specific model with empty voids. Both models are reversible ones and their void volume fraction rates are the lowest of all the models. The specific model with inclusions in the voids follows the same changes as the Gelin model but with lower rates. These rates could be increased with the Rice and Tracey model proposed for perfectly plastic cell material. Further developments of the specific model will integrate this assumption, as around voids there is a plastic strain concentration. The results of the different damage models have to be compared with experimental measurements of the void volume fraction. A research program is in progress with IRSID laboratories for the observation of the relative density changes of 42CD4 grade steel under tensile and torsion loading as well as in different areas of the workpiece of the workability test.

Acknowledgements

The authors are grateful to Fortech Company and Giat Industries for their financial support.

References

[BOY 97] BOYER J.-C., STAUB C., "An orthotropic ductile damage model including shear stress effect", *Third International Conference on Materials Processing Defects*, M. Predeleanu, P.Gilormini, Eds., Elsevier Sciences B.V, Amsterdam, pp. 13–22, 1997.

[BOY 99] BOYER J.-C., C. STAUB C., E. VIDAL-SALLE, "A shear stress dependent ductile damage model", *Plasticity'99: the seventh international symposium on plasticity and its current applications*, Khan A.S., Ed., Neat Press, Fulton Maryland, pp. 697–700, 1999.

[GEL 95] GELIN J.C., "Theoretical and numerical modelling of isotropic and anisotropic ductile damage in metal forming processes", *Second International Conference on Materials Processing Defects*, S.K. Gosh and M. Predeleanu, Eds., Elsevier Sciences B.V., Amsterdam, pp. 123–140, 1995.

[GUR 77] GURSON A. L., "Continuum theory of ductile rupture by void nucleation and growth", *Trans. ASME, J. Eng. Mat. Tech.*, Vol. 99, 1, pp. 2–15, 1977.

[OUD 95] OUDIN J., BENNANI B., PICART P., "Constitutive models for microvoid nucleation, growth and coalescence in elastoplasticity, finite element reference modelling", *Second International Conference on Materials Processing Defects*, S.K. Gosh and M. Predeleanu, Eds., Elsevier Sciences B.V., Amsterdam, pp. 107–122, 1995.

[RIC 81] RICE R., TRACEY D. M., "On the ductile enlargement of voids in triaxial stress fields", *J. Mech. Phys. Solids*, 17, pp. 210–217, 1981.

[THO 90] THOMASON P.F., *Ductile fracture of metals*, Pergamon Press, 1990.

[TVE 81] TVERGAARD V., "Influence of voids on shear band instabilities under plane strain conditions", *Int. J. Fracture*, 17, pp. 389–395, 1981.

Chapter 8

Experimental and Numerical Analysis of Blanking for Thin Sheet Metal Parts

Vincent Lemiale and Philippe Picart
Laboratoire de Mécanique Appliquée R.CHALEAT – UMR CNRS 6604, Besançon, France

Sébastien Meunier
AUGE Découpage, Besançon, France

1. Introduction

Blanking is one of the most used processes to obtain small and thin components in electronic and mechanical industries. Currently, the designing and manufacturing of these components are mainly based on know-how and empirical rules. The quality of the component is mainly determined by examination of the shape of the cut edge. That shape is related to the process parameters and the material behaviour. The industrial requirements need to produce smaller components with high controlled geometry. In many cases, the practical approach is not yet sufficient.

During last years, a large number of publications about blanking have been proposed. Experiments were carried out and most of them are associated with numerical simulations [MAI 91], [TAU 96], [GOI 00]. These experiments are focussed on the influence of the process parameters, such as the geometry of the punch and die, or the punch-die clearance [GOL 99], [TAU 96]. Some investigations also concern the analysis of the deformation through the thickness in situ using a CCD camera [GOI 00], [TAK 96], [YOS 99]. These works are limited to sheet metal parts with a thickness greater than 1 mm.

Presently the industrial knowledge concerning the metal blanking process is still empirical. Therefore, scientific approaches have been proposed to model blanking. Most of them use the finite element formulation [GOO 00], [KOM 99], [SAA 99]. The blanking process implies localized high plastic strains as in other sheet metal forming processes, stamping or bending. On the other hand, in blanking the plastic strain is used to initiate the fracture mechanism that must be avoided in other processes. So the main mechanical problem is related to the prediction of the initiation of the crack and the propagation across the sheet thickness.

The present paper proposes first an experimental investigation of the blanking process. The experimental device presented here permits together the measurement of the punch force versus penetration and the recording of the displacement field through the thickness. The thickness of the tested specimens is 0.2 mm. The purpose of the experiments is to test the capability of material for blanking process. Secondly, considering that the initiation of the crack can be related to the damage evolution in the sheet thickness along a shear band, a numerical modeling is presented. The Gurson's model is adopted to describe the damage evolution by nucleation and growth of microscopic voids in the matrix.. The numerical simulations are performed using the finite element code POLYFORM developed in our laboratory. A simulation with the same geometrical conditions than the experiment is proposed to validate the modeling. Finally, the capabilities of the experimental device to analyse the material behaviour in blanking are exposed and discussed.

2. Experimental device

In this section the design of the experimental device is presented. The main characteristics are described so that the different aspects of the instrumentation.

2.1. *Specifications*

The geometry of the blanked part is very simple. A rectangular part with 9 mm length and 2 mm large is chosen in order to be able to record the displacement field through the sheet thickness (see Figure 1). The retained configuration is very closed to the manufacturing conditions. The thickness of the sheet can vary from 0.05 mm to 0.5 mm, that corresponds to the current thickness for electronic components. The instrumentation of the device have been defined to investigate the material behaviour during blanking. A load sensor located on the punch measures the punch force while a displacement sensor measures the punch displacement. Moreover, the displacement field is recorded in situ using a CCD camera. The die had to be modified on one side to allow the examination of the sheet thickness during blanking.

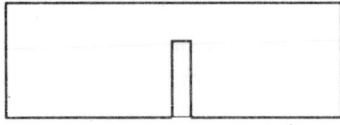

Figure 1. *Geometry of the blanked part*

2.2. *Presentation of the set-up*

Taking into account the previous specifications, a blanking device is designed (see Figure 2). The apparatus is installed on a testing machine that provides the axial force to the upper boby (1). The punch (7) is fixed on the upper body. While the upper body is going down, the sheet specimen is maintained on the die by the blankholder (6) using four springs fixed between the upper body (2) and the center body (3). The blankholder is also used to insure the guidance of the punch. To estimate the spring forces, a blanking test has been done without sheet specimen. The resulting force is obtained using the load cell measurement of the testing machine. The value of the force is 3.5 kN. So the blankholder imposes a pressure of 40 MPa on the sheet specimen.

The radius of the punch corner is about 10 μm. The value is similar for the die radius. Finally, the clearance between the punch and the die is 20 μm, that corresponds to 10% of the sheet thickness (0.2 mm) for the specimens used. A real photograph of the device is also presented in Figure 4a.

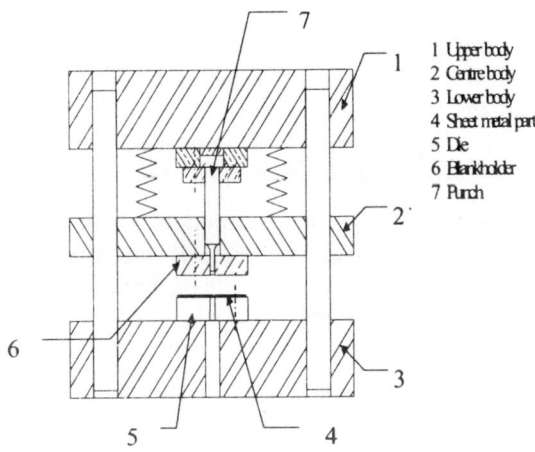

1 Upper body
2 Centre body
3 Lower body
4 Sheet metal part
5 Die
6 Blankholder
7 Punch

Figure 2. *Blanking device*

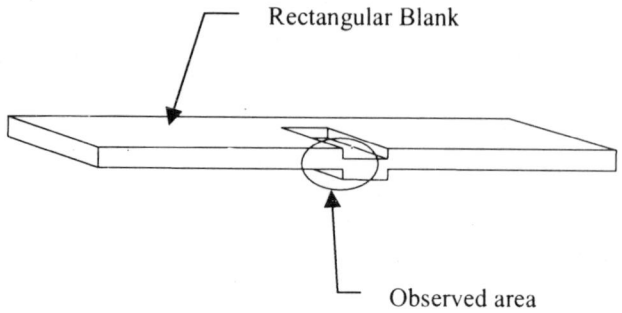

Figure 3. *Location of the observed area*

2.3. *Measurement*

To measure the punch force, four strain gauges are fixed directly on the punch. An inductive displacement transducer located between the upper body and the centre body is used to measure the punch displacement. The displacement field through the sheet thickness is recorded in situ using a CCD camera equiped with special optics and placed in front of the specimen during the test (see Figure 3 and Figure 4b).

The magnification must be sufficient to accurately examine the displacement field. Due to the overall dimensions of the apparatus, the digital camera must be installed at a certain distance of the sheet part. With the optical system fixed in front of the camera (see Figure 4b), the retained distance between the optics and the sheet specimen is 110 mm. The maximum magnification is about 1.5 µm per pixel.

The displacement field and then the strain field are computed by analysing the contrast of the images from the grey level of each pixel. That method is very useful here because the surface aspect of the sheet provides a adequate contrast. No specific preparation of the specimen is required that would be difficult because of the very small dimensions of the observed area (1mm x 1mm). The resulting displacement and strain fields are not presented in this paper because they need further investigations.

Figure 4. *Photographs, a-Blanking device, b-CCD camera and optical system*

3. Experimental investigation of material behaviour in blanking

3.1. *Load-penetration curve for the punch*

Due to the capacity of the CCD camera, 25 images per second, the blanking tests are performed at a low speed of 20 $\mu m.s^{-1}$. The experiments are made on three different copper alloys with 0.2 mm thickness. The corresponding load-penetration curves for the punch are proposed in Figure 5. The main steps of the material behaviour of the sheet during blanking are clearly observed on each curve: first, a elastic part, second, a plastic part with damage effects and finally the propagation of the crack through the sheet thickness. Nevertheless, the material behaviour is rather different for each one.

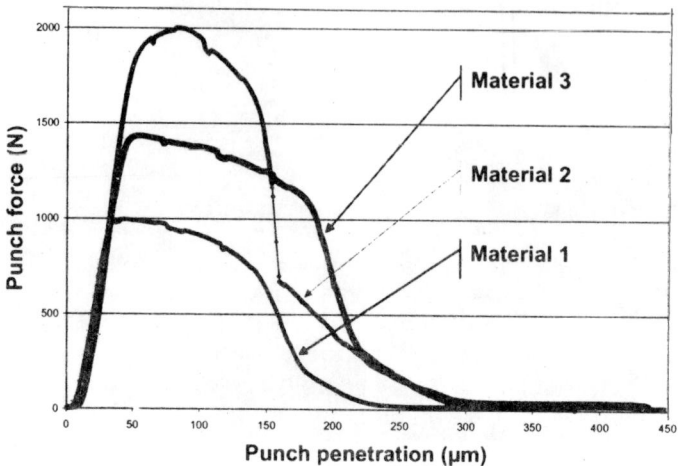

Figure 5. *Experimental load-penetration curves for the punch*

The maximum value for the punch load varies from 1000 N for material 1 to 2000 N for material 2. This result is coherent with the corresponding yield tensile stresses (from 400 MPa for material 1 to 600 MPa for material 2). The aspect of these curves, especially for the non linear area, shows that the plastic deformation for these materials is rather different between materials 1 and 3 and material 2. Consequently, the characteristics of the cut edge (roll over, shear zone, fractured zone and burr) vary. The corresponding values are given in Table 1. Especially the fractured zone is more important for material 1 and 2 than for material 3. It can be easily associated to the load-penetration curves. The slow and long decreasing of the punch load for material 3 explains the difference concerning the fracture zone. On the other hand, the shear zone is the most important for material 3.

Table 1. *Analysis of the cut edge geometry for the 3 materials*

Length of the specific zones for the cut edge	Percentage of the thickness		
	Material 1	Material 2	Material 3
Roll over	11.9	21.5	5
Shear zone	34.7	38.3	80
Fractured zone + burr	53.4	40.2	15

To complete the load-displacement curves, to investigate the strain field through the thickness and to detect the initiation of the crack, the analysis of the recording images in situ is essential.

3.2. *Recording of the displacement field in situ*

The combination of the blanking speed (20 μm.s^{-1}) and the acquisition speed of the camera (25 images per second) allows to record up to 250 images during a test. An example of image for the sheet thickness obtained with CCD camera is proposed in Figure 6. The magnification is about 1.5 μm per pixel. A set of images is also presented on Figure 7 for material 1 for different values of the punch penetration (10%, 32% and 55%) with the corresponding point on the load-displacement curve. On image a (10% penetration), the specimen is bent. At this stage, the strain is still elastic. On image b (32% penetration), we observe a plastic strain, mainly due to shearing, with a progressive damage. The punch load decreases. Finally, on image c (55% penetration), the crack has been initiated and propagates through the specimen. On the corresponding part of the curve, we notice a small decreasing of the punch load, which can be associated to the transition between the shearing phase and the fracture phase. In Figure 8a, for material 1 the crack initiation is observed for 32% of punch penetration. In Figure 8b, the sheet is totally fractured.

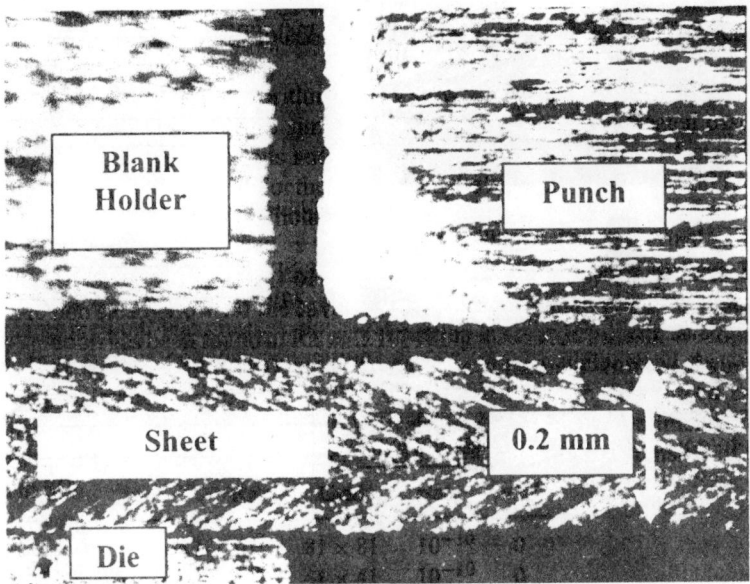

Figure 6. *CCD image of the blanking elements*

The blanking tests presented here show that the measurement system is able to correctly investigate the material behaviour during blanking. The characteristic zones of the cut edge, the strain field, the initiation and propagation of the crack through the thickness can be study in situ using that experimental device.

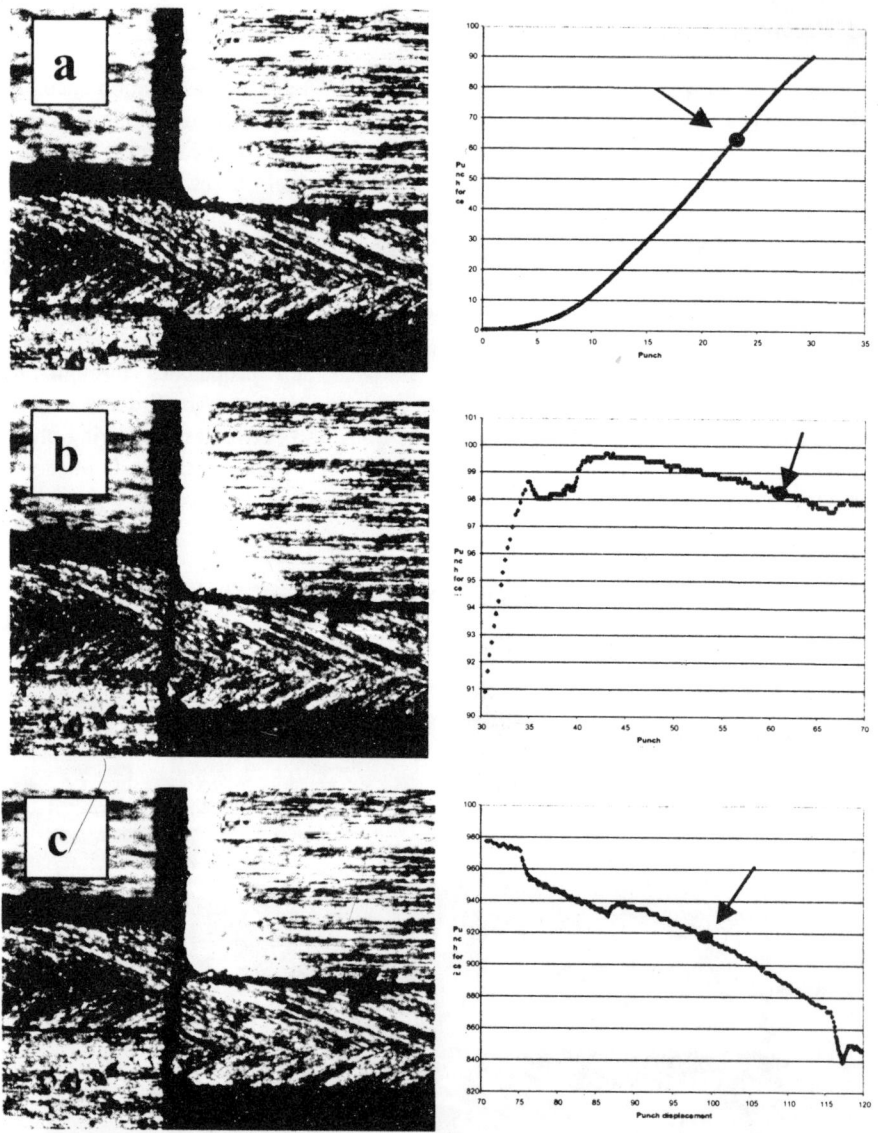

Figure 7. *Images at three steps during the blanking test and corresponding positions on the load-penetration curves for material 1: a-10% of punch penetration, b-32%, c-55%*

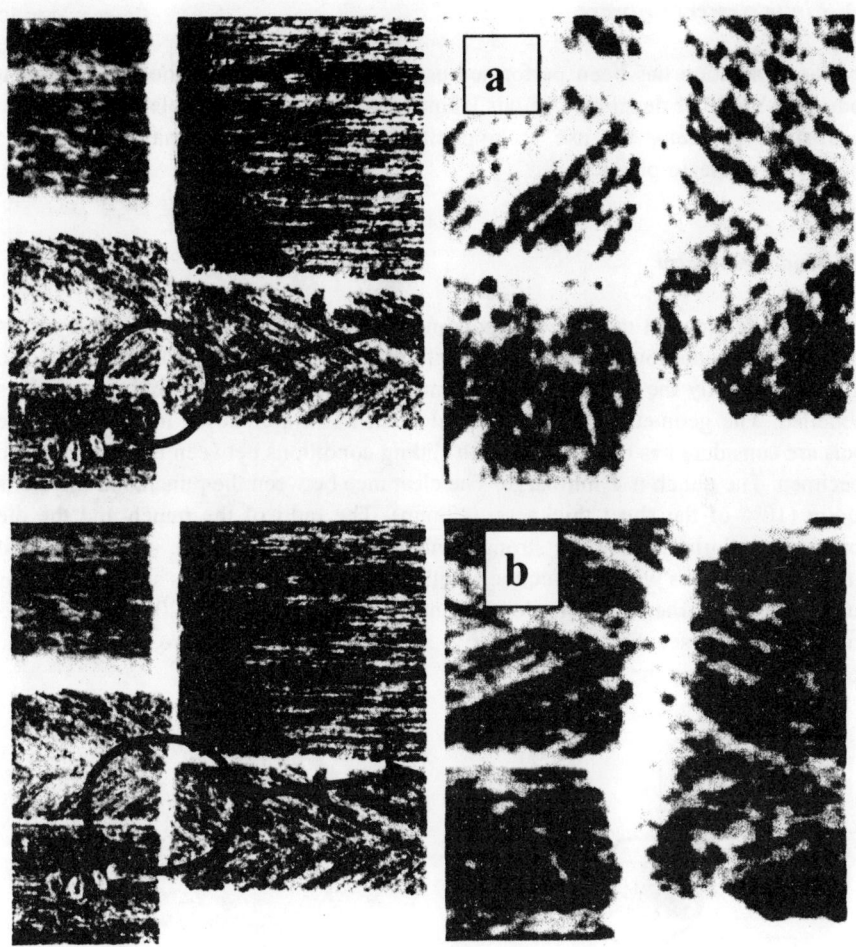

Figure 8. *Zooms of the fracture zone at two steps during blanking for material 1, a-32% of punch penetration, b-55%*

4. Simulation of the blanking process

The finite element method is well adapted to simulate blanking. In this section, the main characteristics of the finite element formulation used are presented. The material and geometric parameters used for the simulation of the blanking test are given. Finally, the results of the simulation are exposed in term of damage distribution.

4.1. *Finite element software*

The simulation has been performed using the 2D version of the finite element code POLYFORM developed in our laboratory. Considering the blanking test as a quasi-static problem, we use an implicit algorithm. The material behaviour is considered as elasto-plastic.

4.2. *Model geometry*

The finite element model corresponding to the geometry of the blanking test is used for the simulation. A 2D analysis with plane strains assumptions is adopted to describe correctly the mechanical problem. Due to the symmetry, only one half is modelled. The geometry and the finite element mesh are shown in Figure 9. The tools are considered as rigid bodies with sliding conditions between the tools and the specimen. The punch is 2 mm large. The clearance between the punch and the die is 20 µm (10% of the sheet thickness, 0.2 mm). The radii of the punch and the die corners are 10 µm. This value strongly influences the mesh of the sheet. To avoid numerical problems with contact, the length of the finite elements is set to 3 µm near punch and die corners. The finite element mesh is divided into 2020 triangular 3 nodes elements (see Figure 9). Finally, symmetric conditions are applied to the nodes located along the symmetry axis.

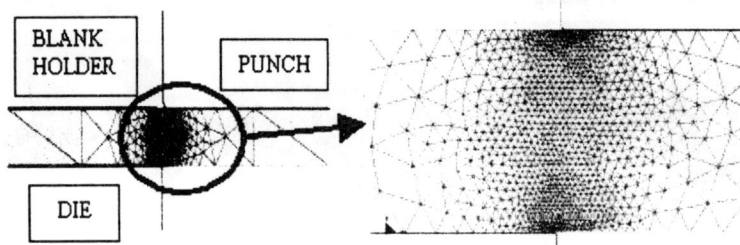

Figure 9. *Finite element model*

4.2. *Material behaviour*

The specimen is a copper alloy. The Young's Modulus is equal to $1.30\ 10^5$ MPa, the Poisson's ratio is 0.32. The material is assumed to be isotropic, and the flow stress is given by the following equation:

$$\sigma_M = 440(1 + 6.25\overline{\varepsilon}^P)(MPa) \qquad [1]$$

where:

σ_M is the effective flow stress

$\bar{\varepsilon}^P$ is the effective plastic strain

4.4. *Damage modeling*

To describe the damage phenomenon during blanking, we adopt the porous ductile model proposed by Gurson and resumed by Tvergaard [TVE 90] and Needleman. In this model, the development of damage is divided in three steps : the growth, the nucleation and the coalescence of the microscopic voids in the material matrix. The approximate yield condition has the following form:

$$\Phi = \frac{\sigma_{eq}^2}{\sigma_M^2} + 2q_1 f^* \cosh\left(\frac{3}{2}q_2 \frac{\sigma_m}{\sigma_M}\right) - 1 - \left(q_1 f^*\right)^2 = 0 \qquad [2]$$

where σ_{eq} represents the equivalent von Mises stress, σ_M the current flow stress and σ_m the hydrostatic stress in the matrix material. q_1 and q_2 are the material parameters introduced by Tvergaard. The function f^* was introduced by Tvergaard and Needleman to model the complete loss of material stress carrying capacity at a realistic void volume fraction. This function is defined by

$$f^*(f) = \begin{cases} f & \text{for } f \le f_c \\ f_c + \dfrac{f_u^* - f_c}{f_F - f_c}(f - f_c) & \text{for } f > f_c \end{cases} \qquad [3]$$

where f is the current void volume fraction, the void volume fraction at final fracture is denoted by f_F, $f_u^* = f^*(f_F) = 1/q_1$ and f_c is the critical value of f at the beginning of coalescence. Finally, the growth of the current void volume fraction is related to the plastic deformation b:

$$\dot{f} = (1 - f)tr(\dot{\varepsilon}^P) + A\dot{\bar{\varepsilon}}^P \qquad [4]$$

where $\dot{\varepsilon}^P$ is the plastic part of the macroscopic strain rate tensor and $\dot{\bar{\varepsilon}}^P$ is the effective plastic strain rate representing the microscopic strain state in the matrix material. A normal distribution is considered for nucleation, so A has the following form:

$$A = \frac{f_N}{s_N \sqrt{2\pi}} \exp\left[-\frac{1}{2}\left(\frac{\bar{\varepsilon}^P - \varepsilon_N}{s_N} \right)^2 \right]$$ [5]

where ε_N is the mean strain for nucleation, s_N is the corresponding standard deviation and f_N is the volume fraction of void nucleating particles.

Table 2 summarizes the numerical values for the material damage parameters.

Table 2. *Damage material parameters for the FE simulation*

Material parameter	Value
f_0	0
q_1	1.5
q_2	1
f_N	0.04
s_N	0.1
ε_N	0.2
f_C	0.15
f_F	0.25

4.5. Prediction of the damage distribution

In Figure 10, the distribution of the void volume fraction for a punch displacement of 44 µm (22% of punch penetration) is plotted. The maximum value is located near the punch and die corners. The damage occurs in a narrow shear band between the punch and the die. Considering that the damage material parameters used for the simulation are not really adapted to the tested materials. This repartition is only qualitative. The material damage parameters will be identified to obtained more realistic quantitative values of the damage.

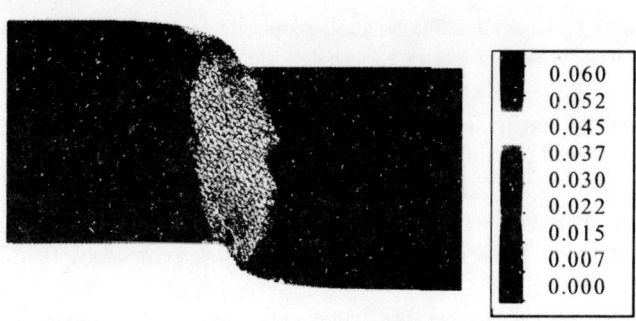

| 0.060 |
| 0.052 |
| 0.045 |
| 0.037 |
| 0.030 |
| 0.022 |
| 0.015 |
| 0.007 |
| 0.000 |

Figure 10. *Distribution of the void volume fraction at 22% of punch penetration*

4. Concluding remarks

A specific blanking apparatus adapted for thin sheet metal parts has been developed. This apparatus is instrumented to measure the load-displacement curve for the punch and to record in situ the displacement field through the thickness with a CCD camera. Load-displacement curves are presented for three copper alloys specimens with 0.2 mm thickness. A set of recorded images through the sheet thickness during blanking are proposed and permit to identify and quantify the three main phases of the blanking, the elastic phase, the shear phase and the fracture phase. Using this device, the experimental investigation of the material behaviour in blanking for very thin sheet is now possible. Currently, a set of experiments on several copper alloys has started. Further investigations about the influence of the process parameters as the punch-die clearance and the punch geometry are undertaken. The exploitation of the CCD images will be pursue to improve the detection of the shear and fracture phase.

Concerning the numerical simulation, the strategy used to take into account damage has been presented. Qualitative results have been obtained that are in good agreement with those expected. However, these results have to be improved by confrontation with the expérimental one. Moreover, it is obvious that the behaviour of the sheet strongly depends on the material properties, so the material damage parameters must be identify by inverse method using the experimental results.

Acknowledgements

This work is supported by Auge Decoupage company, Besançon, France, specialized in the manufacturing of lead-frames for electronic industries.

References

[BEL 98] BELLENGER E., BUSSY P., "Plastic and viscoplastic damage models with numerical treatment for metal forming processes", *Journal of Materials Processing Technology*, 80–81, 1998, pp. 591–596.

[BRO 00] BROKKEN D., BREKELMANS W.A.M., BAAIJENS F.P.T., "Discrete ductile fracture modelling for the metal blanking process", *Computational Mechanics*, 26, 2000, pp. 104–114.

[BRO 00] BROKKEN D., BREKELMANS W.A.M.., "Predicting the shape of blanked products: a finite element approach", *Journal of Materials Processing Technology*, 103, 2000, pp. 51–56.

[FAU 98] FAURA F., Garcia A., ESTREMS M., "Finite element analysis of optimum clearance in the blanking process", *Journal of Materials Processing Technology*, 80–81, 1998, pp. 121–125.

[GOI 00] GOIJAERTS A.M., STEGEMAN Y.W. *et al.*, "Can a new experimental and numerical study improve metal blanking?", *Journal of Materials Processing Technology*, 103, 2000, pp. 44–50.

[GOI 01] GOIJAERTS A.M., GOVAERT L.E., BAAIJENS F.P.T., "Evaluation of ductile fracture models for different metals in blanking", *Journal of Materials Processing Technology*, 110, 2001, pp. 312–323.

[GOL 99] GOLOVASHCHENKO S.F., "Numerical and experimental analysis of the trimming process", *Proc. of the 4th Int. Conf. and Workshop on Numerical Simulation of 3D Sheet Forming Processes,* Numisheet, 1999.

[HAM 01] HAMBLI R., "Blanking tool wear modelling using the finite element method", *International Journal of Machine Tools & Manufacture*, 41, 2001, pp. 1815–1829.

[KOM 99] KOMORI K., "Simulation of shearing by node separation method: effect of parameter in fracture criterion", *Proc. of the 4th Int. Conf. and Workshop on Numerical Simulation of 3D Sheet Forming Processes,* Numisheet, 1999.

[MAI 91] MAILLARD A., "Etude expérimentale et théorique du découpage", Master Thesis, University of Compiègne, 1991.

[MAI 00] MAITI S.K., AMBEKAR A.A., SINGH U.P., DATE P.P, NARASIMHAN K., "Assessment of influence of some process parameters on set metal blanking", *Journal of Materials Processing Technology*, 102, 2000, pp. 249–256.

[PYT 00] PYTTEL T., JOHN R., HOOGEN M., "A finite element based model for the description of aluminium sheet blanking", *International Journal of Machine Tools and Manufacture*, 40, 2000, pp. 1993–2002.

[SAA 99] SAANOUNI K., FRANQUEVILLE Y., "Numerical prediction of damage during metal forming processes", *Proc. of the 4th Int. Conf. and Workshop on Numerical Simulation of 3D Sheet Forming Processes,* Numisheet, 1999.

[TAK 96] TAKAHASHI T., AOKI I., "Development of Analysing System Applicable for Large Plastic Deformation", *Proc. of the 5th ICTP*, vol. 3, 1996, pp. 583–590.

[TAU 96] TAUPIN E., BREITLING J., WU W.T., ALTAN T., "Material fracture and burr formation in blanking results of FEM simulations and comparison with experiments", Journal of Materials Processing Technology, 59, 1996, pp. 68–78.

[TVE 90] TVERGAARD V., "Material failure by void growth to coalescence", *Advances in Applied Mechanics*, Vol. 27, 1990, pp. 83–151.

[YOS 99] YOSHIDA Y., YUKAWA N., ISHIKAWA T., "Deformation analysis of shearing process using rigid plastic finite element method considering fracture phenomenon", *Proc. of the 6th ICTP*, Vol. 3, 1999, pp. 2245–2250.

Chapter 9

Development in Finite Element Simulations of Aluminum Extrusion

J. Lof, K. Valkering and J. Huétink

Dept of Mechanical Engineering, University of Twente, The Netherlands

1. Introduction

At present, the design of extrusion dies and operations in extrusion companies is primarily based on trial and error. The subjective experience of the die designer, the press operator and the die corrector to a large extend determine the performance of the extrusion die and the efficiency of the process. To improve on this situation it is necessary to obtain more objective knowledge about the extrusion process.

The knowledge available on this process has been gathered primarily through experimental work and analytical calculations [JOH 62, LAU 81]. It is only during the last ten years that numerical simulations of aluminium extrusion have been reported in literature. This is largely due to the high computational demands that are associated with this kind of simulation. Early work was mainly concerned with 2D extrusion problems [GRA 92, DEV 92, KAN 94] or simple 3D geometries with low extrusion ratios [MOR 92, KIU 95, KAN 96]. With the increase of computer power more complex extrusion problems have been modelled. Simple porthole dies have been modelled in 3D by Tong [TON 95] and by Mooi [MOO 96]. For these kinds of extrusion problems, pre-processing issues start to emerge. For the simulation of more complex profiles it is essential to automate large parts of the pre-processing. Van Rens [REN 99] developed a meshing algorithm specifically aimed at the geometry associated with extrusion of thin-walled sections, enabling simulations of this type of profile.

The primary focus of the research reported here is directed at solving two general problems that are found in extrusion practice. The first problem is the control of the aluminium flow through the die, especially at the exit area where inbalances in velocity over the profile can lead to distorted or otherwise unacceptable profiles. In practice very unpredictable behaviour is observed, necessitating trial-pressings and die correction. The second problem is the strength and stiffness of the extrusion die. The large thermo-mechanical loads imposed on the die during the process cause deflection and sometimes plastic deformation of the die. This problem occurs especially when using porthole dies. Both problems undermine the reliability of the die, which results in an inefficient use of the available press capacity. To improve on the current situation, the dies should become stronger, more reliable and more predictable.

In this paper a short overview of the results of the simulations will be given. The numerical model is discussed first, in Section 2. In order to accurately predict the processes that occur inside the bearing channel of the extrusion process, detailed simulations of the bearing area are made which yielded very interesting results. These results are discussed in Section 3. In order to make simulations of the extrusion of a complex profile, pre-processing time and analysis time become major issues. To enable practical use to be made of these simulations, a pre-processing application was made which automates most of the pre-processing and provides a direct coupling between the design of the dies in a CAD application and the analysis code. This is discussed in Section 4.

2. Numerical model

The simulations of the extrusion process were carried out using the finite element code DiekA. This is an Arbitrary Lagrangian Eulerian (ALE) code developed by Huétink et al. [HUE 86]. This code is specialized in the simulation of forming processes like deep drawing, rolling, cutting and extrusion.

The deformation resistance of hot aluminium during extrusion is usually described by a viscoplastic constitutive relation. In the simulations presented in this work, special attention is paid to the aluminium flow in the bearing area illustrated in Figure 1. Inside the bearing channel, plastic deformation is very limited and elastic effects play an important role in determining the contact pressure between aluminium and die. Therefore these effects have an important influence on the resistance in the bearing channel. To accurately describe the behaviour of the material in the bearing area, an elasto-viscoplastic model is used for the simulations presented in this paper. An elaborate description of this model is presented in [LOF 00/1, LOF 00/2].

The constitutive behaviour of the specific aluminium alloy is represented as a relation between the flow stress and the strain rate. For the simulations presented in this work a modified Sellars-Tegart law (hyperbolic sine law) is used [JON 69, LAN 00]:

$$\sigma_f(\dot{\kappa},T) = s_m \operatorname{arcsinh}\left(\left(\frac{\dot{\kappa}+\dot{\kappa}_0}{A}\exp\left(\frac{Q}{RT}\right)\right)^{\frac{1}{m}}\right)$$

Here σ_f is the flow stress, which is dependent on the plastic strain rate $\dot{\kappa}$ and the temperature T. The parameters s_m, A, m, Q are experimentally determined constants, R is the universal gas constant and the parameter $\dot{\kappa}_0$ is used to introduce a small elastic region in the constitutive relation. The parameters in this model are fitted to data obtained experimentally for an AA6063 alloy.

In order to verify the accuracy of the numerical model, verification of experimentally obtained results is necessary. This is done by comparison of simulation results with extrusion experiments performed independently at the University of Delft. An elaborate description of this verification is presented in [LOF 00/1, LAN 00]. Extrusion pressure was predicted with an accuracy of 10% compared to the experimental results. It must be stressed that no fitting parameters were used to obtain this accuracy, proving that it is possible to get good results with extrusion simulations. However, the experiments were conducted under laboratory conditions and the material data were determined very precisely. Therefore it cannot be expected that simulations of industrial extrusion will yield equal accuracy.

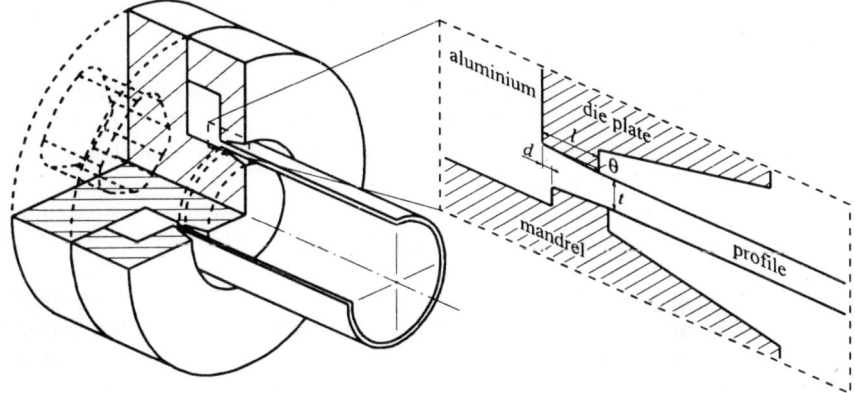

Figure 1. *Schematic picture of an extrusion of a circular tube and a blow-up of the bearing area*

3. Simulating the bearing

The bearing, illustrated in Figure 1, is the most critical part in the process. The bearing determines the final shape of the profile and it is also used to control the local exit velocity of the material. This is done by varying the bearing length l. Simulating the bearing area constitutes a major challenge. The scale of the bearing area is very small compared to the container. In the area just in front of the bearing high deformation rates occur, while inside the bearing channel plastic deformation is minimal or non-existent. In addition to this, the contact behaviour between aluminium and die switches from stick to a slipping friction somewhere in this area.

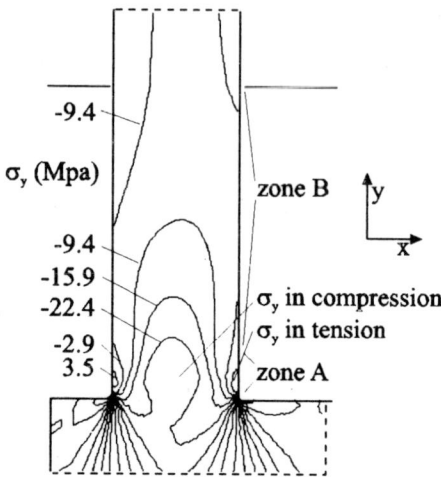

Figure 2. *Contact in the bearing channel as depicted in Figure 1*

To get a good understanding of the processes that occur inside and just in front of the bearing, very detailed 2D simulations of the bearing have been made. With these simulations the influence of design parameters like profile thickness *t* and bearing length *l* have been examined. Also, the influence of non-alignment of the two bearing surfaces *d* and bearing angle θ has been examined. An elaborate description of these simulations can be found in [LOF 00a, LOF 00b].

One result of these simulations is of particular importance to extrusion practice. From the simulations it followed that in a parallel bearing ($\theta = 0$) contact between bearing channel and aluminium is not always present. For short bearing lengths, the solution with contact in the bearing is not stable and will eventually move to a situation without contact. This effect is explained in more detail with the help of Figure 2. Two zones (A and B) are indicated in this figure. In zone A stresses in the y-direction will be in tension (positive values). In this area the material has the tendency to lose contact with the bearing wall, because the material is stretched in the y-direction. This tendency is counteracted by the pressure that is built up inside the bearing channel by the frictional forces at the wall (zone B). If these frictional forces are sufficient, which is the case for long bearings, the contact will remain. If the forces are too small, contact will be lost at zone A and eventually along the entire bearing length.

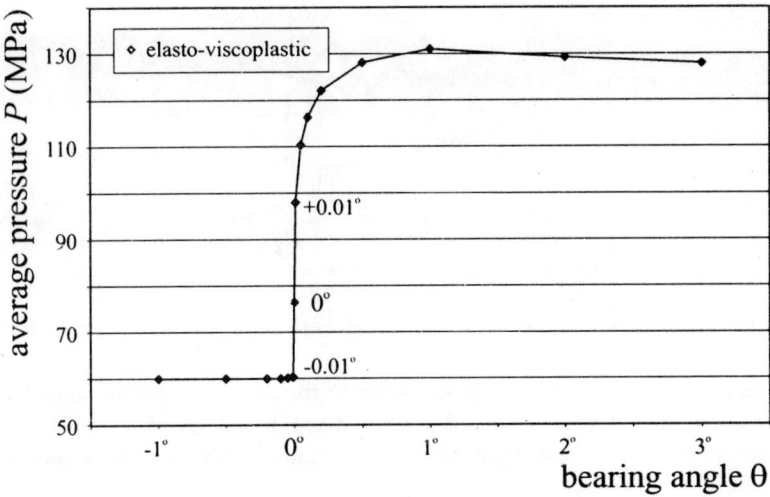

Figure 3. *Relation between the bearing angle and the resistance in the bearing channel*

The stability problem can be further explained based on Figure 3. Here, the bearing angle θ is plotted against the average pressure at the inflow side of the model. This average pressure is used as a measure for the resistance in the bearing channel. When the angle changes from negative to positive, the solution jumps from a situation without contact in the bearing channel to a situation with contact. This causes a very abrupt jump in the resistance in the bearing channel. The result is that for a parallel bearing it is impossible to predict the resistance. This brings to light a fundamental problem in extrusion practice, where use of a parallel bearing is common. In the opinion of the authors this problem is the most important contributing factor to the unpredictable behaviour of extrusion dies. The direct result of this unpredictability is the fact that the extrusion process has never left the trial and error stage because no commonly applicable design rules can be developed.

To remove this unpredictability from the process, either a negative or a positive bearing angle can be used. A negative angle has the distinct advantage that it significantly reduces the resistance in the bearing. However, the bearing length can no longer be used to control the exit velocity of the profile. For this reason a design as illustrated in Figure 4 can be used. A small chamber is placed in front of the bearing. The dimensions of this chamber can be used to control the local resistance of the bearing.

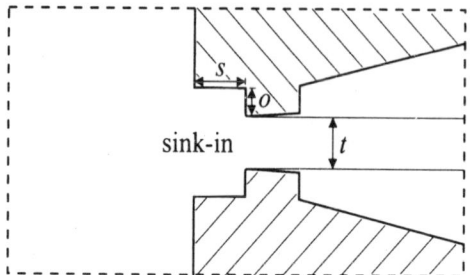

Figure 4. *Alternative bearing design with sink-in to control the local exit velocity*

With additional simulations the influence of the design parameters o and s have been charted and a design rule has been made based on the results of these simulations. This rule has been tested in practice with very satisfying results [LOF 00c].

4. Simulating the entire process

To analyse the aluminium flow through a complex extrusion die, it is necessary to make 3D simulations of the entire process. For the practical application of these simulations in an industrial environment a number of demands have to be met. First of all, the creation of the model must be easy and fast. Secondly the analysis time has to be reasonable. An elaborate description of the steps taken to ensure that these conditions are met is presented in [LOF 00a].

To make the simulations more efficient, the bearing is modelled with a so-called equivalent bearing model. This is a very coarse mesh of the bearing, which is constructed in such a way that it has the same characteristics as a much more detailed model. The radius in front of the bearing is modelled as a sharp corner. However, such a sharp corner exhibits locking effects, because of the boundary conditions on the perpendicular surfaces. This is illustrated in Figure 5 on the left. When only a limited number of elements are present in the bearing, this locking will affect the resistance in the bearing, effectively blocking a significant part of it. To avoid this a triple node is used. The node at the corner is replaced by three separate nodes, which are connected to the elements as illustrated in Figure 5 (middle). The degrees of freedom of these nodes are connected to each other as indicated. The resulting displacement field is indicated on the right in Figure 5. The equivalent bearing model was compared to a 2D reference model, in which the small rounding in the entrance of the bearing channel was meshed smoothly. Good agreement between the two models was found [LOF 00a]. By using the equivalent bearing model, a very suitable compromise between accuracy and analysis time is obtained. This enables simulations of complex dies to be carried out in a reasonable time frame (~10 hours).

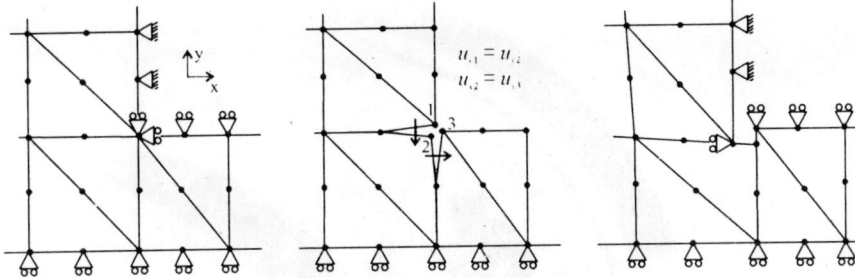

Figure 5. *Triple node construction to model the corner of the bearing (left: locking corner node; middle: triple node construction; right: resulting displacement field)*

To make the pre-processing as efficient as possible, an application was developed that automatically generates the numerical model. This application is integrated in Pro/Engineer, which is the CAD system used by PHOENIX to design extrusion dies for manufacturing. An illustration of a typical extrusion die is given in Figure 6. This geometry is obtained directly from the die maker. From this geometry, the geometry of the aluminium domain can be easily constructed. An automatic meshing algorithm is then used to mesh this aluminium domain (Figure 7). Boundary conditions and the equivalent bearing model are automatically applied to the model. To construct the triple nodes at the bearing entrance, the aluminium domain is split in three parts as can be seen in Figure 8. Using Pro/Engineer built in routines the triple nodes can automatically be generated. In addition to the aluminium model, a FEM model for the die is also constructed. This model can be used at a later stage to calculate the deformations and stresses that occur in the die. All these steps are performed within Pro/Engineer, with minimal input of the user. This application provides a very powerful and direct coupling between the die design in Pro/Engineer and the FEM analysis in DiekA.

Figure 6. *Typical extrusion die (quarter because of symmetry) with associated profile*

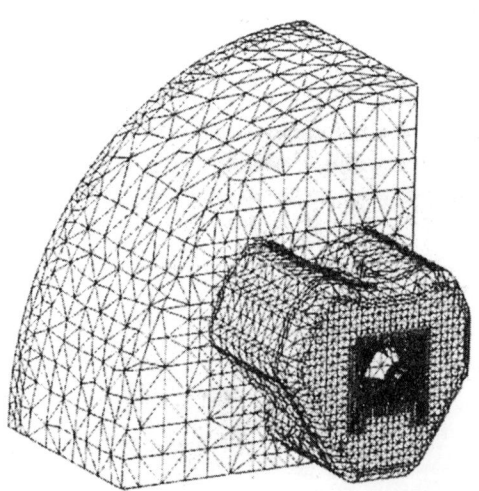

Figure 7. *Finite element mesh used for the aluminium flow simulation*

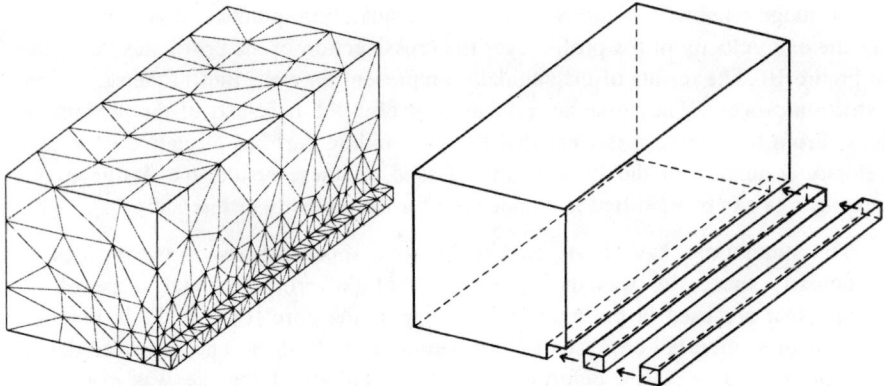

Figure 8. *Exploded view of the parts necessary for the construction of triple nodes*

A large variety of results can be obtained from the aluminium flow calculation. These include the velocities, stresses, equivalent plastic strain rate and reaction forces. The reaction forces are used in the stress analysis of the die. Also interesting are the velocities and the pressure distribution in the aluminium. The pressure on a cross-section through the die is shown in Figure 9. It clearly shows the pressure build-up from the bearing through the feeder holes to the billet.

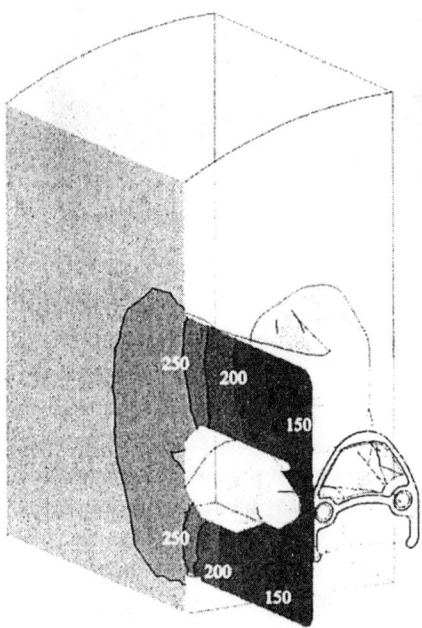

Figure 9. *Pressure distribution on a cross-section of the aluminium (MPa)*

To judge whether this die will perform adequately in practice, it is important to see the exit velocity of the profile over the cross-section of the profile, as illustrated in Figure 10. The results of the simulation represent the velocities at the start of the extrusion process. They give an indication of how the nose-end of the profile will look. From this figure it is clear that the thin middle part has a relatively low exit velocity compared to the thicker parts. Based on these results the design of the bearing area can be modified to compensate for this velocity difference.

The simulations have been used to improve specific parts of the die design. Particularly successful was the optimization of the cross-sectional shape of the bridges that are used in porthole dies to support the core [LOF 00d]. In practice, porthole dies often fail because of deformation of the bridges. The improved design did not show any plastic deformation and the failure of the die was eventually caused by wear on the bearing surface. This increased the average lifespan of the dies by a factor of three and also decreased the necessity for frequent die changes during production, resulting in a significant cost reduction.

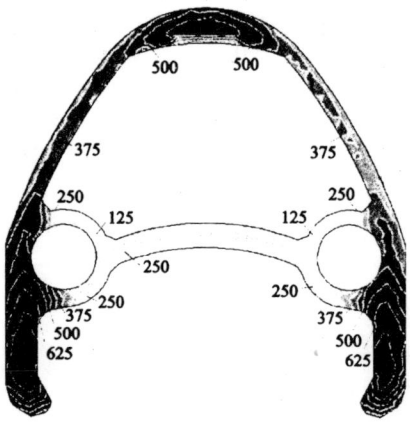

Figure 10. *Exit velocity of the profile (mm/s)*

5. Conclusion

With the help of numerical simulations we have gained important new insights into the mechanics behind the extrusion process. With these insights, we have been able to improve significantly on the design of extrusion dies. In addition, it has been possible to construct design rules [LOF 00c] based on the results of the simulation which remove to a large extend the subjective influence of die designers on the initial design of the extrusion die and replace this with an objective design method. The result is an improved performance of the dies and increased efficiency of the extrusion process.

In addition to this a powerful tool for simulation of extrusion dies, directly coupled to the die designers CAD system, is available. With this tool, other problems can be analysed and it is possible to investigate individual die designs with relative ease. The use of the equivalent bearing model ensures that analysis time is within a reasonable time limit. In the future, further improvements in the analysis code and increase of computer capacity can reduce the pre-processing time and the analysis time even further. This makes the simulations interesting for a growing number of situations.

Acknowledgements

This research was carried out within the framework of the Eureka project IPECTIES. This project was initiated by BOAL B.V. in De Lier, The Netherlands, and PHOENIX S.P.A. in Paderno, Italy. The support is gratefully acknowledged.

References

[DEV 92]. DEVADAS C and. CELLIERS O., "Metal flow during the extrusion process", in *Proc. 5th International Extrusion Technology Seminar*, pages 367–376, Wauconda, Illinois, 1992. Aluminum Association and Aluminum Extruders Council.

[GRA 92] GRASMO G., HOLTHE K., STØREN S., VALBERG H., FLATVAL R., HANSSEN L., LEFSTAD M., LOHNE O., WELO T., ØRSUND R., and HERBERG J., "Modelling of two-dimensional extrusion", in *Proc. 5th International Extrusion Technology Seminar*, pp. 367–376, Wauconda, Illinois, 1992. Aluminum Association and Aluminum Extruders Council.

[HUE 86] HUÉTINK J., *On the simulation of thermomechanical forming processes*, PhD Thesis, University of Twente, The Netherlands, 1986.

[JOH 62] JOHNSON W. and KUDO H., *The mechanics of metal extrusion*, Manchester University Press, 1962.

[JON 69] J.J. JONAS, C.M. SELLARS and McG TEGART W.J., "Strength and structure under hot-working conditions", *Met. Review*, Volume 14, pp. 1–24, 1969.

[KAN 94] KANG B.S., KIM B.M., and CHOI J.C., "Preform design in extrusion by FEM and its experimental confirmation", *J. Mater. Proc. Tech.*, 41:237–248, 1994.

[KAN 96] KANG Y.S. and YANG D.Y., "Investigation into the thermo-viscoplastic finite element analysis of square die extrusion of square section with Langrangian description", *Int. J. Mach Tools Manufact.*, 36:907, 1996.

[KUI 95] KIUCHI M., YANAGIMOTO J., and MENDOZA V., "Flow of solid metal during extrusion: 3D simulations by FE method", Shen and Dawson ed., *Simulation of Materials Processing: Theory, Methods and Applications*, pp. 847–852, Rotterdam, 1995. Balkema.

[LAN 00] VAN DE LANGKRUIS J., LOF J., KOOL W.H., VAN DER ZWAAG S. and Huétink J., "Comparison of experimental AA6063 extrusion trials to 3D numerical simulations using a general solute-dependent constitutive model", *Comp. Mater. Sci.*, Volume 18, pp. 381–392, 2000.

[LAU 81] LAUE K. and STENGER H., "Extrusion", *American Society for Metals*, 1981.

[LOF 00a] LOF J., *Developments in finite element simulations of aluminium extrusion*, PhD Thesis, University of Twente, The Netherlands, 2000.

[LOF 00b] LOF J. and VAN DEN BOOGAARD A.H., "Adaptive return mapping algorithms for J_2 elasto-viscoplastic flow", *Int. J. Numer. Meth. Engng.*, 2000 (accepted for publication).

[LOF 00c] LOF J., HUÉTINK J.and NILSEN K.E., "FEM simulations of the material flow in the bearing area of the aluminium extrusion process", *in Proc 7th international Extrusion Technology Seminar*, Vol. 2, pp. 211–222, 2000.

[LOF 00d] LOF J., KLASEBOER G., HUÉTINK J. and KOENIS P.T.G., "FEM simulations of aluminium extrusion using an elasto-viscoplastic material model", *Proc. 7th International Extrusion Technology Seminar*, Vol. 2, pp. 157–168, 2000.

[MOO 96] H.G. MOOI, *Finite element simulations of aluminium extrusion*, PhD thesis, University of Twente, The Netherlands, 1996.

[MOR 92] MORI K.I., OSAKADA K., and YAMAGUCHI H., "FE simulations of 3D extrusion of sections", in D.R.J. Owen ed., *3rd Int. Conf. Comp. Plast.*, pp. 1139–1149, Pineridge Press, 1992.

[REN 99] B.J.E. VAN RENS., *Finite element simulations of the aluminium extrusion process*, PhD Thesis, Eindhoven University of Technology, The Netherlands, 1999.

[TON 95] L. TONG., *FE simulation of bulk forming processes with a mixed Eulerian-Lagrangian formulation*, PhD Thesis, Swiss Federal Institute of Technology Zürich, Switzerland, 1995.

Chapter 10

Investigation of Springback Using Two Different Testing Methods

Martin Rohleder and Karl Roll
DaimlerChrysler AG, Production Planning Mercedes-Benz Passenger Cars, Sindelfingen, Germany

Alexander Brosius and Matthias Kleiner
University of Dortmund – Chair of Forming Technology, Germany

1. Introduction

For the manufacture of complex body, frame and running gear parts in the automotive industry various techniques of sheet metal forming are used. Typical for these complex sheet metal forming parts is inhomogeneous deformation, which leads, together with elastic-plastic material properties and the form drag caused by the geometry of the part, to deviations from the desired shape after unloading. This phenomenon is called *springback*.

Because of automated production and demand for high dimensional and shape accuracy nominal geometry deviation due to springback becomes a central concern (Roll *et al.*, 2000); (Wagoner *et al.*, 1996); (Roll, 1997). In the context of lightweight vehicle construction the problem is amplified by the increasing use of high-strength steels and aluminum alloys. These alloys show in general larger deviations due to springback compared with conventional mild steel (Beth, 1993). Despite intensive efforts throughout the last years a reliable prediction of springback deviations by means of finite element simulation is still not possible.

Process Conditions

➤ Forming Process
➤ Process Velocities
➤ Process Forces
➤ Tribology
➤ Temperature

Geometry

➤ Sheet Thickness
➤ Sheet Dimensions
➤ Tool Geometry

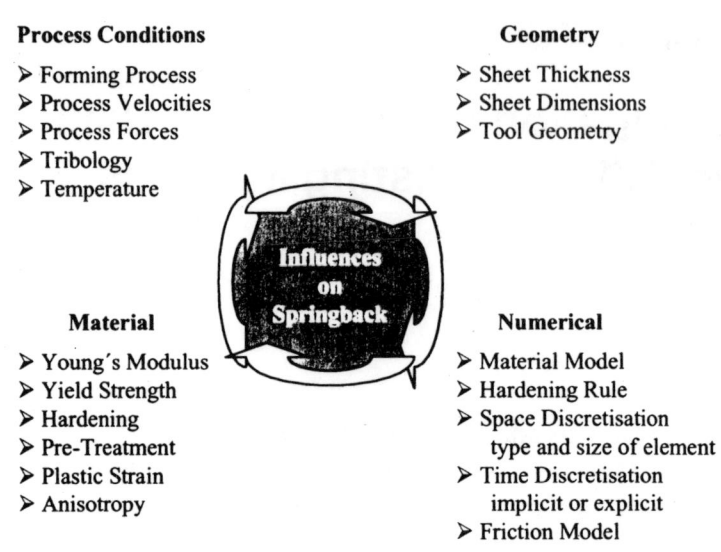

Material

➤ Young's Modulus
➤ Yield Strength
➤ Hardening
➤ Pre-Treatment
➤ Plastic Strain
➤ Anisotropy

Numerical

➤ Material Model
➤ Hardening Rule
➤ Space Discretisation
   type and size of element
➤ Time Discretisation
   implicit or explicit
➤ Friction Model

Figure 1. *Variables influencing springback predictions using finite element analysis [5]*

Springback results of sheet metal forming simulations are influenced by a large number of variables (Wagoner *et al.*, 1996). Physical and process specific variables influencing the amount of springback are already determined (Beth, 1993). If springback is calculated numerically in addition to the variables mentioned above, numerical variables have an influence, see Figure 1.

2. Draw-bend test

The draw-bend test (Li *et al.*, 1999) was used to investigate the capability of simulating springback due to bending loads with four commercial finite element codes, which make use of implicit as well as explicit time integration schemes and the Hill 1948 material description (N.N., 1999a; b; c); (N.N., 2000).

The principle of the testing method implies that bending and un-bending occur in the forming regions. This load reversal can be considered in the simulation by employing a material law with kinematic hardening.

2.1. Testing method

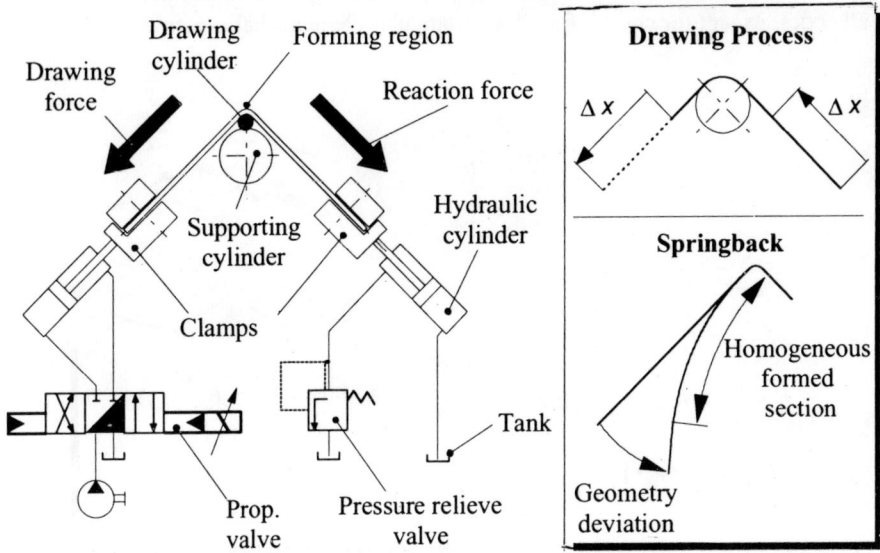

Figure 2. *Draw-bend test, principle and set-up*

First, a sheet metal strip was pre-bent on a CNC-press brake to a 90°-angle. In a second step a special device was used to draw the strips over a cylinder to realize bending and unbending conditions. For this, the endings of the strip were fixed with clamps while a defined drawing force was applied. The stress induced by the force acting on the sheet metal cross section was lower than the yield strength to ensure that only plastic deformations occurred in the forming regions. With the proportional valve the movement and the acting force of the clamps could be controlled. To achieve a homogeneous formed section the clamps were moved 2.5 times the cylinder circumference. After unloading, the deviation from the desired 90°-angle is considered to be the effect of springback (Figure 2).

The influence of friction on springback behavior has been excluded in this study.

2.2. Material, geometry and load parameters

For the investigations a mild steel DC04 with an initial sheet thickness $s_0 = 1.0$ mm was used. Due to the different radii of the drawing cylinders, the length of the strips varied to ensure the required movement of the clamps Δx. Figure 3 shows the influencing parameters as well as the applied drawing forces and the element lengths used. The drawing force F has to be lower than the yielding strength to avoid plastic deformation outside the forming region. Therefore a scaling factor f_i has been introduced in the equation below. Four-node shell elements have been used in all codes, except for code I, for which triangular elements have been used.

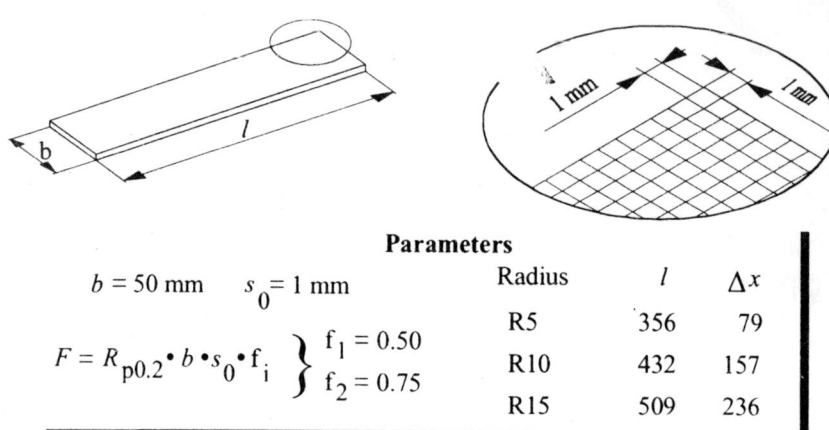

Parameters

$b = 50$ mm $s_0 = 1$ mm

$$F = R_{p0.2} \cdot b \cdot s_0 \cdot f_i \left.\begin{array}{c} f_1 = 0.50 \\ f_2 = 0.75 \end{array}\right\}$$

Radius	l	Δx
R5	356	79
R10	432	157
R15	509	236

Figure 3. *Parameters of experiments and simulations*

2.3. Evaluation strategy

To compare numerical and experimental results, the drawn parts have been measured with a 3D-coordinate-measuring machine (3D-CMM) along the line as shown in Figure 4.

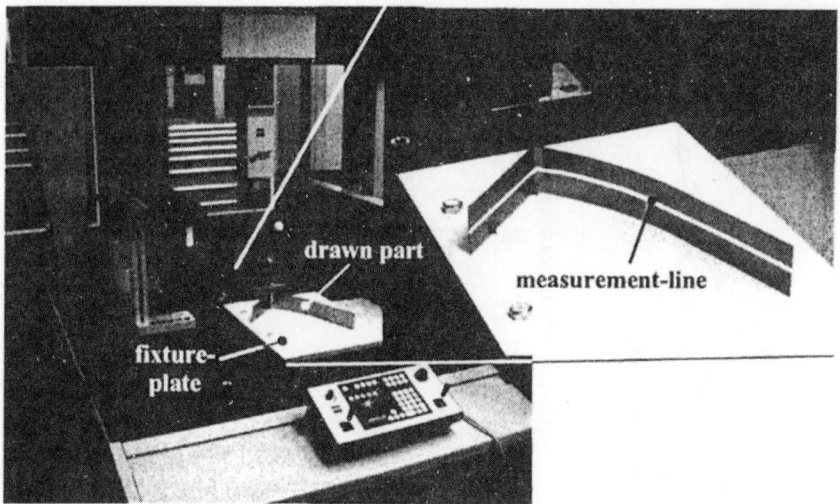

Figure 4. *Measurement line at the drawn part*

To ensure the validity of the comparison between simulation and experiment the mesh has been transformed to the fixture conditions used during the scan process with the 3D-CMM. A SIMPLEX-optimization algorithm (Schwefel, 1995) was applied to find the best fit into the fixture conditions. Moreover, the algorithm has been used to determine the distance from the measured geometry to the calculated mesh in the direction of the associated normal vectors.

2.4. Results of the draw-bend test

Figure 5 shows exemplarily for a radius R = 10 mm the result of the springback analysis at the scan-line. The springback behavior of the other considered radii is similar to the one displayed. On the left a contour-plot of the scan-line from the measurements is compared with the meshes. On the right the deviations from the measured data are plotted versus the abscissa. A positive value indicates a deviation in the direction of the positive normal vector.

With the exception of code I all programs compute a lower springback than the experimental one. This phenomenon can be explained by means of Figure 6.

The diagram shows the calculated deformation of the cross-section under load-conditions as well as the resultant geometrical moment of inertia. An increasing geometric moment of inertia leads to a greater geometric stiffness, which impedes the deformation after unloading. A comparison between the numerical results and measurements is still not possible, because only unloaded geometry can be measured by a CMM.

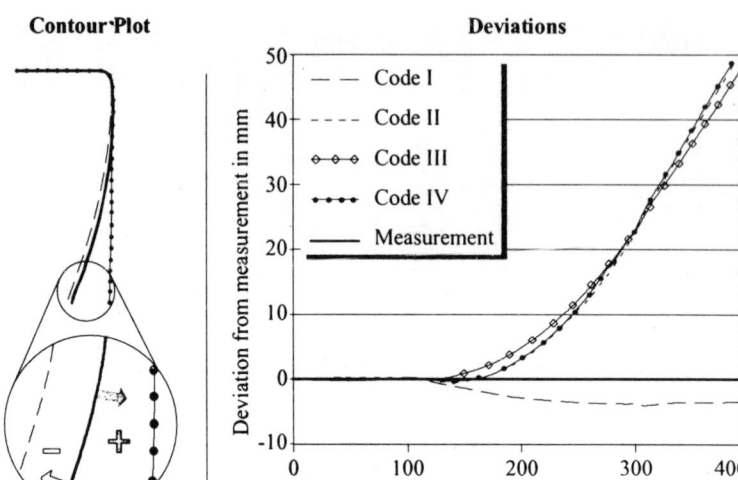

Figure 4. *Results of the draw-bend test*

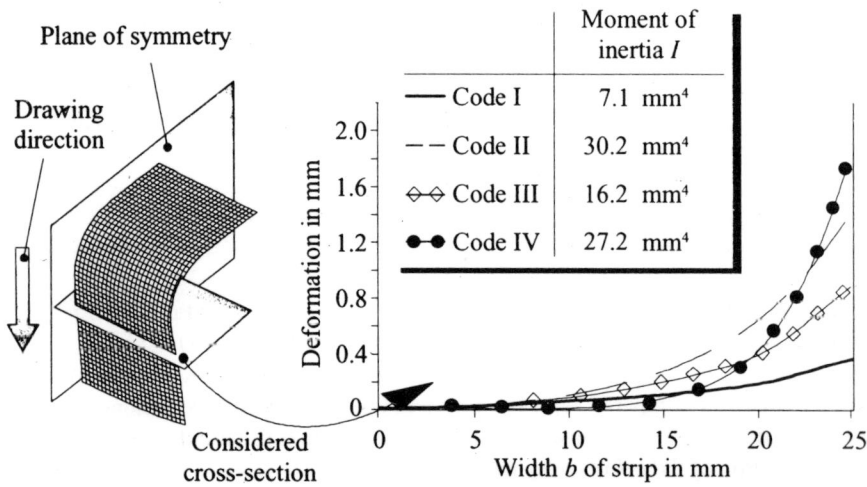

Figure 5. *Cross section of the strip*

However, the computed deformation due to springback is a consequence of the false calculated geometry during the forming simulation in form of the displayed curvature. The better results calculated with code I have to be verified, because the implemented shell is known as a stiff element to prevent the "shear-locking effect". In the present state of the investigation, it is assumed that the reasons for the warping can be found in the hardening law or element formulation. Therefore a sensitivity analysis of parameters to the kinematic hardening effect will be carried out next.

3. Deep drawing of a cup

The second test consisted of taking a ring out of the wall of a deep drawn cup and splitting it with a single axial cut to measure the residual radius (N.N., 1999).

3.1. Procedure

The deep drawing of the cup was carried out on a single acting hydraulic press with draw cushion. The experimental parameters during deep drawing were as follows: material mild steel DC04, initial sheet thickness $s_0 = 1.0$ mm, punch diameter $d_P = 147.6$ mm, die diameter $d_D = 152.0$ mm, punch radius $R_P = 8.0$ mm, die radius $R_D = 6.5$ mm, blankholder force $F_B = 134.0$ kN, drawing depth $h = 90.0$ mm.

The die clearance of $u_Z = 2.2$ mm was used to prevent ironing effects during the forming process. These effects can have a big influence on the forming result. However, the shell elements, used for the simulations, are not able to handle these physical effects.

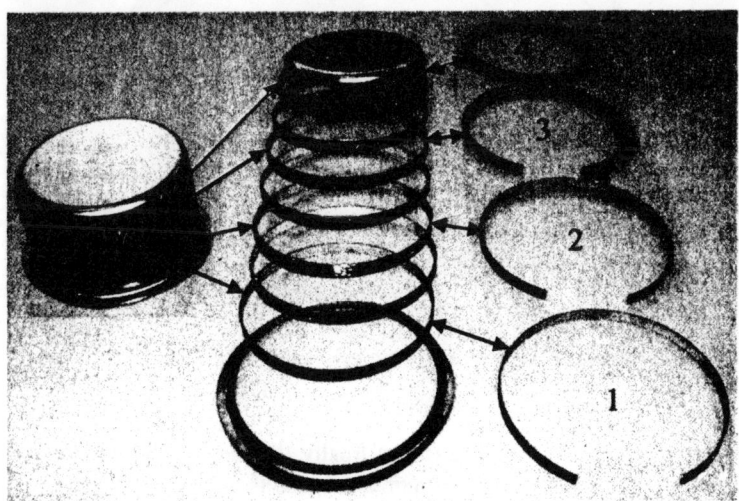

Figure 7. *Cutting and splitting procedure*

During deep drawing the stamping and blankholder force as well as the draw-in were measured. Thus, the forming conditions could be reproduced exactly in the finite element simulation and a reliable comparison of the results is guaranteed.

After deep drawing four rings were taken out of the wall of each cup using an electrical discharge machining to prevent the influence of heat and mechanical cutting methods. After that the rings were split with a single axial cut and positioned on a disk. The geometry of the rings was measured on a 3-dimensional measuring machine and the resultant radius was calculated along the curvilinear. One result is presented in Figure 7.

3.2. *Experimental results*

Figure 8 displays the experimentally determined springback of DC04. The curves represent the mean value of three performed experiments. Ring 1 and 2 showed the biggest amount of springback. The springback of ring 3 was less and ring 4 retained almost its initial shape.

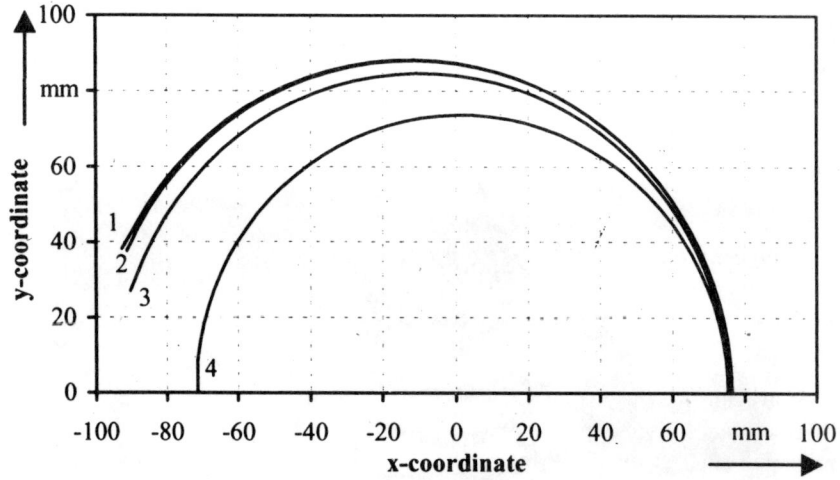

Figure 8. *Experimentally determined ring geometry*

The radius along the curvilinear and finally the mean radius of the rings were calculated for the evaluation of the numerical results. The determination of a mean radius was possible, because the radius of each ring varied only in a narrow range over the curvilinear. The maximum variation was 1.6%.

3.3. *Numerical results*

Initially, the calculations were performed using three different finite element codes, which mainly differ in the discretization of the time domain. As input data determined material, measured process and numerical parameters, recommended by each code developer, were used (N.N. 1999a, b, c). The friction coefficient was chosen to 0.07.

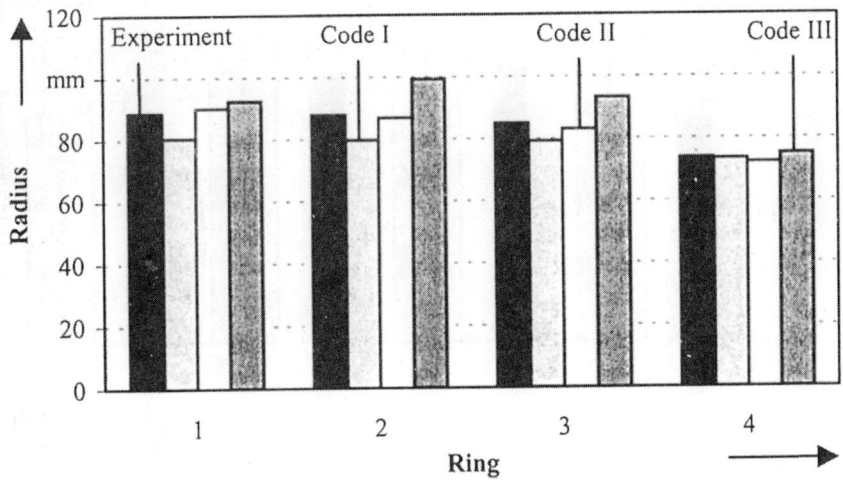

Figure 9. *Numerical results using different codes*

In Figure 9 the calculated and measured mean radii of all four rings are shown. A satisfactory result is that all three codes calculated the correct direction of springback and that the maximum deviation between simulation and experiment was less than 13.0%.

However, the results of the three codes compared to each other are quite different. As Figure 9 illustrates, the explicit forming simulation using a fully integrated shell element followed by an implicit springback calculation led to best results. The residual mean radius of ring 2 was nearly identical to the experimental result. Regarding ring 1 and 3 this code correlated best with the experimental data. The only exception represented ring 4, where the implicit simulation provided the best result. In general the fully implicit simulation underestimated and the fully explicit simulation overestimated the amount of springback.

The calculated springback deviation was very sensitive to numerical stamping results. Therefore simulations using three different friction coefficients and two different material models were performed. The numerical results at the end of the stamping as well as the calculated springback were compared with experimental results. This sensitivity study was carried out using code III, because it is easy to use and in general no convergence problems of the solutions occur.

In the present investigation a higher friction coefficient led to a better correlation between numerical and experimental springback results, see Figure 10.

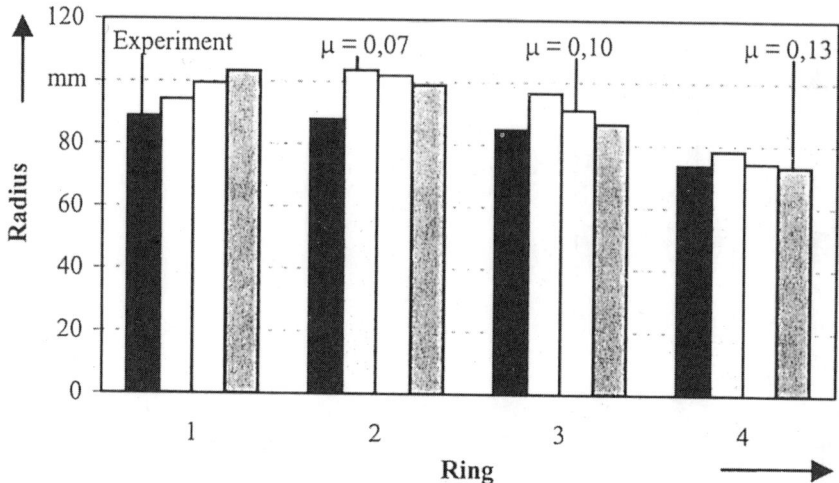

Figure 10. *Radius using different friction coefficients*

The reason for that is that better correspondence of numerical to experimental stamping results. For example draw-in and wall thickness at the end of the stamping are given in Figure 11 and Figure 12.

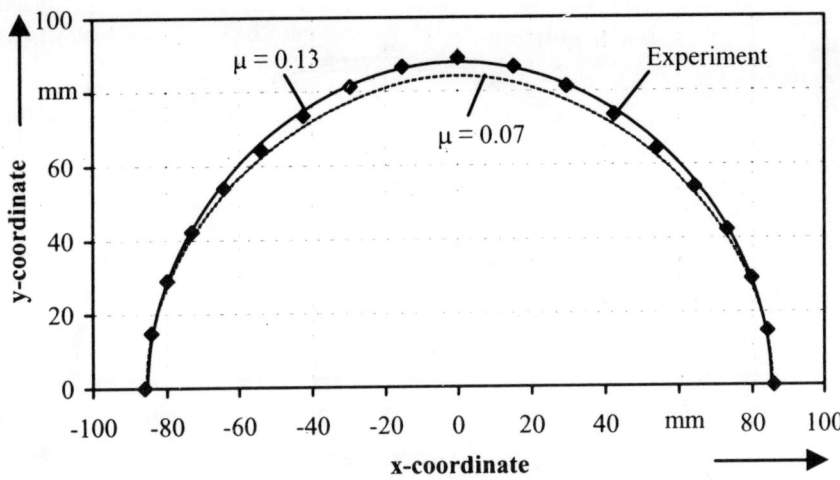

Figure 11. *Draw-in (flange) using different friction coefficients*

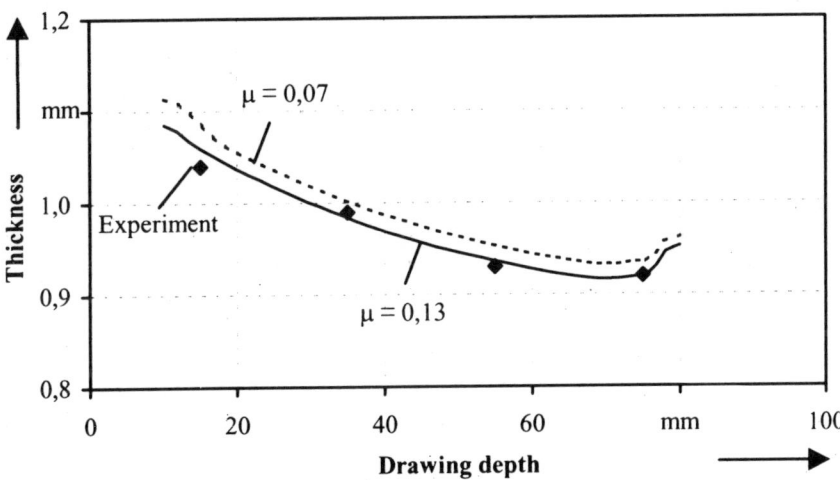

Figure 12. *Wall thickness using different friction coefficients*

Similar results were obtained using different material models. The three-dimensional Barlat, Lege and Brem 1991 model provided better results than Hill 1948 (Barlat *et al.*, 1991); (Hill, 1948). Figures 13 and 14 present the results.

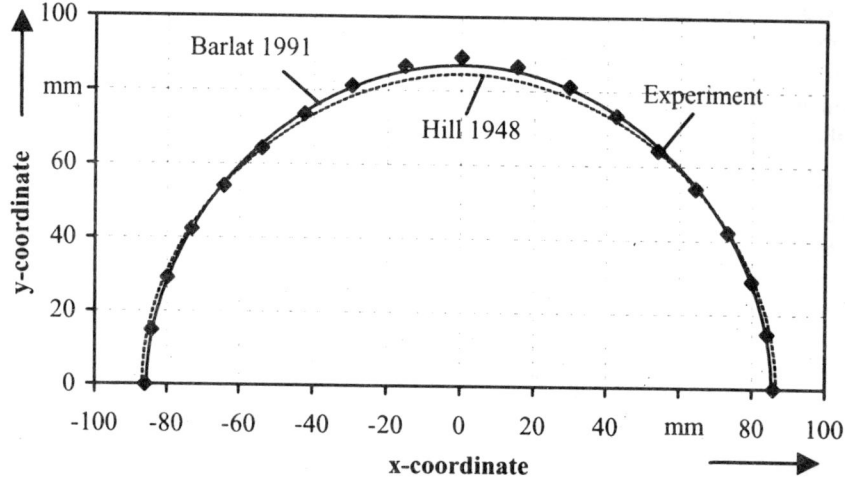

Figure 13. *Draw-in (flange) using different material models*

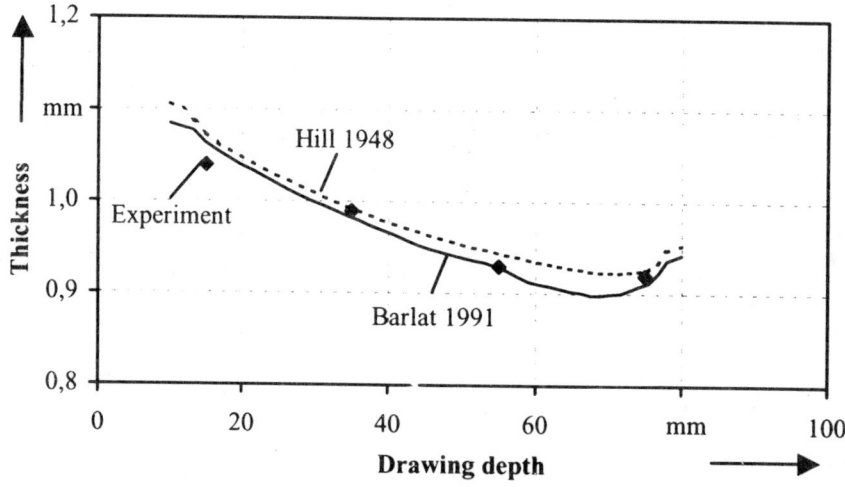

Figure 14. *Wall thickness using different material models*

4. Conclusion

The draw-bend test has been studied in order to test the capability of four different commercial finite element codes to predict springback. A comparison of the results showed that with the codes it is not possible to compute the deformations caused by springback satisfactorily. The reason for this seems to be related to the wrong calculated geometry in form of a large curvature region. This leads to an increased stiffness which prevents the springback deformation. The cause for this geometrical defect is assumed to result from the element formulation or parameters of the hardening law.

However, numerical springback results of taking a ring out of the wall of a deep drawn cup were in good correlation to experimental data. Concerning three different finite element codes, explicit forming followed by an implicit springback calculation led to best results. The two other codes over- and underestimate the amount of springback. The friction coefficient and material model had a large influence. To calculate springback deviations correctly, realistic stamping results are considered to be essential.

References

Barlat F., Lege D.J. and Brem J.C., "A Six-Component Yield Function For Anisotropic Materials", *International Journal of Plasticity*, Vol. 7, pp. 693–712, 1991.

Beth M., "Untersuchungen zum Rückfederungsverhalten von Feinblechen bei Tief- und Streckziehvorgängen", Dr.-Ing. Dissertation Technische Universität Darmstadt, 1993.

Hill R., "A Theory of Yielding and Plastic Flow of Anisotropic Metals", *Proceedings of the Royal Society*, Vol. 193, pp. 281–297, 1948.

Li K., Geng L. and Wagoner R. H., "Simulation of Springback: Choice of Element", *Proceedings of the 6th International Conference on Technology of Plasticity*, pp. 2091–2098, 1999.

N.N., "Benutzerhandbuch INDEED Version 7.3", *Innovationsgesellschaft für fortgeschrittene Produktionssysteme in der Fahrzeugindustrie mbH*, 1999.

N.N., "LS-DYNA3D, User's Manual Version 950", Livermore Software Technology Corporation, 1999.

N. N., "OPTRIS The Virtual Press, User's Guide Version 6.0", Dynamic Software, 1999.

N. N., "PAM-STAMP, Solver Notes Manual", Pam Systems International, 2000.

Rohleder M. and Roll K., "Springback Prediction on a Ring Taken from a Deep Drawn Cup", *Proceedings of the 9th International Conference on Sheet Metal*, pp. 595–604, 2001.

Roll K., Altan T., Tekkaya A. E. and Herrmann M., "Virtuelle Umformtechnik", Umformtechnik 2000 PLUS, pp. 255–274, 1999.

Roll K., "Simulationssysteme im Vergleich - Handlungsschwerpunkte für die Zukunft", 4. Sächsische Fachtagung Umformtechnik, Fraunhofer-Institut für Werkzeugmaschinen und Umformtechnik IWU Chemnitz, pp. 315–334, 1997.

Schwefel H.P., *Evolution and Optimum Seeking*, John Wiley & Sons, Inc., 1995.

Tang S. C. and Ren F., "On Solutions by Different Numerical Methods in Simulation of Sheet Metal Forming Process", *Proceedings of the 6th International Conference on Technology of Plasticity*, pp. 2149–2154, 1999.

Wagoner R.H. and He N., "Springback Simulation in Sheet Metal Forming", *Proceedings of the 3rd International Conference and Workshop on Numerical Simulation of 3-D Sheet Metal Forming Processes*, pp. 308–315, 1996.

Chapter 11

Modeling of AlMg Sheet Forming at Elevated Temperatures

A. H. van den Boogaard
Department of Mechanical Engineering, University of Twente, The Netherlands

P. J. Bolt and R. J. Werkhoven
TNO Industrial Technology, Eindhoven, The Netherlands

1. Introduction

Because of the need for weight reduction in the transport industry, the use of aluminum is taken increasingly into consideration [RWT 98]. But although aluminum is lighter than steel, the formability is less, hence press operations are often more critical. Press operations can be improved by optimizing the mechanical properties of the sheet by local heating or cooling [SHE 78, WIL 88, SCH 95a, SCH 95b, BOL 00, BOL 01]. An extra benefit of warm forming is that the stretcher lines that occur when 5xxx alloys are deformed at room temperature do not appear at elevated temperatures.

In deep drawing of an AA 5754-O aluminum cylindrical cup, the limiting drawing ratio could be increased from 2.1 to 2.6 by heating the flange up to 250ºC and by cooling the punch at room temperature as demonstrated in an experiment performed by the authors (see Figure 1). In this paper experiments with the AA 5754-O alloy are analyzed, to determine whether a numerical analysis can predict the punch force-displacement curves and the thickness distribution of the final product. Uniaxial tensile tests were performed at 4 different temperatures and at 2 strain rates. With the data from these experiments, the parameters for two material models were fitted. Finally, the cylindrical deep drawing experiments were simulated using these models. The results are presented in Section 3.

Figure 1. *Maximum cup heights achieved at 20°C (left) and with the flange at 250°C (right)*

2. Material model

2.1. *Experiments*

Two different material models were used for the analyses. First, a phenomenological model was used and, secondly, a so-called physically based model. The physically based model still has a number of parameters that are difficult to measure and hence are used as fitting parameters. The choice of parameters and state variables is, however, based on physical quantities like the dislocation densities, in contrast to the purely phenomenological models.

Both models give a flow stress as a function of the deformation path, temperature and strain rate. The translation of this (equivalent) stress to the general stress space is performed by an isotropic Von Mises yield surface.

Uniaxial tensile test experiments were performed at temperatures of 25°C, 100°C, 175°C and 250°C at strain rates of 0.002 and 0.02 s^{-1}. The resulting engineering stress–strain curves are presented in Figure 2.

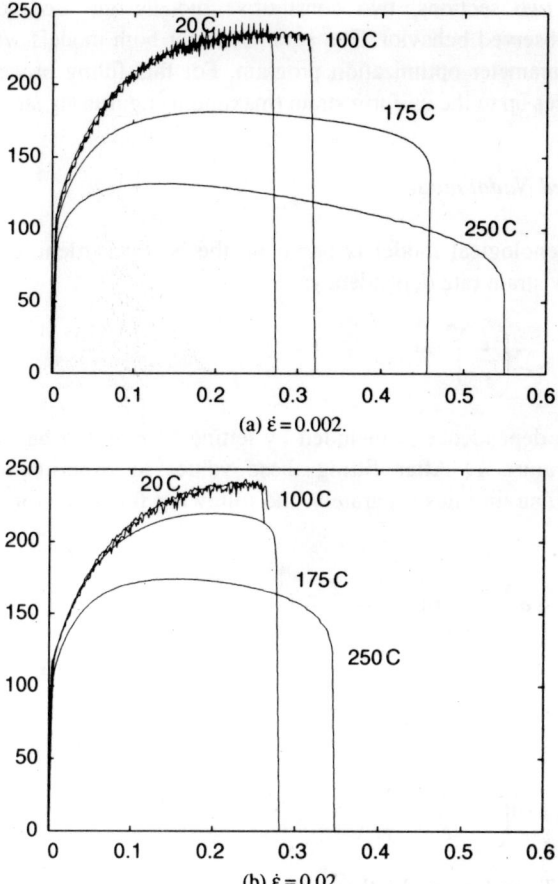

Figure 2. *Measured engineering stress–strain curves*

As can be seen from the stress–strain curves the typical serrations, attributed to the interaction between solute atoms and dislocations, disappear in the tests with a temperature at or above 175°C. It is noteworthy that up to a temperature of 100°C the influence of temperature and strain rate is marginal. Furthermore, the initial yield stress at 175°C is almost equal to the yield stress at room temperature, and at 250°C it is only 10 to 15 MPa less. At higher temperatures and especially at a low strain rate, the hardening decreases significantly leading to lower flow stresses, but also to a decreasing uniform strain (strain where the engineering stress is at a maximum). In spite of a decreasing uniform strain, the ultimate strain increases significantly, especially at low strain rate indicating a relatively stabile necking stage. This can be attributed to the relatively high strain rate dependency at higher temperatures.

In the next two sections, two constitutive models are used to describe the experimentally observed behavior. The parameters for both models were determined by a MATLAB parameter optimization program. For this fitting procedure, only the stress–strain values up to the uniform strain (maximum engineering stress) were used.

2.2. The extended Nadai model

The phenomenological model is based on the Nadai hardening law combined with a power law strain rate dependency:

$$\sigma = C(\varepsilon + \varepsilon_0)^n \left(\frac{\dot{\varepsilon}}{\dot{\varepsilon}_0} \right)^m \qquad [1]$$

The temperature dependence is included by letting C, n and m be functions of the absolute temperature T. After fitting these values to experiments at different temperatures and strain rates separately, the following functions for the parameters were chosen.

$$C(T) = C_0 + a_1 \left[1 - \exp\left(a_2 \frac{T - 273}{T_m} \right) \right] \qquad [2a]$$

$$n(T) = n_0 + b_1 \left[1 - \exp\left(b_2 \frac{T - 273}{T_m} \right) \right] \qquad [2b]$$

$$m(T) = m_0 \exp\left(c \frac{T - 273}{T_m} \right) \qquad [2c]$$

The parameters T_m and $\dot{\varepsilon}_0$ make the respective parts of the formulas dimensionless but are not independent of the other parameters, hence they can be given a suitable constant value. The remaining 9 parameters were fitted to the 8 uniaxial tensile tests simultaneously, resulting in the values presented in Table 1.

Table 1. *Parameters for the extended Nadai model*

T_m	800 K	a_1	132 MPa	n_0	0.32
ε_0	0.006585	a_2	3.5	m_0	0.009
$\dot{\varepsilon}_0$	0.002 s^{-1}	b_1	0.13	c	7.5
C_0	474 MPa	b_2	2.5		

2.3. *The Bergström model*

The physically based model used in this paper is a model described by Bergström and later adapted by Van Liempt [BER 69, BER 83, VL 94]. The model incorporates the influence of the temperature on the yield stress and on the hardening rate and includes recovery aspects. This model was initially used for the simulation of hot deformation of steel [RIE 99].

Basically the model determines the (equivalent) stress as:

$$\sigma = g(T)\,(\sigma_0 + \alpha G_{ref} b \sqrt{\rho} + \sigma^*(\dot{\varepsilon},T)) \qquad [3]$$

where the function $g(T)$ was originally defined by the ratio between the elastic shear modulus at temperature T and at the reference temperature T_{ref}: $G(T)/G_{ref}$. This choice is motivated by the idea that each mechanism of dislocation movement is related to the stress field in the lattice and hence to the actual shear modulus. The second part on the right-hand-side is the familiar Taylor equation, see e.g. [EST 96, MEY 99].

The essential part is the evolution of dislocation density ρ. This will give a temperature and strain rate influence on the hardening, while σ^* yields an instantaneous temperature and strain rate influence on the flow stress.

The dynamic stress σ^* is commonly defined as

$$\sigma^*(\dot{\varepsilon},T) = \sigma_0^* \left\{ 1 + \frac{kT}{\Delta G_0} \ln\left(\frac{\dot{\varepsilon}}{\dot{\varepsilon}_0} \right) \right\}^p \qquad [4]$$

for $\dot{\varepsilon}_0 \exp(-\Delta G_0 / kT) < \dot{\varepsilon} < \dot{\varepsilon}_0$ and with k the Boltzmann number. This formula induces a translation of the stress–strain curves, with a magnitude that decreases with increasing temperature. In the experimental stress–strain curves (Figure 2) it was observed that the small influence on the initial yield stress present at low temperatures does not decrease at higher temperatures. Therefore, the contribution of σ^* is neglected altogether in this paper. This means that all the influence of the temperature on the flow stress is introduced indirectly by the influence on the hardening rate. For fcc alloys, this behavior is also noted in literature e.g. [YAO 00]. Note that in the present extended Nadai model the strain-rate and temperature influence acts directly on the flow stress.

The evolution of dislocation density is formulated as a differential equation:

$$\frac{d\rho}{d\varepsilon} = U(\rho) - \Omega(\dot{\varepsilon},T)\rho \qquad [5a]$$

with

$$U = U_0 \sqrt{\rho} \qquad [5b]$$

$$\Omega = \Omega_0 + C \exp\left(-\frac{mQ_v}{RT} \right) \dot{\varepsilon}^m \qquad [5c]$$

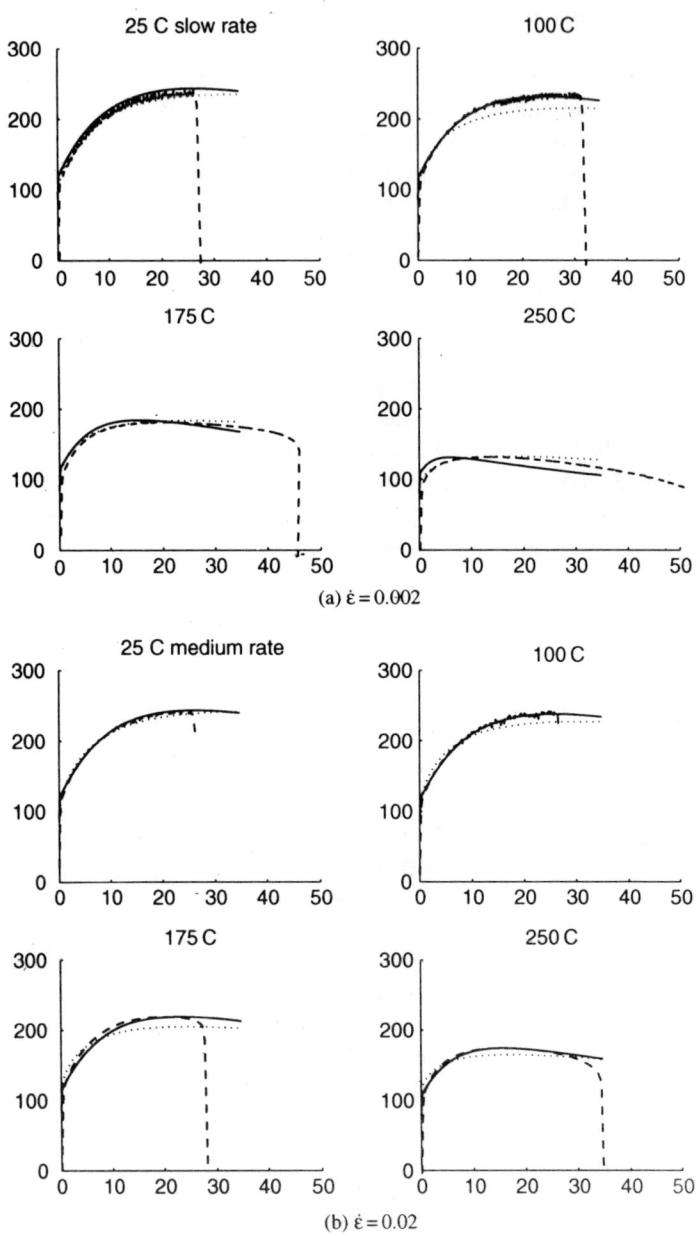

Figure 3. *Stress–strain curves at different temperatures, experiments (dashed), Bergström model (solid) and Nadai model (dotted)*

The function U represents storage of mobile dislocations (making them immobile), and Ω represents remobilization or dynamic recovery. The functions U, and especially the function Ω determine the shape of the hardening curve at different temperatures and strain rates.

The initial dislocation density was chosen to be 10^{11} m^{-2} which seems to be a reasonable value for annealed aluminum. A ten times lower or higher value only had a small influence on the initial stages of plastic deformation. The magnitude of the Burger's vector b was taken from literature and a value of $\alpha = 1.0$ was chosen.

The function $g(T)$ was fitted as a third order polynomial with the initial yield stress at a strain rate of 0.02 because the original scaling with $G(T)$ yielded a too strong decrease of the initial yield stress.

The remaining 6 parameters were fitted to the 8 tensile tests simultaneously and resulted in the values presented in Table 2.

Table 2. *Parameters for the Bergström model*

σ_0	115.8 MPa	m	-0.715	$\rho(t=0)$	$10^{11} m^{-2}$
α	1.0	U_0	$4.591 \cdot 10^8$ m^{-1}	σ^*	0 MPa
b	$2.857 \cdot 10^{-10}$ m	Ω_0	14.78		
C	12311	Q_v	58950 J/mol		

In Figure 3 the simulated engineering stress–strain curves are plotted, together with the experimental data. It can be seen that both models are more or less capable of describing the experiments. It should be noted that the comparison is only valid for uniform strain, which means up to the maximum engineering stress.

For the extended Nadai model the flow stress at 100°C is underestimated in order to reach a low enough flow stress at 175 and 250°C. The Bergström model does not perform very well if the initial yield stress is overestimated as in the low strain rate case at 250°C.

Although the models yield more or less similar stress–strain curves for constant strain rate simulations, large differences appear if a jump in the strain rate is applied. In Figure 4 the stress–strain curves are plotted for deformation at 175°C and for strain rates of 0.002 s^{-1} (lower curve) and 0.02 s^{-1} (upper curve). If a strain rate change from 0.002 to 0.02 s^{-1} or from 0.02 to 0.002 s^{-1} is applied after a strain of 5%, the Nadai model immediately follows the curve corresponding to a constant strain rate. With the Bergström model the constant strain rate curve is only slowly approached after continuous straining. Some initial experiments showed that real material behavior is something in between and should be investigated more thoroughly.

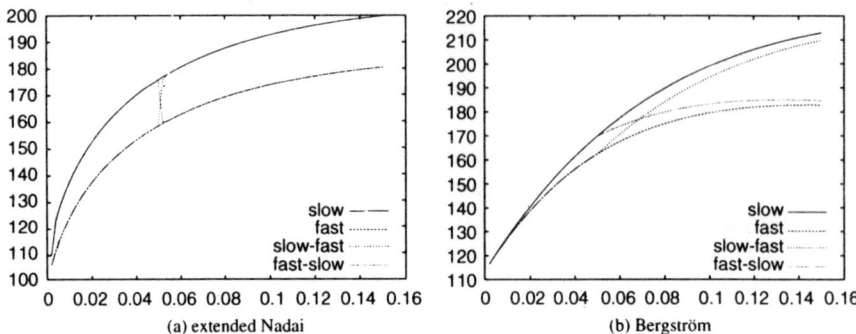

Figure 4. *Stress–strain curves with and without strain-rate jumps*

3. Deep drawing experiments and simulations

A number of cylindrical cups were deep drawn at different temperatures as shown in Figure 1. The principal compressive or tensile strains are schematically indicated in Figure 5 for the 3 distinctive areas in the cup. These experiments were simulated with the two material models, described in the previous section.

All experiments and simulations were performed with blanks of 230 mm diameter and a punch stroke of 80 mm. The punch diameter was 110 mm and the punch radius was 10 mm. The die hole diameter was 113 mm and the die radius was 15 mm. The blank holder force was equivalent to an initial pressure of 1.0 MPa on the blank holder/sheet contact area. A water based lubricating paste was used which still gave sufficient lubrication at 250°C.

Three experiments were performed with a punch velocity of 120 mm/min. The nominal die and blank holder temperatures were 20°C, 175°C and 250°C, while the punch was kept at 20°C. The die and blank holder were heated with internal heat rods, and the punch was cooled with water through internal channels. The temperatures of the punch and die were measured with thermo couples.

A model with 4 node constant dilatation axi-symmetric continuum elements was used for the simulation of the deep drawing experiments. The sheets of 1.2 mm thickness were modeled with 2 elements in thickness direction and an element size of 1 mm in radial direction. The extended Nadai material model was implemented as a user routine in MSC.MARC. In this model also a part of the punch, die and blank holder were modeled, including heat rods and cooling channels. From these analyses it appeared that the sheet in contact with the punch or the die/blank holder takes the temperature of that tool very rapidly. The die shoulder and punch radius showed only a small cooling, respectively heating, during the analysis. The simulation with the Bergström model was performed with the in-house code DIEKA. With the experience from the complete thermal analysis, the tools were now modeled as rigid contours with a prescribed temperature.

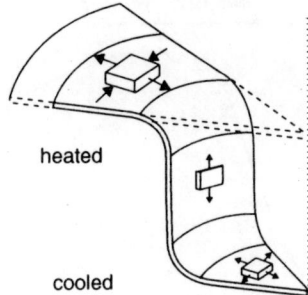

Figure 5. *Different stress states in a deep drawn cup*

The friction between tool and workpiece is one of the less known factors in the simulation. In the simulations a Coulomb friction coefficient of 0.06 is assumed between tool and workpiece. This value was measured experimentally at room temperature. It was recently shown that for this combination of materials and lubricants, at 175°C and 250°C, the friction coefficient can be two to three times as high as at room temperature.

In Figures 6 and 7, the force-displacement diagrams of the punch and the thickness distributions of the cup at a depth of 80 mm are plotted for the experiments and the simulation with extended Nadai and Bergström model, respectively.

4. Discussion

In the extended Nadai model the flow stress directly depends on the equivalent strain, strain-rate and temperature. In the Bergström model an evolution equation for the dislocation density is solved. As a result, the Nadai model will show a different flow stress upon strain rate change directly, while the Bergström model will reach a new flow stress only after some additional strain. It was expected that this difference would be clearly visible in the simulation of the deep drawing experiments, since strain-rate and temperature are not constant in that process.

Comparing the different punch force-displacement curves, it can be seen that both numerical models underestimate the experimental curves. The underestimation is more severe with the extended Nadai model than with the Bergström model. The wiggles in the numerical curves are due to not fully converged increments.

The trends with changing temperature are predicted well, but the difference between 20°C and 175°C is overestimated. The influence of the temperature on the thickness after 80 mm punch stroke is most pronounced in the bottom of the cup. The thickness reduction in the bottom of the cup is exaggerated in the simulations with both models.

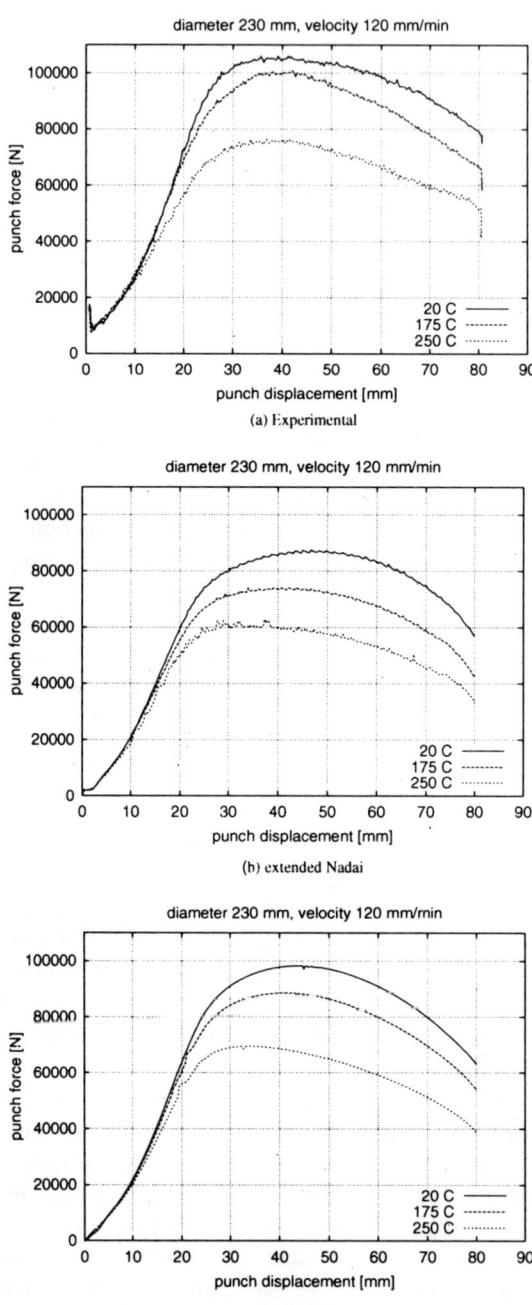

Figure 6. *Punch load-displacement curves*

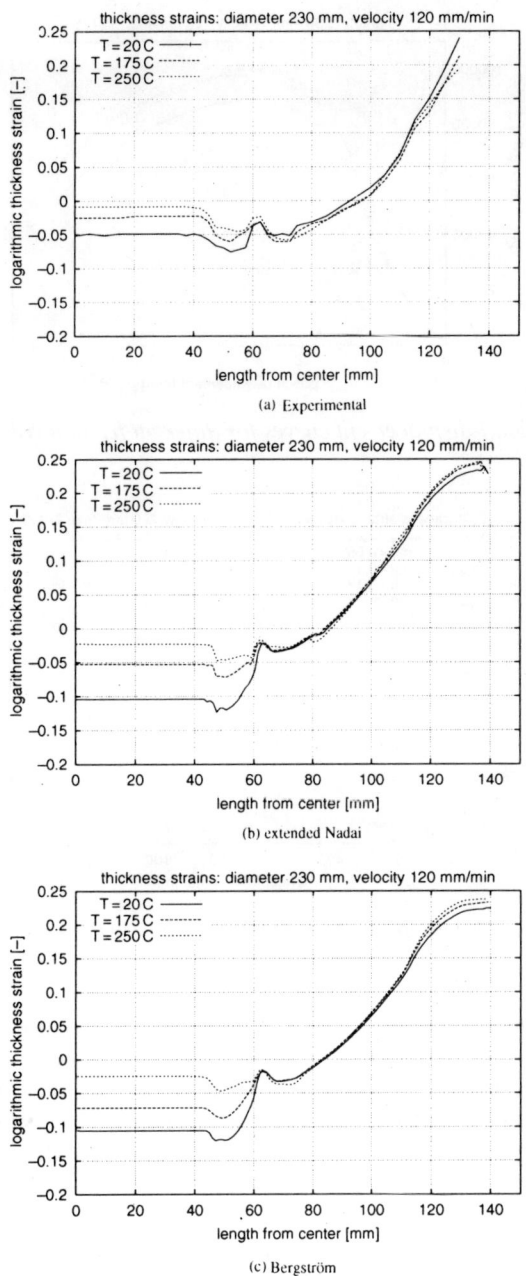

Figure 7. *Thickness distributions*

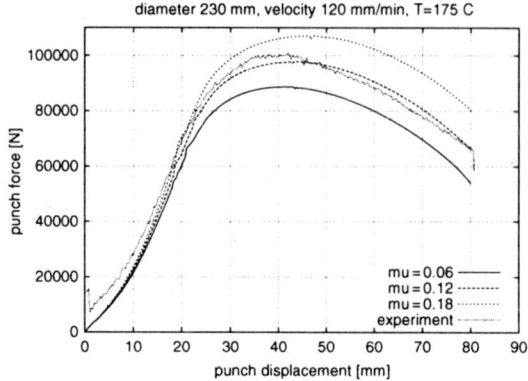

Figure 8. *Punch load-displacement curves for different friction coefficients*

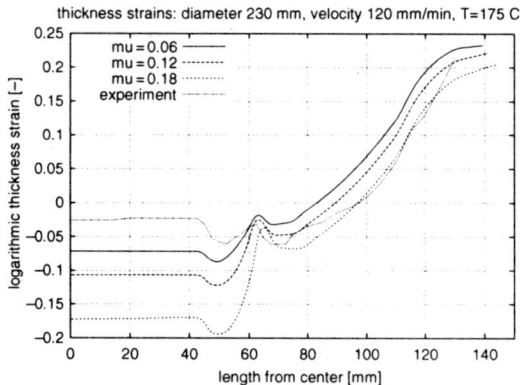

Figure 9. *Thickness distributions for different friction coefficients*

In the simulations of the deep drawing experiments the friction between tool and workpiece is one of the fundamental unknowns. Values, based on room temperature experience were used, but it is clear that these values are likely to change as the temperature increases. High temperature experimental data was only obtained recently. For temperatures of 175°C and 250°C, friction coefficients between 0.12 and 0.18 were measured. In Figure 8 the influence of friction is shown for an analysis with the Bergström model. The friction coefficient between punch and blank was kept at 0.06, but the friction coefficient between blank holder and blank and between die and blank was varied between 0.06, 0.12 and 0.18. With a friction coefficient of 0.12 the calculated punch force–displacement diagram resembles the experimental one quite well. Assuming that the friction coefficient at room temperature is indeed around 0.06, this also explains the overestimation of the difference between punch force at 20°C and at 175°C.

As a result of the increased friction however, the prediction of the thickness strain deteriorates even more (Figure 9). Probably the most important reason for the overestimation of the thickness strain is the adoption of an isotropic Von Mises yield surface. The yield surface of aluminum in principal stress space has a much more corner-like shape in the equi-biaxial area. Initial numerical tests with room-temperature values showed that by adopting a more appropriate yield function the thickness strain prediction improved, however, not yet up to the experimental values.

It is interesting to note that application of anisotropy alone is not enough, due to the so-called anomalous behavior of aluminum. By using a quadratic Hill yield function with realistic Lankford R-values (less than one), the thickness decreases even more than with the Von Mises yield surface. This was to be expected, because in that case the stresses at the equi-biaxial point decrease compared to the stress at the plane strain point. As can be seen in Figure 5, the bottom is expected to strain more in that case. Only by using a more appropriate shape of yield surface the predicted thickness strain in the bottom of the cup improves, even with R-values less than one. The influence of the yield surface will be investigated in the near future with 3D shell models.

With the deviations between simulation and experiment, the differences between the extended Nadai and the Bergström model cannot be decisively interpreted as an advantage of one over the other.

Acknowledgement

The work presented in this paper was performed as part of project ME 97033 of the Netherlands Institute for Metals Research (www.nimr.nl).

References

[BER 69] Y. BERGSTRÖM. "Dislocation model for the stress-strain behaviour of polycrystalline α-Fe with special emphasis on the variation of the densities of mobile and immobile dislocations". *Mater. Sci. Eng.*, vol. 5, 193–200, 1969.

[BER 83] Y. BERGSTRÖM. "The plastic deformation of metals - a dislocation model and its applicability". *Reviews on Powder Metalurgy and Physical Ceramics*, vol. 2, 105–115, 1983.

[BOL 00] P. J. BOLT, N. A. P. M. LAMBOO, J. F. C. VAN LEEUWEN, R. J. WERKHOVEN. "Warm drawing of aluminium components". In *Proceedings of the 7th Saxon Conference on Forming Technology, Lightweight Construction by Forming Technology*, 101–118. Chemnitz, 2000.

[BOL 01] P. J. BOLT, N. A. P. M. LAMBOO, P. J. C. M. ROZIER. "Feasibility of warm drawing of aluminium products". *J. Mater. Process. Technol.*, vol. 115, 118–121, 2001.

[EST 96] Y. ESTRIN. "Dislocation-density–related constitutive modeling". In A. S. KRAUSZ, K. KRAUSZ, eds., *Unified Constitutive Laws of Plastic Deformation*, 69–104. Academic Press, San Diego, 1996.

[MEY 99] M. A. MEYERS. "Dynamic deformation and failure". In M. A. MEYERS, R. W. ARMSTRONG, H. O. K. KIRCHNER, eds., *Mechanics and Materials; Fundamentals and Linkages*, 489–594. John Wiley & sons, Inc., New York etc., 1999.

[RIE 99] A. D. RIETMAN. *Numerical Analysis of Inhomogeneous Deformation in Plane Strain Compression*. Ph.D. thesis, University of Twente, 1999.

[RWT 98] "Innovativer werkstoffeinsatz in kraftfahrzeugen, kolloquium RWTH Aachen". vol. Tagungsband 719. Verlag und Vertriebsgesellschaft mbH, Düsseldorf. 1998.

[SCH 95a] D. SCHMOECKEL, B. C. LIEBLER, F.-D. SPECK. "Temperaturgeführter stoff-flußbeim tieziehen von Al-blech - grundlagen und modellversuche". *Bänder Bleche Rohre*, vol. 36(6), 14–21, 1995.

[SCH 95b] D. SCHMOECKEL, B. C. LIEBLER, F.-D. SPECK. "Temperaturgeführter stoff-flußbeim tieziehen von Al-blech - realversuche". *Bänder Bleche Rohre*, vol. 36(7/8), 24–27, 1995.

[SHE 78] F. SHEHATA, M. J. PAINTER, R. PEARCE. "Warm forming of aluminium/magnesium alloy sheet". *Journal of Mech. Work. Techn.*, vol. 2, 279–291, 1978.

[VL 94] P. VAN LIEMPT. "Workhardening and substructural geometry of metals". *J. Mater. Process. Technol.*, vol. 45, 459–464, 1994.

[WIL 88] D. V. WILSON. "Aluminium versus steel in the family car - the formability factor". *J. of Mech. Working Technol.*, vol. 16, 257–277, 1988.

[YAO 00] X. YAO, S. ZAJAC. "The strain-rate dependence of flow stress and work-hardening rate in three Al-Mg alloys". *Scandinavian Journal of Metallurgy*, vol. 29, 101–107, 2000.

Chapter 12

Hydroforming Processes for Tubular Parts: Optimization by means of Adaptive and Iterative FEM Simulation

Wim H. Sillekens and Robert J. Werkhoven
TNO Industrial Technology, Department of Manufacturing Development, HE Eindhoven, The Netherlands

1. Introduction

Hydroforming is a manufacturing technique for metals that is currently in the spotlight. In this process – employed for tubular but also for sheet-metal parts – the product is being plastically shaped by means of hydraulic pressure, usually assisted by mechanically applied forces. Those processes that feature a combination of mutually independent working loads enable the process to be actively controlled, meaning that defects may be averted by adapting the stress situation in the material. The quality of the product can thus be optimized within relatively wide boundaries. Figure 1 gives some examples for tubular parts. These include bulging of pre-bent tube (top), T-shape bulging with a counter punch (middle), and expansion in an open die (bottom). Most processes use internal fluid pressure as in these examples, but some variants (*e.g.,* for joining a tube to a mandrel) act with external pressure. Commonly, the working loads on the tube gradually rise as the process proceeds but some steep increase in fluid pressure is enforced once that the die is filled. The latter is called the *calibration step*.

Research and technological development work in this field, as well as industrial application, has boomed during the last century; see for instance [DOH 97], [AHM 00] and [SIE 00] for a review. For an increasing number of products, the assets (concerning such aspects as designing freedom, accuracy and reproducibility, component integrity, process integration and flexibility) are well recognized and prevail over any possible drawbacks (press and tooling costs, production rate, pre- and post-treatments). Applications are in lightweight products, integral parts and in products with high added value; *e.g.,* components for cars (exhaust parts, engine cradles, etc.), office furniture and sanitary appliances. Product materials include stainless steel, mild steel, high-strength steel, and aluminium as well as other non-ferrous metals in wrought alloys.

Where the use of multiple working loads adds to the versatility of the process, the accomplishment of a satisfying parameter setting (or *loading path*) is still a matter of concern during the start-up phase of a specific production. These process conditions determine not only whether the product is sound – that is, free of fractures and the like – but also affect such further quality demands as product dimensions, mechanical properties and surface condition.

The article at hand addresses some options of attaining the "optimal" loading path for hydro-mechanical forming of tubular parts by means of finite-element process simulation. After outlining the adopted methods, results for some typical cases are presented and discussed. These test cases concern a tube-expansion and a T-shape bulging situation (both die-bound). Calculations are evaluated by considering the obtained loading path, the wall-thickness distribution and the principal plastic strains, as well as a comparison with some experimental data. To start with, however, the distinct aspects concerning the quality of a hydroformed product are recapitulated in the next section.

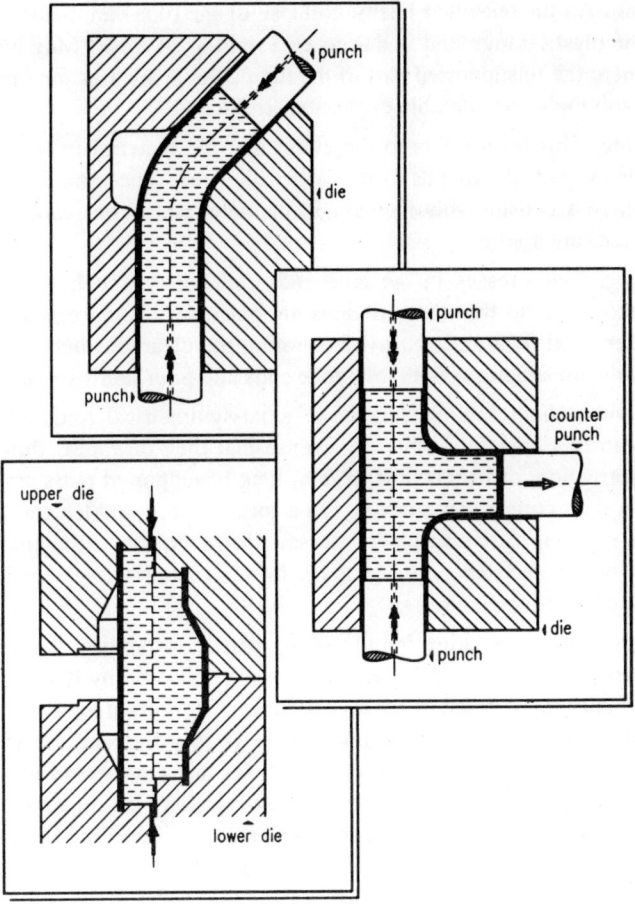

Figure 1. *Some hydro-mechanical forming processes*

2. Quality considerations

The outcome of a metal-forming operation is determined by a large number of parameters. Quality aspects are numerous too, but can roughly be classified according to whether they relate to *process limits* or *product performance*. The next listing is of interest to the hydroforming of tubes.

2.1. *Process limits*

A forming process must in the first instance be designed in a way that it yields products with no apparent flaws. Typical defects are the following:

– Buckling. An uncontrolled lateral collapse of the tube can occur in the elastic as well as the plastic range and is due to excess in axial force. May be critical for situations where the unsupported part of the tube is long, such as for free bulging or processes involving a considerable expansion length.

– Reversing. This term refers to the effect that the material in the feed-in zone folds inwards instead of expands to the larger diameter. The limit is of interest for situations where a considerable increase in diameter has to be realized and where high axial forces are applied.

– Upsetting. The stresses in the tube ends can become so high that the yield strength is exceeded, so that these regions are plastically compressed and possibly impede further axial feeding; *e.g.,* by penetrating the clearance between punch and die. Susceptible are situations with long tube ends and poor lubrication.

– Folding/wrinkling. The appearance of axial-symmetrical folds embracing the whole tube can be either reversible, in a sense that they disappear during the next stages of deformation, or irreversible. Again, long unsupported parts are susceptible in connection with compressive forces. On a local scale, wrinkles can occur under the influence of compressive in-plane stresses. However, the origination of these is of a complex nature and depending on many factors (material, wall thickness, stress and strain state, die contact, etc.).

– Necking/bursting. Necking is induced by plastic instability. It means that the deformation localizes in a confined area (thinning), followed by bursting (fracture). Necking is usually represented in the so-called forming-limit diagram, depicting the two principal plastic in-plane strains on its axes. In this diagram, the forming-limit curve or FLC quantifies those strain combinations that correspond to the onset of necking for a given material, and are thus critical. Necking/bursting is a typical concern for situations where a large local expansion in surface area is involved (severe stretch bulging, filling of tight die corners).

2.2. Product performance

Apart from the fact that the product must be free of defects, it has also to fulfil a number of requirements regarding functional or in-service behaviour. These product-specific demands can be quite diverse, but commonly fall within these categories:

– Geometrical tolerances. Usually, the acceptable deviations from the nominal shape and dimensions are specified. Some control over geometrical aberrations exists during the calibration step, since this affects elastic spring-back as well as conformity to (tight) die corners and other geometrical details. The wall-thickness distribution in particular depends on the loading path and may be of importance in connection with assembly, structural integrity and overall weight.

– Structural integrity. This covers several issues on the mechanical behaviour of the finished part, including its strength and stiffness, crashworthiness, and fatigue.

These depend on the tube material and the product geometry, but also on the process since this determines such matters as wall-thickness and strain distribution.

– *Surface characteristics*. The surface roughness generally increases for a sheet metal subjected to plastic deformation (orange-peel effect). Although difficult to quantify, the resulting surface topography will depend on initial roughness, grain size and texture, amount of strain and strain path, as well as on die contact. However, it is apparent that surface roughening in hydroforming can be more severe than in conventional sheet-metal forming since strains are generally greater and mainly come about in a free deformation mode. Apart from this, surface damage such as scores (scratches) and tool marks (impressions) can be inflicted on the part due to the high contact pressures and relative movement between tube and tooling. Lubrication is a key factor in this respect.

Before embarking on the subject of process optimization, the aim of such a procedure should be specified. Restricting ourselves to the loading path, "optimal" process conditions are defined here as the combination of and relationship between the working loads that effectively avert any process limits and renders the best product performance. Clearly, this solution depends on any of the above-mentioned parameters and is therefore specific for each particular hydroforming job.

3. Approach adopted for process optimization

In the modeling of forming processes, the finite-element method (FEM) has established itself as the prevailing tool for analysis, with a variety of software codes available. The most straightforward approach is to use an FEM model with a predetermined set of control parameters and evaluate the process simulation with respect to the resulting geometry, plastic strains, tool loads and so on. Some recent studies into tube hydroforming deal with coupled analysis incorporating an elasto-plastic damage model in order to predict ductile fracture [SAA 01], and sensitivity analysis regarding material parameters and loading path [MAS 01].

Concerning the FEM optimization of such forming processes, one avenue is to incorporate a feedback algorithm into the model that is able of (semi)-automatically assessing the preferred parameter settings for a given process. Generally, this can be done by including some criterion that evaluates the previous step by monitoring if some undesired phenomenon occurs and an optimization procedure that yields an appropriate parameter response to be used as input for the next step. For hydro-mechanical forming, such a process optimization concerns the relation between the individual loading components; *e.g.*, axial feeding versus internal pressure.

The current study investigates two such strategies and focuses on process optimization with regard to the avoidance of defects (mainly wrinkling). The distinction between both strategies is that the control parameters are either adjusted during the simulation of the process (the *adaptive method*) or between consecutive simulations (the *iterative method*). Details on this are as follows.

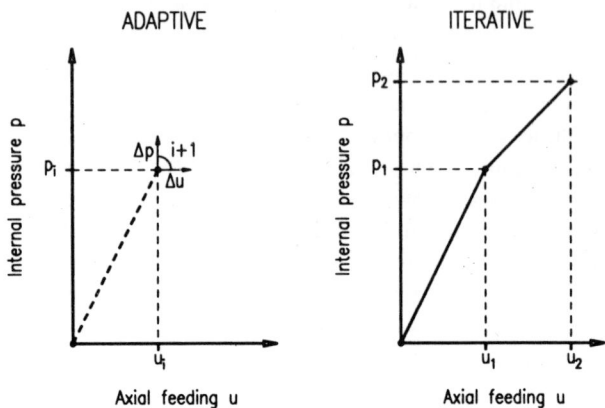

Figure 2. *The two explored strategies for FEM optimization*

3.1. *The adaptive method*

This optimization method has previously been suggested in [DOE 98] and [ALT 99]. The basic idea is that the hydro-mechanical forming process is modelled in a single simulation by successive additions of individual load increments, depending on what is considered instantly appropriate. In the former paper, the algorithm presented prevents failure of the tube by incrementally adjusting the axial forces and pressure boundary conditions by the use of fuzzy-logic control. Failure is defined here as free wrinkling of the tube and detected in the simulation by monitoring the strain differences between inner and outer side of the tube wall (as a measure of bending) at any location. The latter paper proposes to determine "good" loading profiles – i.e., internal pressure versus time and axial feeding versus time – in the same adaptive fashion. However, wrinkling is defined here as the occurrence of a negative node velocity vector (that is, movement towards the centreline of the tube) at any location along the axis.

In the current implementation, the process is modelled by incrementally adding an axial-feeding step Δu or an internal-pressure step Δp, depending on the outcome of the previous increment. The concept is visualized in the graph on the left-hand side of Figure 2. The criterion to assess which measure to be taken is (the initiation of) wrinkling.

For the symmetrical situation, wrinkling is defined simply as the occurrence of a local maximum in radial node location along the axis. Wrinkling is thus described as a geometrical condition (and actually a form of reversible folding). For the non-symmetrical situation, however, this approach cannot be simply implemented. Here, the Nordlund wrinkle indicator – which is proposed in [NOR 97] and [NOR 98] – is used. This scalar quantity is based on the local value of the second-order increment of internal work (or more exactly the rate of internal power) for capturing the initiation of wrinkling, and is assessed for single finite elements. By scaling, it is ensured that it is always in the range $[-1,1]$. Potential areas for wrinkles will be found where this quantity becomes negative. In the current implementation, the further condition of a positive thickness strain was added in order to exclude any effects that may be attributed to necking (which is another effect that triggers the indicator).

The adaptive algorithm is designed such that axial feeding is maintained until wrinkling is detected; then, pressure is increased so that wrinkling reduces until the height (in the symmetrical situation) or the indicator (in the non-symmetrical situation) crosses a pre-set critical value again. This algorithm is implemented in the FEM programme MSC.MARC as a user-subroutine.

3.2. The iterative method

This optimization method has been examined earlier in [GHO 99] and [GEL 01]. Basically, the method involves a number of subsequent simulations of the whole process. These iterations aim at minimising a so-called objective function, which represents the general quality of the product. In the former paper, the loading path is varied according to a prescribed functional relationship with specific degrees of freedom – namely a Bézier representation for axial feeding and internal pressure in dependence on time. The latter paper introduces a quadratic approximation for the wall thickness versus the process parameters that is solved according to a distinct optimization scheme (which is said to be a more effective way of solving the problem). A similar method is presented in [KRI 01], using Bézier functions for the loads and some adapted description of the objective function that places more emphasis on the suppression of wall thinning.

In the current algorithm, the loading path is depicted by variable internal-pressure values p_1 and p_2 at some fixed axial-feeding positions u_1 and u_2. Hence, it is a two-parameter optimization. This is shown in the graph on the right-hand side of Figure 2. Using $s_{o,k}$ as the original and $s_{e,k}$ as the final wall thickness at the nodal point k (total number of nodes n), the objective function f_{obj} is defined as:

$$f_{obj} = \sum_{k=1}^{n} \left(\frac{s_{e,k} - s_{o,k}}{s_{o,k}} \right)^2 .$$

[1]

The optimal loading path is assumed to correspond with the one that minimizes the objective function. Thus, the method aims at reaching a wall-thickness distribution that is as close as possible to the original one. Excessive thinning and thickening is penalized, leading the deformation in the corridor between necking and wrinkling and keeping it close to the so-called "ideal deformation path".

The iterative algorithm is implemented in MSC.MARC in conjunction with Matlab; the latter acts as a shell programme that controls the previous one by providing a parameter update for each following simulation. This uses the sequential quadratic programming (SQP) method as the optimization procedure.

4. Case I: tube-expansion test

The first example that will be dealt with involves the die-bound expansion of a tube under the concurrent action of internal hydraulic pressure and mechanical axial feeding from one end. Both the adaptive and the iterative algorithm were applied in order to establish the preferred loading conditions and compare their results.

4.1. *The model*

The original tube is an AA6061–T4 aluminium alloy of 30 mm external diameter and 1.5 mm wall thickness. The expansion zone extends over a length of 100 mm with a maximum die diameter of 45 mm. In Figure 3, the configuration is shown. Plastic behaviour (flow stress σ_f as a function of equivalent strain ε_{eq}) is described by the rigid-plastic power law:

$$\sigma_f = K \cdot (\varepsilon_{eq} + \varepsilon_0)^n. \tag{2}$$

For this particular tube material, the strength coefficient $K = 310$ MPa, the strain-hardening exponent $n = 0.24$, and the pre-strain $\varepsilon_0 = 0.024$. Anisotropy is left out of consideration. Friction between the tube and the die is accounted for with Coulomb's law, using a friction coefficient of $\mu = 0.1$.

Using axial symmetry, the actual FEM model is two-dimensional. The tube is meshed with 100×2 four-node volume elements (type 10), and the die is modelled as a rigid body. Axial feeding is prescribed as a displacement of the longer tube end; internal pressure is applied as an equally distributed normal load on the inside of the tube. Boundary condition for the end of the process is that the tube contacts the die at its maximum diameter.

The **adaptive algorithm** was run with an axial-feeding step of $\Delta u = 0.15$ mm and an internal-pressure step of $\Delta p = 0.5$ MPa, using the geometrical criterion to monitor wrinkling. The wrinkle height was set at some critical (or allowed) values h_c to study its influence.

The **iterative algorithm** was run with initial values of $p_1 = 1$ MPa and $p_2 = 40$ MPa. Two settings for the axial-feeding partition were used: $u_1/u_2 = 0.25$ and $u_1/u_2 = 0.50$ (with $u_2 = 25$ mm in each case). Axial feeding per increment was 0.05 mm.

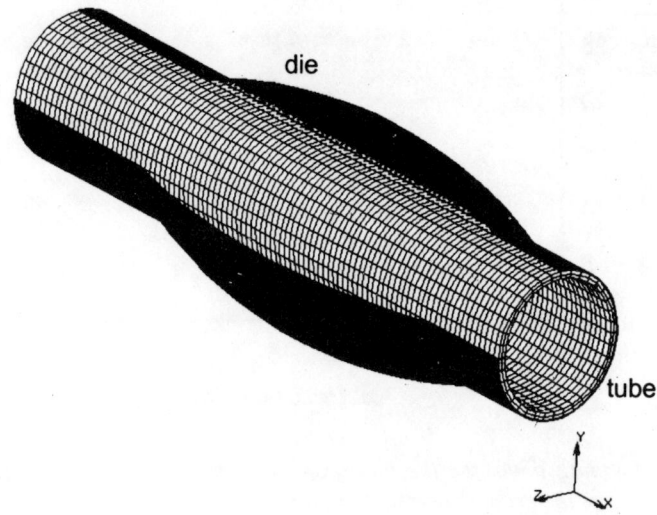

Figure 3. *The tube-expansion test: original geometry (exploded view of the die)*

4.2. Results and discussion

Calculated loading paths are given in Figure 4 for the adaptive algorithm and in Figure 5 for the iterative algorithm. All these simulations were successful in the sense that the die was eventually filled, that no obvious wrinkles remained in the product, and that no excessive upsetting occurred at the pushed tube end.

The adaptive solutions basically consist of three phases: it starts with axial feeding only, then a steep increase in pressure, and it ends with mainly axial feeding. The axial position at which the pressure starts rising depends on the allowed wrinkle height. End points are different in both axial-feeding and internal-pressure values. Practically, such a course could be implemented by approximating it with three straight parts.

For the iterative solutions, the die is filled before the pre-set end value u_2 is reached. An actual bi-linear optimization is only achieved for the lower partition ratio. The imposed end values for the axial feeding are considerably lower than for the adaptive solutions. Typically, the calculations involve several dozens of iterations (depending somewhat on the convergence condition in the optimization procedure), making it a time-consuming exercise.

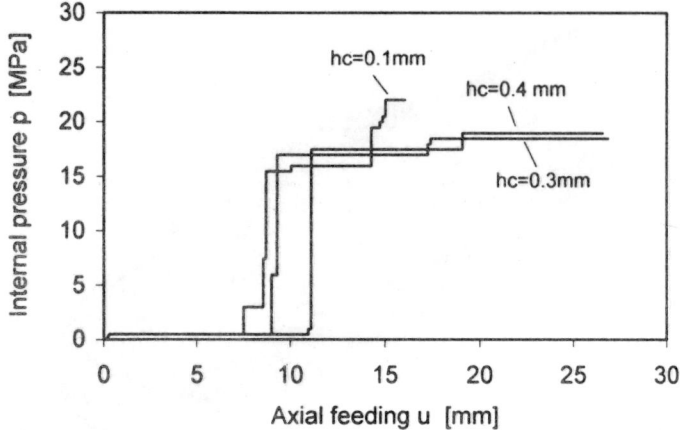

Figure 4. *Loading paths for the tube-expansion test: adaptive solutions for some distinct values of the critical wrinkle height h_c*

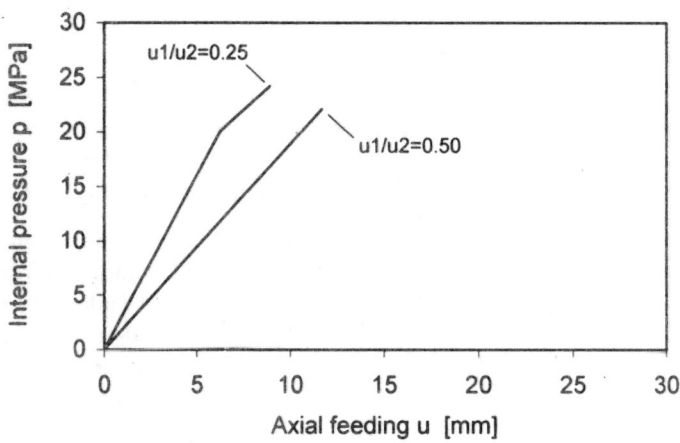

Figure 5. *Loading paths for the tube-expansion test: iterative solutions for some distinct values of the axial-feeding partition u_1 / u_2*

A comparison between all acquired values for the objective function according to equation [1] is given in Table 1. Noteworthy, the lowest value is achieved for the adaptive algorithm, although the minimization of this function is not an explicit goal in this case. It confirms the idea that the present loading-path description for the iterative algorithm needs to be improved.

Table 1. *Objective functions for the tube-expansion test: solutions for the distinct optimization strategies and parameter settings*

ADAPTIVE			ITERATIVE	
critical wrinkle height h_c [mm]			partition u_1/u_2	
0.1	0.3	0.4	0.25	0.50
2.78	1.82	2.12	2.50	2.75

In Figure 6, wall thickness for the best solution of each method (that is, with the lowest value for the objective function) is compared. Obviously, these profiles are not symmetrical in axial direction due to the unilateral feeding. The higher axial feeding in the adaptive solution translates into a wall-thickness distribution in the product that remains closer to original. The difference between both solutions is technically significant. Also, it shows that some upsetting is exhibited in the pushed tube end for the adaptive solution, whereas this is not the case for the iterative one.

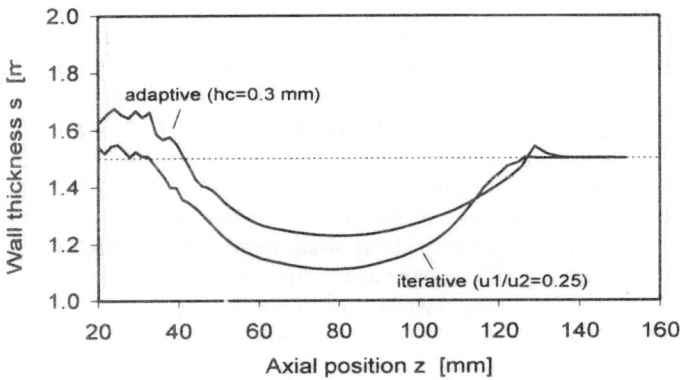

Figure 6. *Wall-thickness distribution in the tube-expansion product: adaptive versus iterative solution for selected parameter settings*

The accompanying plastic deformations for these two solutions are illustrated in Figure 7. In this forming-limit diagram, the major and minor logarithmic strains in the product (ε_1 and ε_2, corresponding with the tangential and the longitudinal strain, respectively) are plotted against each other. Though the overall shape of the product is identical, local deformations are quite different.

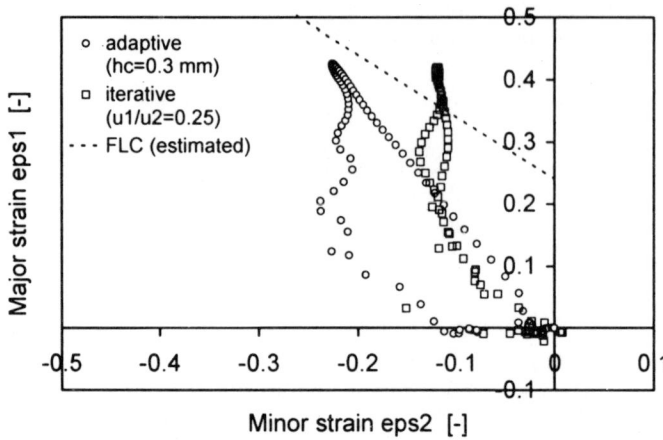

Figure 7. *Principal strains in the tube-expansion product: adaptive versus iterative solution for selected parameter settings*

This representation is commonly used to check if the strains do not exceed the forming capacity of the product material as expressed by the forming-limit curve (FLC). As an approximation, the forming limit set by plastic instability – here, the initiation of localized necking – is a line that follows the description $\varepsilon_1 + \varepsilon_2 = n$, with n as the strain-hardening exponent (Hill's model based on necking theory; see for instance [BAN 00]). By including this estimated FLC for the material under consideration as the dashed line in the graph, it becomes evident that the iterative solution is critical in this respect. Those strain combinations that exceed this FLC are situated in the middle of the expanded section, at the largest perimeter, which is thus prone to necking/bursting. Where the tangential strain ε_1 at a distinct location cannot be altered because it is uniquely determined by the original and the final diameter, the adaptive algorithm succeeds in avoiding this necking limit by introducing as much compressive longitudinal strain ε_2 as allowed.

In order to tackle the problem with necking in the iterative algorithm, the objective function [1] may be adapted by replacing the square-order with a higher-order or asymptotic term. By doing so, more emphasis is placed on preventing strong deviations from the ideal deformation path. Finally, it is to be mentioned that the predictions for this test case need still to be confirmed by experimental means.

5. Case II: T-shape test

The second example that will be discussed is that of the hydro-mechanical forming of a T-shaped tubular test part. In this case, internal hydraulic pressure acts on the tube in conjunction with mechanical axial feeding from both ends. This test –

that is simply referred to as the T-shape test – has established itself as one of the common means for materials characterization and to study process mechanics. The adaptive algorithm was applied to this test case in conjunction with the Nordlund wrinkle criterion. Results will be compared with some experimental data.

5.1. *The model*

The original tube is a mild steel grade FePO4 of 70 mm external diameter, 200 mm length and 1.2 mm wall thickness. The internal diameter of the die is 71.5 mm for both the main and the lateral branch; the transition radius between these is 8 mm. Final bulge height is fixed at 30 mm with a stopper. In Figure 8, the configuration is shown. Plastic behaviour of the tube material is described with the rigid-plastic power law (strength coefficient $K = 490$ MPa, strain-hardening exponent $n = 0.203$, and pre-strain $\varepsilon_0 = 0.014$); anisotropy is left out of consideration. Friction between the tube and the die and stopper is accounted for with Coulomb's law, using a friction coefficient of $\mu = 0.2$.

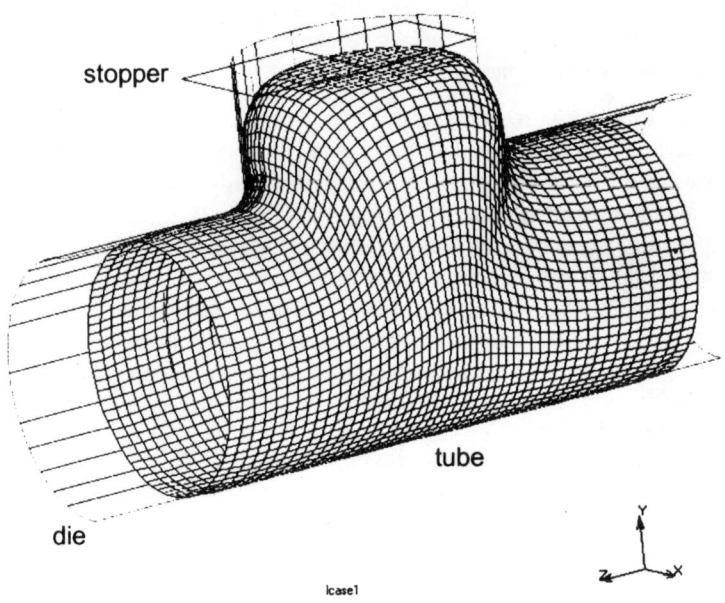

Figure 8. *The T-shape test: final geometry (exploded view of the die)*

Taking advantage of symmetry, the actual FEM model covers only one quarter of the geometry. The tube section is meshed with 1 000 four-node shell elements (type 75); die and stopper are modelled as rigid bodies. Axial feeding is prescribed as a displacement of the tube end, and internal pressure is applied as an equally distributed normal load on the inside of the tube. Since the transition between lateral branch and stopper is sharp (radius zero), full filling of the die cannot be attained so the process is ended at some point during the final stage of corner filling.

The **adaptive algorithm** was run with an axial-feeding step of $\Delta u = 0.10$ mm and an internal-pressure step of $\Delta p = 0.5$ MPa. The Nordlund indicator for wrinkle formation was set at some critical (or allowed) values I_c to study its influence, thereby presuming that its instantaneous value denotes the receptivity to wrinkling.

5.2. *Results and discussion*

In Figure 9, the calculated loading paths are given, starting from the boundary value for the inception of wrinkling ($I_c = 0$) to some value of the indicator close to the lower limit ($I_c = -0.9$). These simulations were successful in the sense that no apparent defects occurred, but only the simulation for the lowest value continued until the lateral branch was completed (the others stopped when a pressure maximum was reached). For this particular value, the process starts with axial feeding only, until the wrinkling limit is reached and further deformation proceeds in an approximately linear loading fashion.

Also, a loading curve is included that was found by experimental optimization. As in the calculations, axial feeding corresponds with the travel of a single punch (so the final reduction in tube length is 60 mm). Basically, the approach consisted of a number of trial-and-error tests in which the loading path was adapted bi-linearly. Defect types that had to be solved were wrinkling in either the lateral branch flanks or in the main branch opposite to the bulge section, or wrinkling and upsetting in the tube ends. Prior to this, the main effects of and interactions between process parameters in a simple proportional loading path were studied by means of a factorial approach [SIL 01]. Criterion for the final evaluation was that a complete product was attained with no visual defects (see also the inserted photograph in Figure 9). Noteworthy, the represented solution – of which the measured loading data are shown – is quite close to a straight line.

When considering the influence of the allowed value of the wrinkle indicator, it becomes clear that I_c has a decisive effect on the calculated loading curve. The solution for $I_c = -0.9$ shows the best resemblance to the experimental curve. This value is quite close to the lower limit, supporting the notion that the Nordlund criterion is actually a conservative estimate for wrinkling.

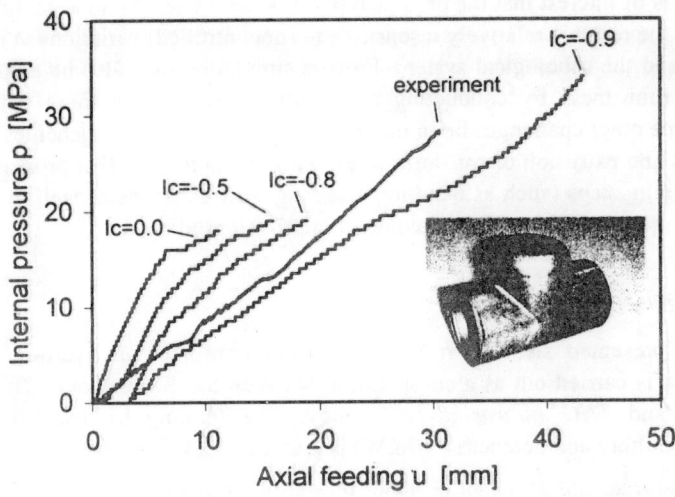

Figure 9. *Loading paths for the T-shape test: adaptive solutions for some distinct values of the critical wrinkle indicator I_c versus experiment (insert: accompanying product)*

6. Conclusions and outlook

FEM simulation with a control algorithm – as is discussed in this article – is a sensible approach towards optimization of hydro-mechanical forming processes with regard to the loading path. Also, the iterative method may be applied for such related optimization issues as tool design (*e.g.*, to evaluate the effect of transition radii).

The findings for the presented algorithms and test cases can be recapitulated as follows.

– The **adaptive method** explores the wrinkling limit of the process only, yet is very quick and flexible concerning the course of the loading path.

– The **iterative method** is essentially capable of screening the full process window, but the current bi-linear optimization is too confined to actually benefit from this.

Further work along this line of development should focus on enhanced loading-path descriptions (for the iterative algorithm, but also by considering volume control as an alternative to pressure control for the fluid), improved materials description (plastic anisotropy) and formability representations (wrinkling, necking), experimental validation and application to more complicated products. With respect to the first issue, however, it should be borne in mind that the proposed solution must also be feasible regarding the industrial implementation, particularly concerning the capabilities of press control.

Also, it is of interest that the predicted process parameters yield a stable process; that is, that the result is relatively insensitive to (uncontrolled) variations in materials behaviour and the tribological system. Process simulation may also be employed to study this robustness by conducting sensitivity analyses with these parameters. Finally, some other challenges lie in incorporating the tube inhomogeneities (such as weld seams and extrusion or roll-form lines, with distinctly different properties) and the pre-forming steps (such as bending, affecting wall thickness as well as material properties) into the FEM models to come to improved predictions.

Acknowledgements

Results presented stem from a Dutch research project on hydro-mechanical forming that is carried out as a co-operation between the *University of Technology Eindhoven* and *TNO Industrial Technology*. The funding by the Ministry of Education, Culture and Sciences (OC&W) is gratefully acknowledged.

The authors would also like to thank P. Nordlund at MSC.MARC in Palo Alto (USA) for providing the user-subroutine for assessment of the wrinkle indicator and valuable advice.

References

[AHM 00] AHMETOGLU M., ALTAN T., "Tube hydroforming: state-of-the-art and future trends", *Journal of Materials Processing Technology*, Vol. 98, 2000, pp. 25–33.

[ALT 99] ALTAN T., KOÇ M., AUE-U-LAN Y., TIBARI K., "Formability and design issues in tube hydroforming", *Hydroforming of Tubes, Extrusions and Sheet Metals – Volume 1*, ed. Siegert K., MAT INFO Werkstoff-Informations gesellschaft, Frankfurt, Germany, 1999, pp. 105–121.

[BAN 00] BANABIC D., BUNGE H.-J., PÖHLANDT K., TEKKAYA A.E., *Formability of Metallic Materials: Plastic Anisotropy, Formability Testing, Forming Limits*, Springer, 2000.

[DOE 98] DOEGE E., KÖSTERS R., ROPERS C., "Determination of optimised control parameters for internal high pressure forming processes with the FEM", *Proceedings of the 6th International Conference on Sheet Metal – Volume II*, Twente, Netherlands, 1998, pp. 119–128.

[DOH 97] DOHMANN F., HARTL CH., "Tube hydroforming – Research and practical application", *Journal of Materials Processing Technology*, vol. 71, 1997, pp. 174–186.

[GEL 01] GELIN J.C., LABERGERE C., "Modelling, optimization and optimal control for hydroforming processes", *Proceedings of the 4th International ESAFORM Conference on Material Forming – Volume 1*, Liège, Belgium, 2001, pp. 377–380.

[GHO 99] GHOUATI O., GELIN J.C., BAIDA M., LENOIR H., "Simulation and control of hydroforming processes for tubes or flanges forming", *Proceedings of the 2nd ESAFORM Conference on Material Forming*, Guimarães, Portugal, 1999, pp. 473–476.

[KRI 01] KRISTENSEN B.E., NIELSEN K.B., "Least-square formulation of the object function, applied on hydro mechanical tube forming", *Proceedings of the 9th International Conference on Sheet Metal*, Leuven, Belgium, 2001, pp. 353–362.

[MAS 01] MASSONI E., "Finite element simulation of tube hydroforming process", *Proceedings of the 4th International ESAFORM Conference on Material Forming – Volume 1*, Liège, Belgium, 2001, pp. 381–384.

[NOR 97] NORDLUND P., HÄGGBLAD B., "Prediction of wrinkle tendencies in explicit sheet metal-forming simulations", *International Journal for Numerical Methods in Engineering*, Vol. 40, No. 22, 1997, pp. 4079–4095.

[NOR 98] NORDLUND P., "Adaptivity and wrinkle indication in sheet metal-forming", *Computer Methods in Applied Mechanics and Engineering*, Vol. 161, 1998, pp. 127–143.

[SAA 01] SAANOUNI K., CHEROUAT A., HAMMI Y., "Optimisation of hydroforming processes with respect to fracture", *Proceedings of the 4th International ESAFORM Conference on Material Forming – Volume 1*, Liège, Belgium, 2001, pp. 361–364.

[SIE 00] SIEGERT K., HÄUSSERMANN M., LÖSCH B., RIEGER R., "Recent developments in hydroforming technology", *Journal of Materials Processing Technology*, Vol. 98, 2000, pp. 251–258.

[SIL 01] SILLEKENS W.H., VELTMANS P.P.H., GUTIERREZ M.A., FERNANDEZ J.I., "Factorial experimentation and simulation of hydroforming processes for tubular parts", *Proceedings of the 9th International Conference on Sheet Metal*, Leuven, Belgium, 2001, pp. 605–612.

Chapter 13

Identification of Non-linear Kinematic Hardening with Bend-reverse Bend Experiments in Anisotropic Sheet-metals

Michel Brunet, Fabrice Morestin and Françis Sabourin
Laboratoire de Mécanique des Solides, INSA, Villeurbanne, France

Stephane Godereaux
Technocentre Renault, Guyancourt, France

1. Introduction

There is a need in sheet metal forming analysis by finite-elements to introduce more realistic constitutive equations. During forming processes, sheet metals can be subjected to a bending-unbending deformation path. That is the case under draw conditions where the sheet metal experiences bending and then unbending as the metal enters the die. It is also the case when the sheet metal passes through drawbeads and more generally for multi-paths forming problems and all bending types of deformation. The final stress state at the end of the forming process is of a crucial importance to predict the spring-back accurately.

The simulation of the spring-back can be seen as a trivial problem compared to the simulation of the forming phase, since it involves no contact and only load and stress relaxation. Despite this fact, results from spring-back simulations have many times shown results, which are far from the level obtained experimentally. The importance of considering the Bauschinger effect by using a kinematic hardening law, even linear, instead of isotropic hardening has been demonstrated by many authors [MAT 95, LI 98, TAN 96].

To this end, the Lemaître and Chaboche [LEM 90] type of non-linear kinematic hardening models are receiving increasingly more attention because of their enhanced performance in predicting cyclic and other transient material behaviour. However, sheet-metals present an initial anisotropy and, using a phenomenological approach, Hill [HIL 48] has developed an anisotropic theory of plasticity, which uses a quadratic yield function. More complex yield functions have been proposed by Hill [HIL 90] and many others researchers and are generally non-quadratic in form.

Zhang and Lee [ZHA 95] have presented a new mathematical model for plane strain bending and spring-back analysis. This model combines effects associated with bending and stretching and considers stress and strain distribution in the thickness direction. Moreover, the Hill non-quadratic yield function is incorporated in the model. Hiwatashi *et al.* [HIW 97] have developed a mixed hardening model in which the shape of the yield locus is generated from the texture and the dislocation structures are responsible for isotropic and kinematic hardening. This micro-structural evolution is modeled with three internal state variables and their evolution equations. Recently, Zhao and Lee [ZHA 99] have presented bend/reverse bend tests but performed with a three-point bend test apparatus and they have not considered the initial anisotropy. Moreover, their optimization is conducted on the punch load-displacement curves instead of the moment-curvature curves and the strain-stress curves of the tensile tests, as in our case. In their paper, FEA is used for the bending calculations whereas an analytical form of the plane strain bending under constant moment is developed here.

In this paper, the quadratic anisotropy of plasticity is extended to include the concept of combined isotropic non-linear kinematic hardening of Lemaître-Chaboche, which is the induced anisotropy. Since we are concerned with sheet-metal forming analysis, models need to be identified up to large strain. Then an inverse identification technique is proposed based both on bending-unbending experiments and tensile tests up to large strain. A four points bending machine has been built in order to promote a constant bending moment. The numerical simulations of the experiments are performed with an integrated optimizer, which gives the four parameters of the mixed non-linear kinematic hardening model.

2. Formulation

In sheet metal forming applications, we are generally concerned with plane stress conditions. Consider x,y to be the "rolling" and "cross" directions in the plane of the sheet, z is the thickness direction. Based on the Hill quadratic yield function, the anisotropic yield function is modified as:

$$q^2 = \{\sigma - \alpha\}^T [M] \{\sigma - \alpha\} \qquad [1a]$$

where:

$$M = \begin{bmatrix} g+h & -h & 0 \\ -h & f+h & 0 \\ 0 & 0 & 2n \end{bmatrix} \qquad [1b]$$

and:

$$\{\sigma - \alpha\} = \begin{Bmatrix} \sigma_x - \alpha_x \\ \sigma_y - \alpha_y \\ \sigma_{xy} - \alpha_{xy} \end{Bmatrix} \qquad [1c]$$

The parameters f, g, h and n are the dimensionless Hill material coefficients. If $f = g = h = 1/2$ and $n = 3/2$, this function is equivalent to Von-Mises' isotropic yield function. The equivalent stress function q gives the current size of the yield surface where the components of the back stresses $\alpha_x, \alpha_y, \alpha_{xy}$ specify the centre of the yield surface and are directly related to the kinematic hardening of the material.

The basic concept of kinematic models is that the yield surface shifts in the stress space so that straining in one direction reduces the yield stress in the opposite direction, simulating the Bauschinger effect and anisotropy induced by work hardening. The simplest model provides linear kinematic hardening and is thus mainly used for low-cycle fatigue evaluations. This model can yield physically reasonable results if the uniaxial behaviour is linearized in the plastic range (a constant work hardening slope). As we are concerned here with large strain, this restriction on the theory's ability to provide reasonable results is not acceptable.

The combined isotropic-kinematic hardening is an extension of the linear model. It provides a much more accurate approximation to the stress-strain relation than the linear model. It also models other phenomena such as ratchetting, relaxation of the mean stress, cyclic hardening, that are typical of materials subjected to cyclic hardening. This model is based on the work of Lemaître and Chaboche [LEM 90]. The size of the elastic range $\bar{\sigma}$ is defined as a function of the equivalent plastic strain $\bar{\varepsilon}^p$ (and other field variables such as temperature if necessary). In this paper, this dependency is stated with a simple exponential law for materials that either cyclically harden or soften as:

$$\bar{\sigma} = \sigma_0 + Q_\infty \left(1 - e^{-b\bar{\varepsilon}^p}\right) \qquad [2]$$

where σ_0 is the yield surface size at zero plastic strain, and Q_∞ and b are material parameters that must be calibrated from cyclic test data. The evolution of the kinematic components of the model is defined as:

$$\{d\alpha\} = C \frac{d\bar{\varepsilon}^p}{\bar{\sigma}} \{\sigma - \alpha\} - \gamma \{\alpha\} d\bar{\varepsilon}^p \qquad [3]$$

where C and γ are additional material parameters to be adjusted. The possible rate of change of these parameters with respect to temperature and field variables is not accounted for here. This equation is the basic Ziegler's law where the last term is the 'recall' term which has been added and which introduces the non-linearity in the evolution law. Using equation.(1a) and the normality rule, the associated flow rule is:

$$\{d\varepsilon^P\} = d\bar{\varepsilon}^P \frac{\partial q}{\partial \{\sigma\}} = \{a\} d\bar{\varepsilon}^P \qquad [4]$$

where the gradient vector is:

$$\{a\} = \frac{1}{q}[M]\{\sigma - \alpha\} \qquad [5]$$

$\{d\varepsilon^P\}$ is the plastic flow increment vector and $d\bar{\varepsilon}^P$ is the equivalent plastic strain increment. By using the consistency condition of yield surface, the following may be obtained:

$$dq = \{d\sigma - d\alpha\}^T \{a\} = \frac{d\bar{\sigma}}{d\bar{\varepsilon}^P} d\bar{\varepsilon}^P \qquad [6]$$

and noticing:

$$H' = \frac{d\bar{\sigma}}{d\bar{\varepsilon}^P} = Q_\infty b e^{-b\bar{\varepsilon}^P} \qquad [7]$$

The plastic strain ratio, R_θ-value, for a tension specimen cut at an angle θ to the rolling direction is defined as the ratio of the plastic strain increment in the width direction to the plastic strain increment in the thickness direction:

$$R_\theta = \frac{d\varepsilon^P_{\theta+\pi/2}}{d\varepsilon^P_z} \qquad [8]$$

The incompressibility assumption of plastic strain gives:

$$d\varepsilon^P_z = \frac{d\bar{\varepsilon}^P}{q}\left[-g(\sigma_x - \alpha_x) - f(\sigma_y - \alpha_y)\right] \qquad [9]$$

Using equation (4) for the plastic strain increments and assuming small elastic strains yields:

$$R_\theta = \frac{s^2\left[(g+h)(\sigma_x - \alpha_x) - h(\sigma_y - \alpha_y)\right] + c^2\left[-h(\sigma_x - \alpha_x) + (f+h)(\sigma_y - \alpha_y)\right] - 4cs(\sigma_{xy} - \alpha_{xy})n}{-g(\sigma_x - \alpha_x) - f(\sigma_y - \alpha_y)} \qquad [10a]$$

noticing: $c = \cos\theta$, $s = \sin\theta$ and where:

$$\sigma_x = \sigma\cos^2\theta, \; \sigma_y = \sigma\sin^2\theta, \; \sigma_{xy} = \sigma\sin\theta\cos\theta \qquad [10b]$$

and σ is the tensile stress applied to the specimen. Without initial back-stresses, the following relations still occur for $\theta = 0°$ and $\theta = 90°$:

$$R_0 = \frac{h}{g} \qquad R_{90} = \frac{h}{f} \qquad\qquad [11]$$

3. Experiments

In order to base the analysis on experimentally measured moment-curvature diagrams, a bending-unbending apparatus has been built. An example of this imposed curvature bending is shown for an aluminium alloy strip in Figure 1. Figure 2 shows the bending-unbending tester, which is controlled by an electric motor associated with a torque-limiter. In order to provide a constant torque and a variable length between to axes, a set of two "Oldham" joints has been used on the axes. Sensors built with gages pasted on the axes measure the bending moment after the necessary calibration. The bend angle ranges from +180° to −180° as shown in Figure 1 in order to reach surface strain up to 10% for sheet thickness of 0.5 to 2 mm. After calibrating the sensors of the apparatus and performing the desired corrections, the moment and curvature are automatically measured as it can be seen in Figure 3 where a few points are selected for the identification technique. On the other hand, the initial anisotropy parameters (the R-values) are first determined independently using the Digital Image Correlation method (D.I.C.), that Brunet *et al.* [BRU 98, BRU 01], applied to uniaxial tests.

Figure 1. *Strip with imposed curvature (maximum)*

Electric motor

Torque limiter

"Oldham" joints

Sensors

Gears

Slide-blocks

Grips

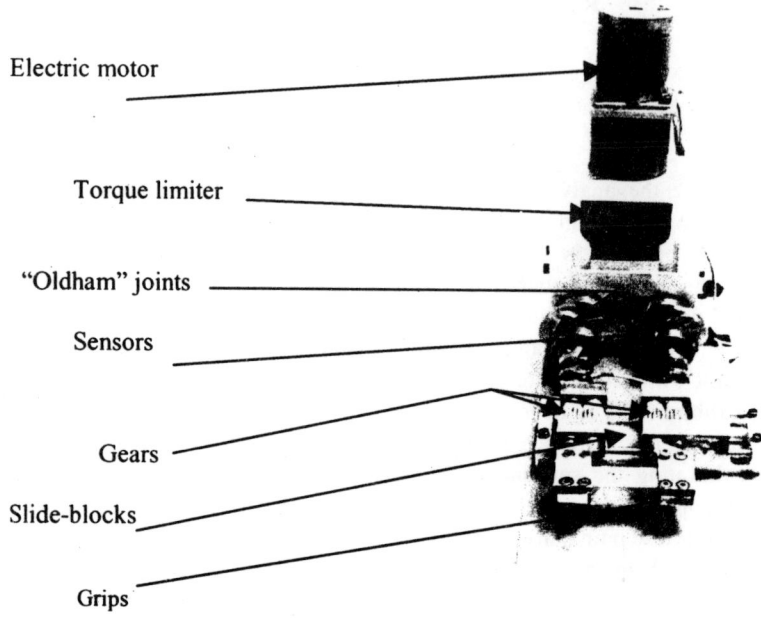

Figure 2. *Bending-unbending test equipment*

The flattening of the curves observed when the sheet experiences zero curvature is due to a load instability or snap-through of the strip. A small curvature appears in the transverse direction during loading due to the plane strain state in the centre of the strip and plane stress state on the edges. All these non-linear geometrical effects due to the shell structure nature of the specimen seem not to significantly affect the experimental results because the strain range studied is large.

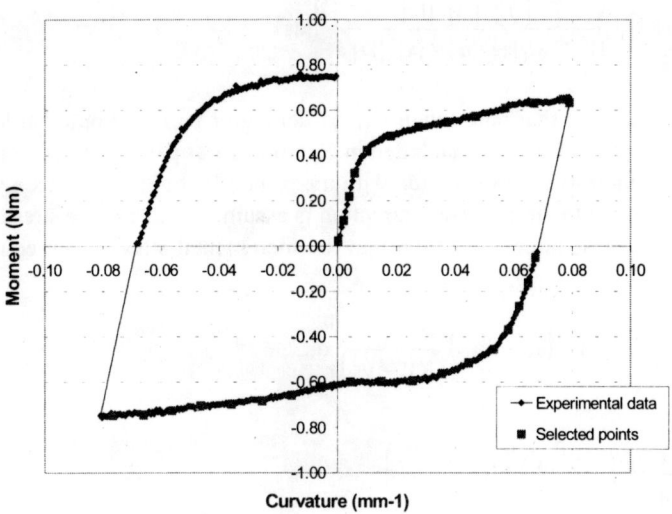

Figure 3. *Experimental moment-curvature curve (Aluminium 1)*

4. Identification

The results of bending-unbending experiments are used to determine the coefficients in the equations 2 and 3 namely σ_0, Q_∞, b, C, γ. The parameters h, g, f of the initial anisotropy are determined independently from tensile tests in the rolling and transverse directions. The strain measurements needed for the determination of R_0 and R_{90} are obtained using the Digital Image Correlation method, [BRU 98].

In order to find the parameters involved in the model, a specific optimization code has been developed. The main advantage of the experiment is that in first approximation, it does not require a Finite Element analysis and then a great save of computational time is obtained. The structural problem to be solved is the plane strain bending of an elasto-plastic sheet as it is developed in the following.

The additive decomposition of the logarithmic strain increment is assumed during active plastic loading:

$$\{d\varepsilon\} = \{d\varepsilon^e\} + \{d\varepsilon^p\} \qquad [12]$$

since the elastic strains remain small in sheet-metals forming analysis.

Accounting for the flow rule, the consistency condition, the evolution of kinematic parameters and introducing the (3x3) matrix [D] of elastic constants, allows the incremental stress-strain relation to be written as:

$$\{d\sigma\} = \left[[D] - \frac{[D]\{a\}\{a\}^T [D]}{H' + C - \gamma\{a\}^T\{\alpha\} + \{a\}^T [D]\{a\}} \right]\{d\epsilon\}$$ [13]

It is worth remarking that pure kinematic hardening or pure isotropic hardening as well as isotropic yield surface is included in equation 13 as particular cases. In the experiment, the orthotropic axes coincide with the principal strain directions x,y and the plane strain state in the transverse y direction is assumed with a plane stress state in the through-thickness direction z such that the incremental stress-strain equation reduces to:

$$d\sigma_x = \frac{E}{1-v^2}\left[1 - (a_x + va_y)^2 \frac{E}{K(1-v^2)} \right] d\epsilon_x$$ [14a]

where:

$$a_x = \frac{1}{q}\left[(\sigma_x - \alpha_x) - h(\sigma_y - \alpha_y) \right]$$ [14b]

$$a_y = \frac{1}{q}\left[-h(\sigma_x - \alpha_x) + (f + h)(\sigma_y - \alpha_y) \right]$$ [14c]

$$K = H' + C - \gamma a_x \alpha_x - \gamma a_y \alpha_y + \frac{E}{1-v^2}(a_x^2 + 2va_x a_y + a_y^2)$$ [14d]

where x is chosen to be the reference direction for the initial anisotropy such that $g + h = 1$.

Let ρ be the radius of curvature, which is imposed incrementally to the strip, the elasto-plastic stress-strain relations are solved for this strain controlled problem with:

$$d\epsilon_x^i = \frac{-z^i d\rho}{\rho(\rho + z^i)}$$ [15]

where z^i is the through-thickness co-ordinate of the integration point considered which allows one to evaluate the bending moment with:

$$M = \sum_{i=2,2}^{N-1} \frac{1}{3}\ell dz\left[\sigma_x^{i-1} z^{i-1} + 4\sigma_x^i z^i + \sigma_x^{i+1} z^{i+1} \right]$$ [16]

The shape of this theoretical curve must fit the M versus $1/\rho$ experimental diagram as the one already shown in Figure 3. The input to the optimization code are the R-values in the directions 0° and 90° to the rolling direction x and the elastic properties E and v. The first step in the optimization routine consists in the calculation of acceptable start values of C, γ, Q_x, b denoted $C_0, \gamma_0, Q_{\infty 0}, b_0$, in the following. This is achieved by considering a pure tensile test and a pure non-linear kinematic hardening function. Effectively, in the case of pure kinematic hardening and uniaxial stress state, it is readily established by analytical integration that:

$$\sigma_x = \sigma_0 + \frac{C_0}{\gamma_0}\left(1 - e^{-\gamma_0 \varepsilon_x^p}\right) \qquad [17]$$

If the function fits the monotonic true stress-natural plastic strain experimental curve of the tensile test, the identification of equation 17 with equation 2 gives the possible starting values for Q_∞ and b.

The next step is the least-squares optimization by minimization of the error in the bending moment. The parameters are determined so that the errors in predicted moments are minimized at some selected points on the experimental moment versus curvature diagram. An iteration procedure ensures a descent of the least-squares function and also feasibility of the parameters on the tensile test (double optimization). A typical feature of the least-squares functional is the occurrence of different local minima instead of the global minimum expected. However, the effect of different starting parameter vectors has been investigated but the initial guesses obtained from monotonic tensile tests seem to be adequate in many cases. The minimization was performed with a Sequential Quadratic Programming (S.Q.P.) algorithm.

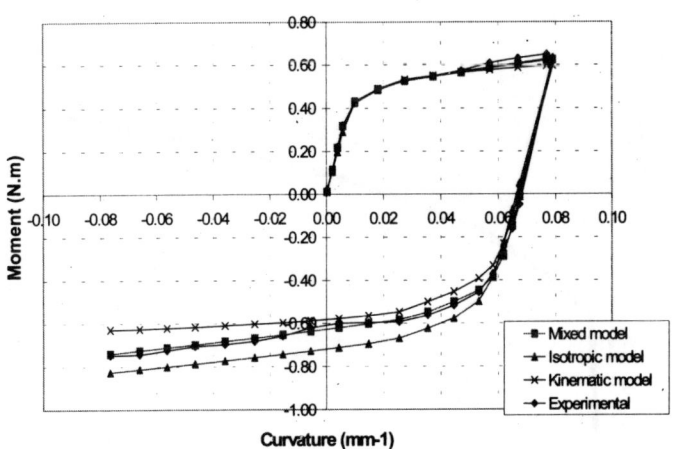

Figure 4. *Experimental and theoretical moment-curvature (Aluminium 1)*

Figure 4 depicts the moment versus curvature for one loading and reverse loading. The material is an aluminium alloy of strip thickness 0.8mm, E = 69000Mpa, σ_0= 137Mpa, R_0= 0.71 and R_{90}= 0.74. The starting parameters have been found to be: C_0= 2887, γ_0 = 120.1 and $Q_{\infty 0}$= 240.3 , b_0 = 9.486. The minimization was conducted as previously explained where the number of observation points was 30.

It can be seen, that very substantial agreement of experimental and simulated data is obtained with the converged values: $C = 740.4$, $\gamma = 4.167$, $Q = 111.6$ and $b = 13.56$. For comparison, the isotropic model and the pure kinematic hardening are also shown in Figure 4. It is clearly observed here that for this aluminium alloy, the isotropic model does not fit the experimental curve in the reverse loading and the Bauschinger effect is better accounted for by the proposed mixed model.

The next Figure 5 compares the theoretical stress-strain curves to the experimental data for the case of the uniaxial monotonic tensile tests. Very good agreement for the stress-strain curves has been obtained due to the fact that the optimization is carried out both on the uniaxial monotonic curve and on the moment-curvature curve. Also, the objective function of the least-squares functional contains quantities of different physical dimensions. This is the case here where the data are given for both in terms of stresses and strains and in terms of moments and curvatures. Then, a reasonable scaling of associated parts in the objective function becomes necessary. The resulting parameters vector of the minimization is called the *weighted least-squares estimate*, Norton [NOR 86], Mahnken and Stein [MAH 96]. It is worth noticing that the data would be incomplete if certain physical effects are not activated during the experiments. It would be the case if data are available only for monotonic loading, thus giving no explicit indication of kinematic hardening.

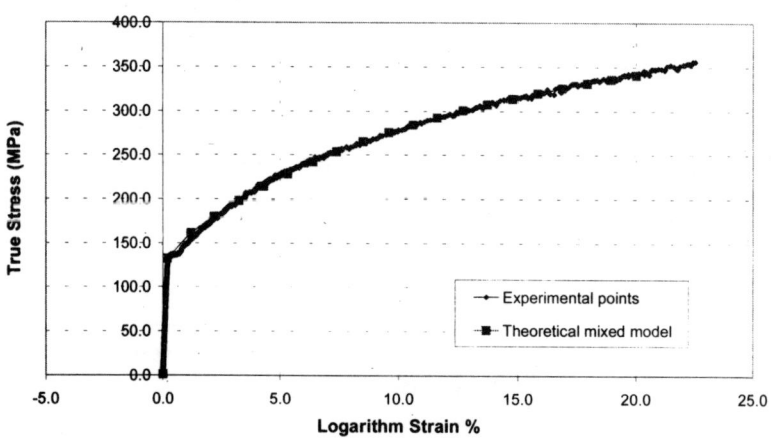

Figure 5. *Experimental and theoretical tensile test curves (Aluminium 1)*

In Figure 6 the theoretical and experimental results for the bending moment versus curvature are compared up to 3 cycles. It can be seen that the reloading is satisfactory after the first unloading. After that, the resulting curves differ slightly up to the second unloading and then discrepancies are observed. The main reason is certainly that the specimen is not in a pure plane-strain bending state after the second reloading due to increasing loading instabilities.

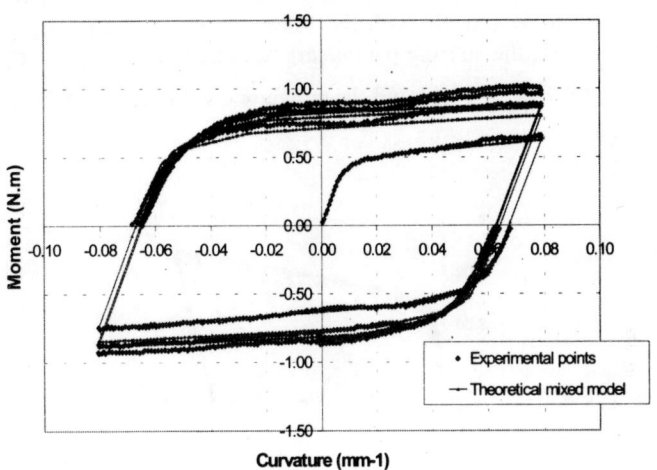

Figure 6. *Three cycles experimental versus theoretical moment-curvature curves (Aluminium 1)*

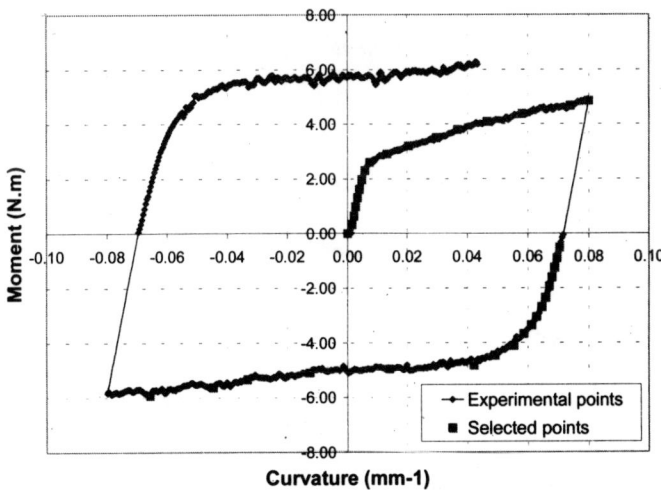

Figure 7. *Experimental moment-curvature (Aluminium 2)*

Figures 7 to 9 show the resulting curves for another kind of aluminium alloy of thickness 2 mm instead of 0.8 mm for the previous one. It is worth noticing that in this case, the selected observation points for the identification technique are not distributed equally along the curvature-moment curve but concentrated in the non-linear part of the curve. The mixed model displays a much better agreement than the pure kinematic or isotropic models with the following set of material parameters: $R_0 = 0.44$ and $R_{90} = 0.61$, the starting parameters have been found to be: $C_0 = 2178$, $\gamma_0 = 104.4$, $Q_{\infty 0} = 208.7$, $b_0 = 10.13$ and the converged values: $C = 1895$, $\gamma = 119.4$, $Q = 334.5$ and $b = 4.062$.

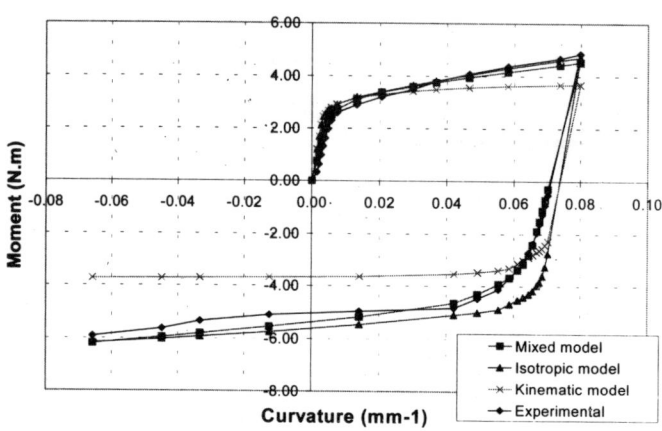

Figure 8. *Experimental and theoretical moment-curvature (Aluminium 2)*

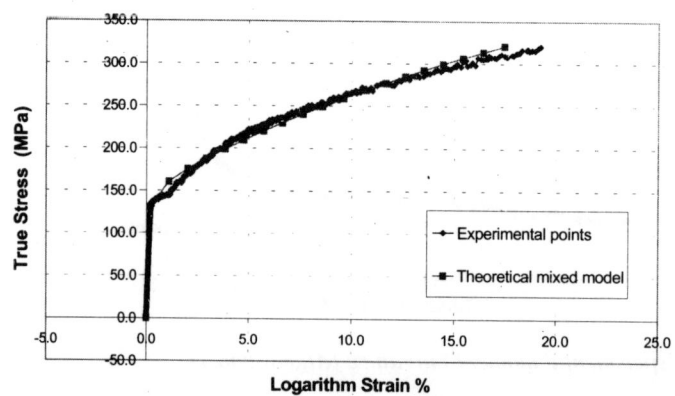

Figure 9. *Experimental and theoretical tensile test curves (Aluminium 2)*

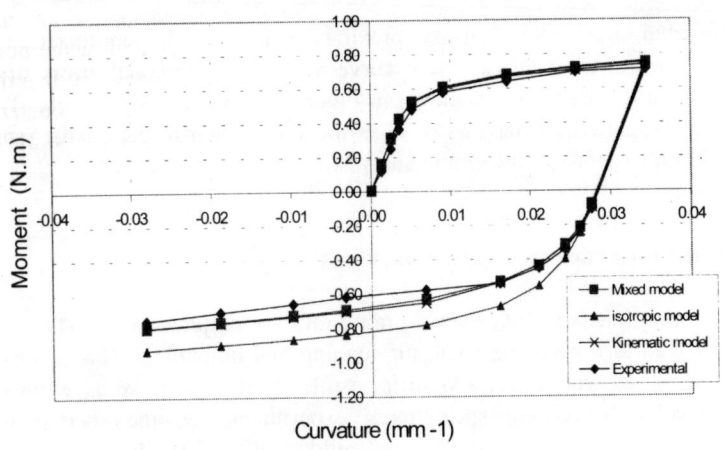

Figure 10. *Experimental and theoretical moment-curvature (Mild-steel)*

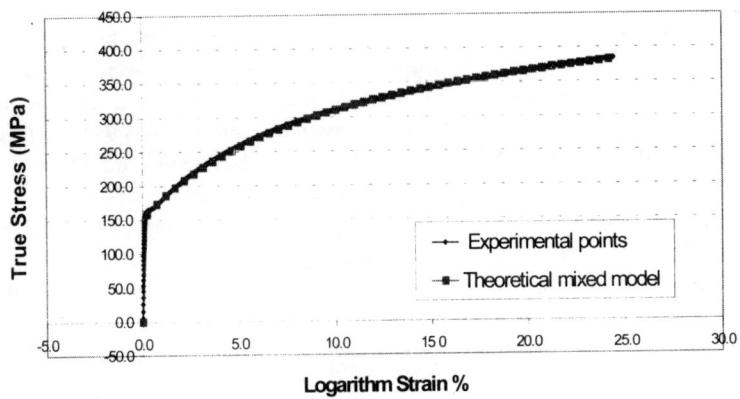

Figure 11. *Experimental and theoretical tensile test curves (Mild-steel)*

Figures 10 to 11 show the resulting curves for a mild-steel of thickness 0.8 mm and elasticity modulus of 206000 Mpa. For this steel sheet, the following set of material parameters has been obtained: $R_0 = 1.77$ and $R_{90} = 2.30$, the starting parameters have been found to be: $C_0 = 2285$, $\gamma_0 = 0.6457$, $Q_{\infty 0} = 228.6$, $b_0 = 12.91$ and the converged values: $C = 2293$, $\gamma = 20.08$, $Q = 214.7$ and $b = 3.264$. It is worth noticing in Figure 10 that this kind of steel sheet is kinematic hardening dominant and the difference between the pure non-linear kinematic hardening curve and the curve of the mixed model is very small.

As for the aluminium alloys of the previous examples, the optimization is performed both on the moment-curvature curve and on the uniaxial stress-strain curve of the tension test as it can be shown in Figures 10 and 11 respectively. This allows one to investigate the behaviors of sheet-metals in order to get a large range of strains suitable for sheet-metal forming analysis.

5. Example of the influence on a spring-back analysis

A sheet material that is sliding over a draw radius is subjected to bending and unbending, which in turn give rise to plastic loading and unloading. This is a case where the mixed model would play a significant role. At this time, we have not yet experimental results concerning spring-back experiments on the sheet-metals previously identified (aluminium 1 and 2 and mild-steel) and we have to content ourselves with a mere numerical comparison between isotropic and mixed models. As the basis for the investigation, a U-shape deep-drawn sheet strip based on the NUMISHEET'93 benchmark geometry has been used [MAK 93].

Generally, it is assumed that the result of spring-back simulated by F.E. analysis may vary greatly according to element size, punch velocity in dynamic explicit analysis, penalty contacts parameters, tool description and constitutive behaviour integration. In this work, our in-house F.E. code has been used whose main characteristic is to give spring-back results slightly influenced by the above parameters, mainly element size as it can be shown in table 1 [SAB 99]. To this end, the 6-node Discrete Kirchhoff Triangular shell element has been used where the tool geometry was described by NURBS patches and the contact conditions, in the explicit dynamic algorithm only for the forming stage, has been prescribed with a dynamic projection method. Once the forming simulation has been carried through, a static implicit algorithm has been applied to the spring-back stage starting from the stress state at the end of the forming process.

Introducing the material parameters (isotropic and mixed models) of the aluminium 1 and the mesh shown in Figures 12 and 13, the radii of the wall side obtained after spring-back are displayed in Table 1. A comparison between the second and third columns in Table 1 reveals the difference between the two hardening models and can give an idea of the error magnitude introduced by neglecting the Bauschinger effect. As expected, the spring-back is greater with isotropic model because the mixed model gives rise to lower stress levels at the end of the forming stage.

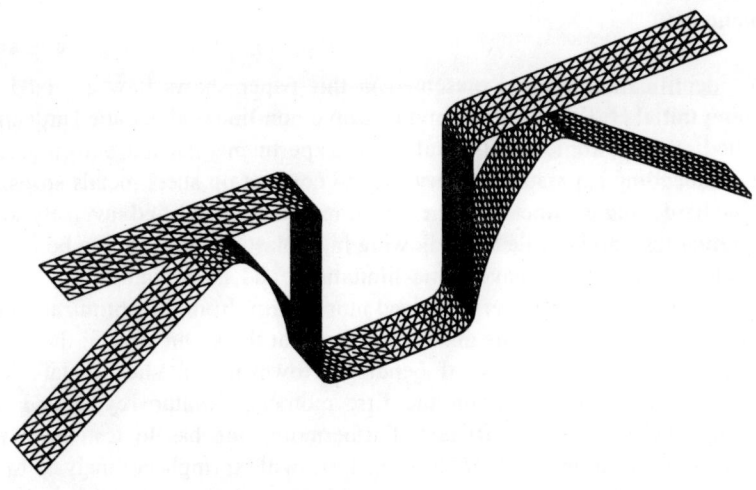

Figure 12. *F.E. mesh of a U-shape deep-drawn sheet strip and spring-back*

Mesh size : 5.0 mm

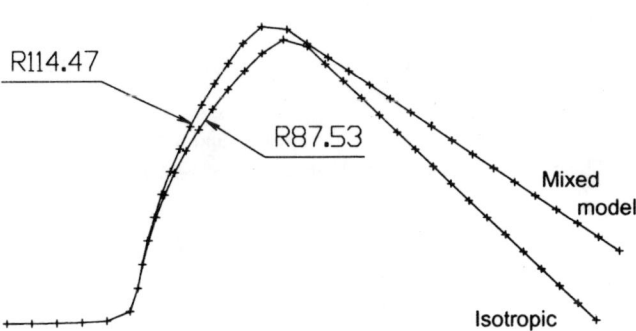

R114.47

R87.53

Mixed model

Isotropic

Figure 13. *Cross section of the mesh after springback*

Table 1. *Side wall radius after springback*

Side wall radius	Isotropic mm	Mixed model mm	Difference %
Mesh size: 2.5 mm	78.01	99.41	27.4
Mesh size: 3.5 mm	79.28	100.63	26.9
Mesh size: 5.0 mm	87.53	114.47	30

6. Conclusion

The identification method presented in this paper shows how a mixed model combining initial Hill's anisotropy and isotropic non-linear kinematic hardening can be applied to anisotropic sheet-metals. The experiments carried out on a specific bending-unbending apparatus are very easy to conduct on sheet-metals strips of any kind. The hardening parameters in the mixed model are identified inversely by using both bending tests and tensile tests allowing high plastic-strain level to be identified. The resulting models still have some limitations and uncertainties due to several reasons: the resulting parameters obtained numerically from the optimization can be affected by measurement errors and by the fact that the strain state in the sample is not exactly a pure strain state of bending. However, for sheet-metals forming simulations, the identification on the first moment-curvature cycle and on the monotonic tensile curve is sufficient. Furthermore, one has to test the model to various F.E. applications such as drawing tests with springback analysis to check whether the performance and prediction are improved or not by comparison with experimental tests. The importance of considering the Bauschinger effect by using a mixed model has been demonstrated on a numerical example.

References

[BRU 98] BRUNET M., MGUIL-TOUCHAL S., MORESTIN F., "Analytical and Experimental Studies of Necking in Sheet Metal Forming Processes", *Journal of Material Processing Technology*, Vol. 80–81, 1998, pp. 40–46.

[BRU 01] BRUNET M., MORESTIN F., "Experimental and Analytical Necking Studies of Anisotropic Sheet-metals", *Journal of Material Processing Technology*, Vol. 112, 2001, p. 214–226.

[CHA 77] CHABOCHE J.L., "Viscoplastic Constitutive Equation for the Description of Cyclic and Anisotropic Behaviour of Metals", *Bulletin Académique Polonais des Sciences*, No. 25, 1977, pp. 33–47.

[HIL 48] HILL R., "A Theory of the Yielding and Plastic Flow of Anisotropic Metals", *Proceedings of the Royal Society London*, 1948, pp. 281–297.

[HIL 90] HILL R., "Constitutive Modelling of Orthotropic Plasticity in Sheet Metals", *Journal of Mechanics and Physics of Solids*, Vol. 38, 1990, pp. 405–417.

[HIV 97] HIWATASHI S., VAN-BAEL A., VAN-HOUTTE P., TEODOSIU C., "Modeling of Plastic Anisotropy based on Texture and Dislocation Structure", *Computational Materials Science*, Vol. 9, 1997, pp. 274–284.

[LEM 90] LEMAITRE J., CHABOCHE J.L., *Mechanics of Solid Materials*, Cambridge University Press, Cambridge, UK, 1990.

[LI 98] LI K, WAGONNER R. H., "Simulation of Springback", *Simulation of Materials Processing: Theory, Methods and Applications, NUMIFORM'98*, (Eds. Huetink and Baaijens), Enschede, The Netherlands, 1998, pp. 21–31.

[MAH 96] MAHNKEN R., STEIN E., "A Unified Approach for Parameter Identification of Inelastic Materials Models in the Frame of the Finite Element Method", *Computer Methods in Applied Mechanics and Engineering*, Vol. 136, 1996, pp. 225–258.

[MAK 93] MAKINOUCHI A., NAKAMACHI E., ONATE E., WAGONNER R. H., "Numerical Simulation of 3D Sheet Metal Forming Process", *Verification of Simulation with Experiments, Proceedings of NUMISHEET'93*, Tokyo, Japan, 1993, pp. 398–400.

[MAT 95] MATTIASSON A., STRANGE P., THILDERKVIST A, SAMUELSSON D., "Simulation of Springback in Sheet Metal Forming", *5th International Conference on Numerical Methods in Industrial Forming Process*, New York, 1995, pp. 115–124.

[NOR 86] NORTON J. P., *An Introduction to Identification*, Academic Press, 1986, San Diego, CA, USA.

[SAB 99] SABOURIN F., BRUNET M., VIVES M., "3D Spring-back Analysis with a Simplified Three Node Triangular Element" *Proceedings of NUMISHEET'99*, Besançon, France, 13–17 September, 1999, pp. 17–22.

[TAN 96] TANG S.C., "Application of an Anisotropic Hardening Rule to Springback Prediction", *Advanced Technology of Plasticity*, 1996, pp. 719–722.

[ZHA 95] ZHANG Z.T., LEE D., "Development of a New Model for Plane Strain Bending and Springback Analysis", *Journal of Materials Engineering and Performance*, Vol. 4, 1995, pp. 291–300.

[ZHA 99] ZHAO K. M., LEE J. K., "On Simulation of Bend/Reverse Bend of Sheet Metals", MED, Vol. 10, *Manufacturing Science and Engineering*, ASME 1999, pp. 929–933.

Chapter 14

On the Simulation of Microhardness at Large Strains Using a Gradient Theory of Plasticity

Thomas Svedberg
Department of Mathematics, Chalmers University of Technology, Göteborg, Sweden

Kenneth Runesson
Department of Applied Mechanics, Chalmers University of Technology, Göteborg, Sweden

1. Introduction

Indentation tests are widely used to determine both elastic and plastic properties of engineering materials. In particular, indentation tests have been used to determine hardness, defined by the indentation load divided by the area of the resulting impression after unloading. Among the different head-shapes of the indenter that have been used, we note a sphere (the Brinell test), a cylinder laying down (the Rockwell C test), a three-sided pyramid (the Berkovich test) and a four-sided pyramid (the Vickers test).

Regular indentation tests, where the size of the indenter is on the scale of millimeters, show no size dependence. However, the smaller indentation probe size used in micro-indentation (and even smaller in so-called nano-indentation) gives results that show a distinct size dependence on the plastic deformation. Experimental results can be found in, e.g. Oliver and Pharr (1992); Stelmashenko et al. (1993); Atkinson (1995); Ma and Clarke (1995); Poole et al. (1996); McElhaney et al. (1998); Elmustafa et al. (2000).

While conventional local plasticity theories do not contain any means to describe such a size-dependence, non-local theories can be used (Eringen (1981)). In using non-local theories, an internal length scale is introduced, and the size-effect shown in experiments can be modeled. Recently, Shu and Fleck (1998) and Begley and Hutchinson (1998) used a strain-gradient theory of plasticity, where not only the strain but also its gradient is present in the yield function, to simulate indentation tests.

In this contribution, we instead use a gradient plasticity formulation, i.e. the gradient is on the internal variable describing the plastic flow and not on the strain itself, as described comprehensively in Svedberg and Runesson (1997, 1998); Svedberg (2000). Alternative formulations of gradient plasticity and its algorithmic treatment have been proposed by a number of authors, e.g. Mühlhaus and Aifantis (1991); Fleck and Hutchinson (1993); Sluys et al. (1993); de Borst and Pamin (1996); Mikkelsen (1997) and Liebe and Steinmann (2001). Non-local methods that employ integral formulations have been discussed by, among others, Bazant and Lin (1988); Pijaudier-Cabot and Huerta (1991); Leblond et al. (1994) and Benallal and Comi (1996).

2. Constitutive model

2.1. *Basic relations*

The constitutive model used is set in the large deformation multiplicative elastoplasticity format, i.e. the deformation gradient is decomposed as $F = \bar{F}^e \cdot F^p$, where $\bar{F}^e$ and F^p are the elastic and plastic parts, respectively. Gradient effects are included in the modeling by enhancing the internal energy with gradients of the internal variables, i.e. $e = e(\bar{F}^e, \bar{s}^e, q_\alpha, \nabla q_\alpha)^1$, where e is the internal energy density, $\bar{s}^e = s - s^p$ is the reversible part of the entropy density and q_α are the internal variables.

Introducing the Helmholtz free energy density ψ through the Legendre transformation $e = \psi + \theta \bar{s}^e$, using the energy equation and the entropy inequality, and using variational arguments, we may establish the following set of basic constitutive relations:

$$S_2 = 2[F^p]^{-1} \cdot \rho_0 \frac{\partial \psi}{\partial \bar{C}^e} \cdot [F^p]^{-T}, \quad \bar{s}^e = -\frac{\partial \psi}{\partial \theta}, \quad \bar{C}^e = [\bar{F}^e]^T \cdot \bar{F}^e \tag{1}$$

$$Q_\alpha = -\rho_0 \frac{\partial \psi}{\partial q_\alpha} + \theta \nabla \cdot \left[\frac{\rho_0}{\theta} \frac{\partial \psi}{\partial (\nabla q_\alpha)} \right] \quad \text{in } \Omega_0 \tag{2}$$

$$Q_\alpha^{(b)} = -N \cdot \rho_0 \frac{\partial \psi}{\partial (\nabla q_\alpha)} \quad \text{on } \partial \Omega_0 \tag{3}$$

where S_2 is the second Piola-Kirchhoff stress, $\bar{C}^e$ is the elastic deformation tensor, ρ_0 is the (initial) density, θ is the (absolute) temperature and $Q\alpha$ are the dissipative stresses. A superscript (b) denotes contributions or entities defined on the boundary only. Here, eq. (1) and the first term in eq. (2) are standard while the second part in eq. (2) and eq. (3) are due to the introduced gradients. For further details, see Svedberg (1999, 2000).

---

[1] Here ∇ denotes the material gradient, if nothing else is explicitly stated.

The specific model studied here is the simplest possible, comprising isothermal hyper-elasticity and plasticity with scalar hardening only. The free energy for this model takes the following form:

$$\Psi = \rho_0 \psi = \Psi^e(\bar{C}^e) + \frac{1}{2}H\kappa^2 + \frac{1}{2}H^g l_0^2[\nabla\kappa^g]^2 \qquad (4)$$

where κ and κ^g are the local and gradient scalar isotropic hardening parameters, H and H^g are the local and gradient hardening parameters and l_0 is the (initial) internal length. The hyper-elastic law defined by Ψ^e is assumed, for example, to be of the logarithmic type (even though the elastic deformations are expected to be small here). Using the appropriate expressions, see Steinmann (1992), we arrive at

$$\bar{T} = \frac{1}{2}\mathbf{E}^e : \ln(\bar{C}^e) \qquad (5)$$

where $\bar{T}$ is the Mandel stress defined as $\bar{T} \stackrel{def}{=} \bar{C}^e \cdot \bar{S}_2$ with $\bar{S}_2 \stackrel{def}{=} 2\rho_0 \, \partial\Psi/\partial\bar{C}^e$, and where $\mathbf{E}^e$ is the elasticity tensor for linear theory (Hooke's law). We note that $\bar{T}$ is symmetrical because of the elastic isotropy.

The hardening stresses are given as

$$K = -H\kappa, \quad K^g = l_0^2 H^g \nabla^2 \kappa^g \text{ in } \Omega_0, \quad \nabla^2 = \text{Laplacian} \qquad (6)$$

$$K^{g(b)} = -l_0^2 H^g N \cdot \nabla\kappa^g \quad \text{on } \partial\Omega_0 \qquad (7)$$

where K and K^g are the "drag-stresses" that represent isotropic hardening (in the simplest possible model).

The accompanying yield function is of the von Mises type, i.e.

$$\Phi = \bar{T}_e - \sigma_y - K - K^g, \quad \bar{T}_e = \sqrt{\frac{3}{2}}|\bar{T}_{dev}| \qquad (8)$$

This model is a simplified version of the model described in Svedberg (2000), where more details can be found.

Comparing this model with the one used by Shu and Fleck (1998) and also with the enhanced model used by Begley and Hutchinson (1998), which are based on small deformation kinematics and where the gradient enhancements are related to the total strain, we believe that the current model offers a simpler implementation and a more realistic description of the real kinematics. Moreover, it is thermodynamically consistent.

2.2. *Rate equations for plastic flow and hardening*

The pertinent rate equations are now introduced as

$$\bar{L}^{p} = \dot{\mu}\bar{M}, \quad \bar{M} \overset{\text{def}}{=} \frac{\partial \Phi}{\partial \bar{T}} = \frac{3\bar{T}_{\text{dev}}}{2\bar{T}_{e}} \tag{9}$$

$$\dot{\kappa} = -\dot{\mu}\left[1 - \frac{K}{K_{x}}\right], \quad \dot{\kappa}^{g} = -\dot{\mu} \tag{10}$$

where we introduced the "plastic velocity gradient" $\bar{L}^{p} = \dot{F}^{p} \cdot [F^{p}]^{-1}$ in (9).

The reason for introducing a separate gradient-hardening variable κ^{g} is that $\dot{\kappa}$ will tend to zero at the state of saturation, as evident from (10). The gradient effect would thus have disappeared if it had been associated with κ.

In order to complete the problem formulation, we need the loading/unloading conditions

$$\dot{\mu} \geq 0, \quad \Phi(\bar{T}, K, K^{g}) \leq 0, \quad \dot{\mu}\Phi(\bar{T}, K, K^{g}) = 0 \tag{11}$$

We shall henceforth assume that $H \geq 0$ and $H^{g} > 0$. It is emphasized that, according to local theory, $H > 0$ is not a sufficient condition to avoid apparent softening (in the sense that the nominal stress shows a peak) and the consequent abrupt localization, since damage as well as geometrical changes will induce a destabilizing effect. The desired stabilization is caused by $H^{g} > 0$, which will be demonstrated more explicitly below.

In order to complete the theory, we must choose a constitutive assumption for $K^{g(b)}$ as to ensure that boundary dissipation is satisfied. Like in the case of small deformations, cf. Svedberg (1996), we set $K^{g(b)} = H^{g} l_{0} c \kappa^{g}$, with $c > 0$, to obtain

$$N \cdot \nabla \dot{\mu} = \frac{c}{l_{0}} \dot{\mu} \quad \text{on } \partial\Omega_{0} \tag{12}$$

which is the adopted natural boundary condition for $\dot{\mu}$ on $\partial\Omega_0$, corresponding to the gradient term. In the literature, cf. Pamin (1994); Mikkelsen (1997), it is common to choose $c = 0$ *ad hoc*, whereas the condition (12) is a possible extension that is thermodynamically admissible. Clearly, when $l_0 = 0$ (local theory), the boundary condition (12) will be irrelevant.

3. Numerical algorithm

3.1. Integration and incremental constitutive relations

Integrating the rate equations above using the (exponential) Backward Euler rule, we can obtain the following incremental equations, cf. Svedberg and Runesson (1998),

$$^{n+1}F^{\mathrm{p}} = {}^{n+1}A \cdot {}^{n}F^{\mathrm{p}} \quad \text{with} \quad {}^{n+1}A = \exp\left(\Delta\mu\, \bar{M}\left(^{n+1}\bar{C}^{\mathrm{e}}\right)\right) \tag{13}$$

$$^{n+1}\kappa = {}^{n}\kappa - \Delta\mu \left[1 - \frac{K(^{n+1}\kappa)}{K_{\infty}}\right], \quad {}^{n+1}\kappa^{\mathrm{g}} = {}^{n}\kappa^{\mathrm{g}} - \Delta\mu \tag{14}$$

In (13), $^{n+1}\bar{C}^{\mathrm{e}}$ is not given by the loading only as for small deformation; rather it is computed as

$$^{n+1}\bar{C}^{\mathrm{e}} \stackrel{\text{def}}{=} [^{n+1}\bar{F}^{\mathrm{e}}]^{\mathrm{T}} \cdot {}^{n+1}\bar{F}^{\mathrm{e}} = {}^{n+1}A^{-\mathrm{T}} \cdot {}^{n+1}\bar{C}^{\mathrm{e,tr}} \cdot {}^{n+1}A^{-1} \tag{15}$$

where the "trial" value of $^{n+1}\bar{C}^{\mathrm{e}}$, denoted $^{n+1}\bar{C}^{\mathrm{e,tr}}$, is given as

$$^{n+1}\bar{C}^{\mathrm{e,tr}} \stackrel{\text{def}}{=} \left[^{n+1}\bar{F}^{\mathrm{e,tr}}\right]^{\mathrm{T}} \cdot {}^{n+1}\bar{F}^{\mathrm{e,tr}} = [^{n}F^{\mathrm{p}}]^{-\mathrm{T}} \cdot {}^{n+1}C \cdot [^{n}F^{\mathrm{p}}]^{-1} \tag{16}$$

It appears that the formulation is essentially that of *local* plasticity theory with the important exception that $\Delta\mu$ is determined from a differential equation (and not merely an algebraic equation). This equation stems from the expression for Φ in (8), which may be rewritten as follows if (14) and (6) are used:

$$^{n+1}\Phi = -f(^{n+1}\bar{C}^{\mathrm{e}}, {}^{n+1}\kappa) + l_0^2 H^{\mathrm{g}} \nabla^2(\Delta\mu) \tag{17}$$

with the notation

$$f(^{n+1}\bar{C}^{\mathrm{e}}, {}^{n+1}\kappa) \stackrel{\text{def}}{=} -\bar{T}_{\mathrm{e}}(^{n+1}\bar{C}^{\mathrm{e}}) + \sigma_{\mathrm{y}} + K(^{n+1}\kappa) + {}^{n}K^{\mathrm{g}} \tag{18}$$

3.2. *Finite element equations*

As noted above, the constitutive problem now represents a full-fledged boundary value problem (compared with an algebraic equation for the local theory). Usually, the constitutive problem and the equilibrium problem are solved as a coupled system, requiring the discretization of the plastic multiplier to be C_1-continuous.

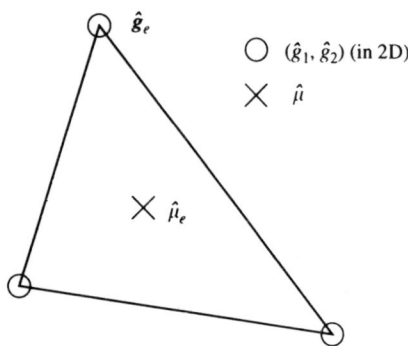

Figure 1. *2D triangular element with linear interpolation of the gradient field and constant multiplier field (element wise)*

However, a different approach is used here: By using a (dual) mixed finite element formulation for the plastic multiplier, where we introduce the gradient of the plastic multiplier $g \stackrel{\text{def}}{=} H^g \nabla(\Delta \mu)$, a C_0-implementation can be used. The simplest possible two-dimensional triangular element is shown in Figure 1. For a more detailed review of the algorithm used, see Svedberg (2000) and Svedberg and Runesson (1998).

4. Computational results

4.1. *Model problem representing surface indentation*

The model problem consists of a rectangular domain in 2D, which is meshed using triangular plane strain gradient enhanced elements. The element mesh is constructed in such a fashion that the element size is small close to the contact area and larger away from it.

The indenter is modeled as rigid, and it is either a sphere (cylinder) of radius r (Brinell type), a prism with half angle 72° (Vickers type), or a flat punch with width $2r$. Different probe sizes are modeled by varying the ratio l/r, and a comparison is made with pure local theory ($l = 0$). The contact between the probe and the material is assumed friction-less.

The simulations are performed using deformation control of the indenter, and the final indentation depth is the same for the different calculations. The mesh consists of 2892 CST elements and 3020 DOF, as shown in Figure 4.

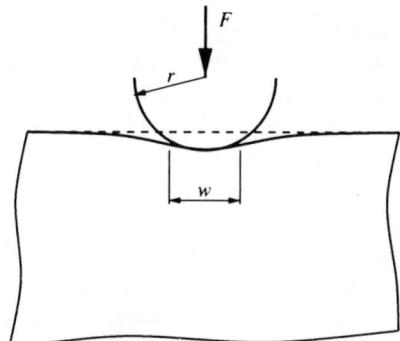

Figure 2. *Geometric setup*

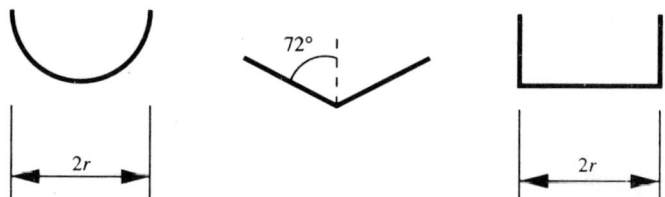

Figure 3. *The different indenter geometries used: Brinell type, Vickers type and flat punch*

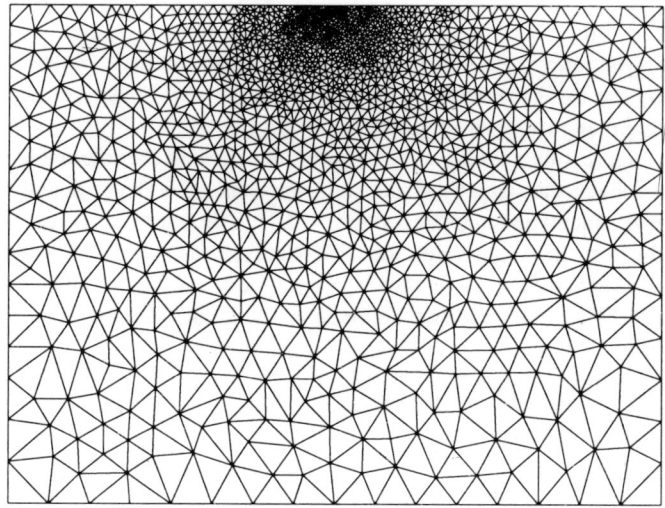

Figure 4. *The mesh used for all simulations*

4.2. *Numerical simulations and discussion*

The first set of figures (Figures 5–7) show the accumulated plastic strain at the end of the loading for local theory and for decreasing indenter size (or, equivalently, increasing internal length). The smoothing effect of the gradient-enhancement is clearly seen for the different indenters; it is most evident for the flat indenter in Figure 7 and for the Brinell type indenter in Figure 5. As the size of the indenter decreases towards (and below) the internal length, the size of the plastic zone increases significantly, cf. the results in Figure 4 in Begley and Hutchinson (1998). It should be noted that the Vickers type indenter is "infinite" and that the ratio r/l in that case is a means of distinguishing the different simulations and to enable comparison with the other indenters.

Comparing the results in Figure 7 with the results in Figure 8a (flat ended circular punch) in Shu and Fleck (1998), we can see that the results are similar. However, a quantitative comparison is not feasible since we assume plane strain while Shu and Fleck (1998) used an axi-symmetric formulation. The results in Figure 5 and Figure 6 can also be compared with those in Figure 9a (sphere) and b (circular cone) in Shu and Fleck (1998). In this case the results are not as close, possibly due to the different settings.

From the extent of the incremental plastic zone for the last load-step in the simulations in Figures 8–10 it is evident that the current plastic zone has its center at the edge of the contact between the indenter and the surface. As above, the size-effect is most evident for the flat indenter and least for the Vickers type indenter.

The next series of figures (Figures 11–13) show the load-displacement curves for the three different indenter types. The difference of the load-displacement curves for different indenter size is a measure of the size-effect on hardness. Note that the hardness H is here defined as the total applied force divided by the "projected" contact area on the initial surface. It can be noted that the increase in hardness for increasing indentation depth is similar for the Brinell type indenter in Figure 11 and the flat punch in 13, while the Vickers type indenter in Figure 12 shows a significantly stronger dependence of the indentation depth. The deviating results for the Vickers type indenter for local (and "nearly" local) theory in Figure 12 seem to be caused by the local mesh density at the contact, possibly exaggerated as a result of the adopted plane strain assumption.

Finally, Figure 14 shows how the hardness depends on the internal length l in the gradient model and the size of the indenter (represented by the radius of the indenter r) at the final indentation depth (depth/r = 0.4). The hardness is normalized with respect to the corresponding value for local theory, i.e. $l = 0$. All the results show a quite linear response (except for the Vickers type results mentioned above). Once again, the results for the Vickers type indenter stand out from the others that behave quite similarly.

5. Concluding remarks

A recently developed thermodynamically consistent theory of gradient plasticity for large strains was employed to investigate the size-effect on micro-indentation. The main objective was to illustrate the differences (and possible similarities) of the deformation field and the hardness, due to different head shape, between different indenter types used in practice. In particular, the morphology of the indented surface was studied. Since the study was restricted to plane strain, it was possible to make a qualitative comparison only. The results of the Brinell type indenter and the flat punch showed similarities, whereas the Vickers type indenter gave different results. For example, the Vickers type indenter showed a significantly stronger size-effect in the sense that the increase of hardness with reduced indenter size was larger. This characteristic has been confirmed elsewhere in the literature, cf. Figure 4a in Shu and Fleck (1998). A possible explanation is that the deformation field produced by the Vickers type of indenter is less smooth with larger local gradients than for the other indenters.

References

M. Atkinson. Examination of reported size effects in ultra-micro-indentation testing. *J. Mater. Science*, 30:1728–1732, 1995.

Z.P. Bazant and F.-B. Lin. Non-local yield limit degradation. *Int. J. Num. Meth. Engng.*, 26:1805–1823, 1988.

M.R. Begley and J.W. Hutchinson. The mechanics of size-dependent indentation. *J. Mech. Phys. Solids*, 46:2049–2068, 1998.

A. Benallal and C. Comi. Localization analysis via a geometrical method. *Int. J. Solids Structures*, 33(1):99–119, 1996.

R. de Borst and J. Pamin. Some novel developments in finite element procedures for gradient-dependent plasticity. *Int. J. Num. Meth. Engng.*, 39:2477–2505, 1996.

A.A. Elmustafa, J.A. Eastman, M.N. Ritter, J.R. Weertman, and D.S. Stone. Indentation size effect: Large grained aluminium versus nanocrystaline aluminium-zirconium alloys. *Script. Mater.*, 43:951–955, 2000.

A.C. Eringen. On nonlocal plasticity. *Int. J. Engng. Sci.*, 19(12):1461–1474, 1981.

N.A. Fleck and J.W. Hutchinson. A phenomenological theory of strain gradient plasticity. Technical report, Division of Applied Sciences, Harvard University, 1993.

J.B. Leblond, G. Perrin, and J. Devaux. Bifurcation effects in ductile metals with nonlocal damage. *J. Appl. Mech.*, 61(2):236–242, 1994.

T. Liebe and P. Steinmann. Theory and numerics of a thermomechanically consistent framework for geometrically linear gradient plasticity. *Int. J. Num. Meth. Engng.*, 51:1437–1467, 2001.

Q. Ma and D.R. Clarke. Size dependent hardness of silver single crystals. *J. Mater. Res.*, 10:853–863, 1995.

K.W. McElhaney, J.J. Vlassak, and W.D Nix. Determination of indenter tip geometry and indentation contact area for depth-sensing indentation experiments. *J. Mater. Res.*, 13:1300–1306, 1998.

L.P. Mikkelsen. Post-necking behavior modelled by a gradient-dependent plasticity theory. *Int. J. Solids Structures*, 34:4531–4546, 1997.

H.-B Mühlhaus and E.C. Aifantis. A variational principle for gradient plasticity. *Int. J. Solids Structures*, 28(7):845–857, 1991.

W.C. Oliver and G.M. Pharr. An improved technique for determining hardness and elastic modulus using load and displacement sensing indentation experiments. *J. Mater. Res.*, 7:1564–1583, 1992.

J. Pamin. *Gradient-dependent plasticity in numerical simulation of localization phenomena.* PhD thesis, Delft University of Technology, 1994.

G. Pijaudier-Cabot and A. Huerta. Finite element analysis of bifurcation in nonlocal strain softening solids. *Comp. Meth. Appl. Mech. Engng.*, 90:905–919, 1991.

W.J. Poole, M.F. Ashby, and N.A. Fleck. Micro-hardness of annealed and work-hardened copper polycrystals. *Script. Mater.*, 34:559–564, 1996.

J.Y. Shu and N.A. Fleck. The prediction of a size effect in micro-indentation. *Int. J. Solids Structures*, 35:1363–1383, 1998.

L.J. Sluys, R. deBorst, and M. Mühlhaus. Wave propagation, localization and dispersion in a gradient-dependent medium. *Int. J. Solids Structures*, 30(9):1153–1171, 1993.

P. Steinmann. *Lokalisierungsprobleme in der Plasto-Mechanik.* PhD thesis, Universität Karlsruhe, 1992.

N.A. Stelmashenko, M.G. Walls, L.M. Brown, and Y.V. Milman. Microindentation on W and Mo oriented single crystals: An STM study. *Acta metall. mater.*, 41:2855–2865, 1993.

T. Svedberg. *A Thermodynamically Consistent Theory of Gradient-Regularized Plasticity Coupled to Damage*. Licentiate thesis, Chalmers University of Technology, 1996.

T. Svedberg. *On the Modelling and Numerics of Gradient-Regularized Plasticity Coupled to Damage*. PhD thesis, Chalmers University of Technology, 1999.

T. Svedberg. Gradient-regularized hyperelasto-plasticity coupled to damage – thermodynamics and numerical algorithm. 2000. Submitted.

T. Svedberg and K. Runesson. A thermodynamically consistent theory of gradient-regularized plasticity coupled to damage. *Int. J. Plasticity*, 13(6-7):669–696, 1997.

T. Svedberg and K. Runesson. An algorithm for gradient-regularized plasticity coupled to damage based on a dual mixed FE-formulation. *Comp. Meth. Appl. Mech. Engng.*, 161:49–65, 1998.

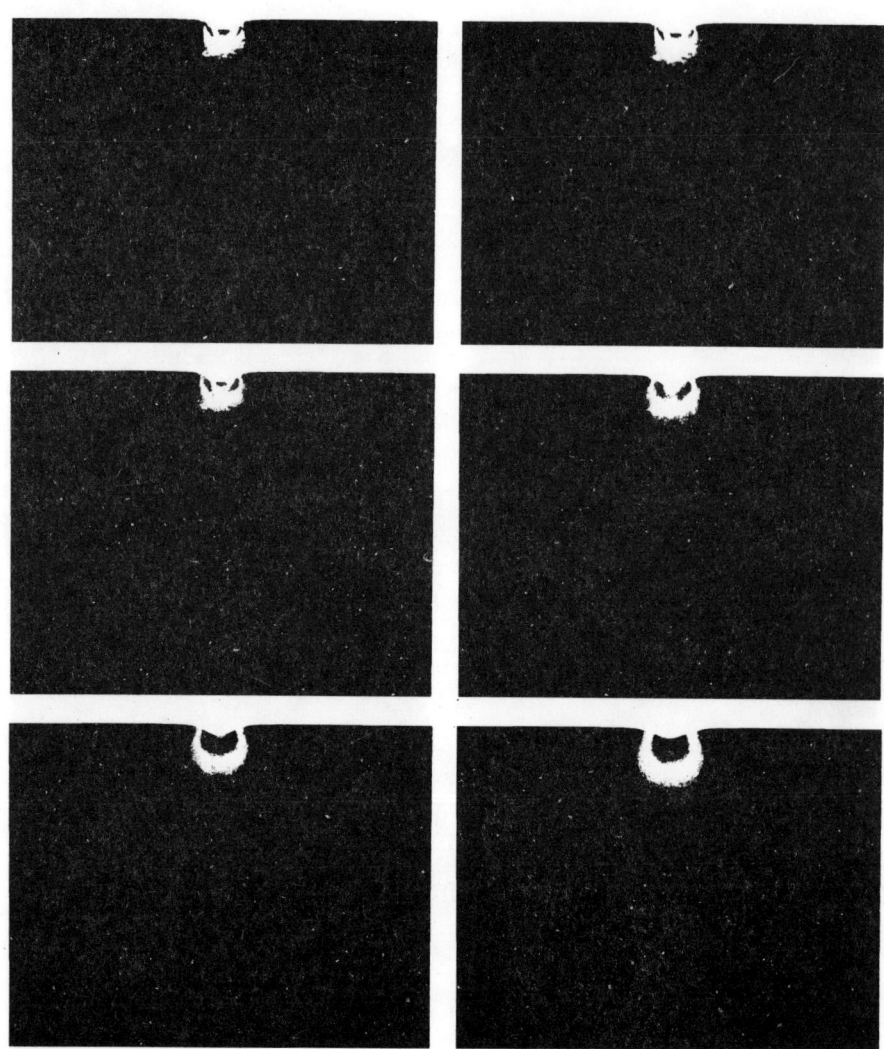

Figure 5. *Distribution of accumulated plastic strain for local theory and for gradient theory with l/r = 1/8, 1/4, 1/2, 1 and 2. Brinell type indenter*

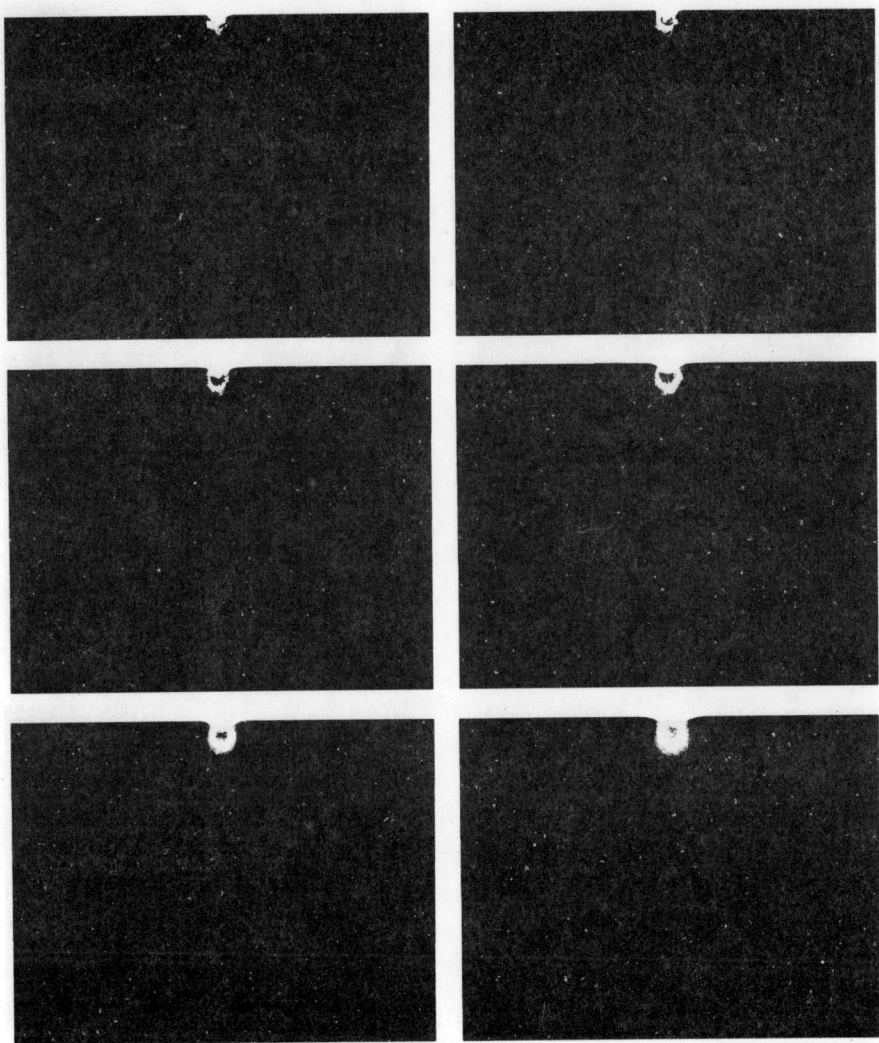

Figure 6. *Distribution of accumulated plastic strain for local theory and for gradient theory with l/r = 1/8, 1/4, 1/2, 1 and 2. Vickers type indenter*

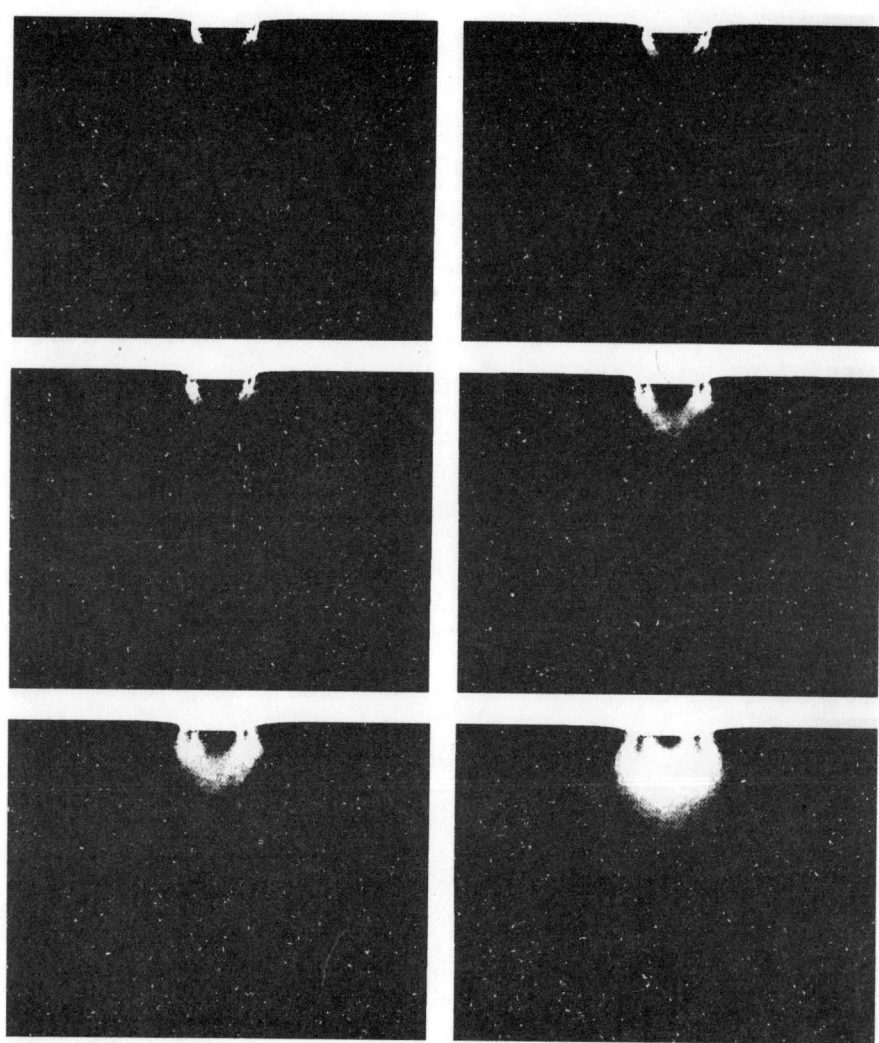

Figure 7. *Distribution of accumulated plastic strain for local theory and for gradient theory with l/r = 1/8, 1/4, 1/2, 1 and 2. Flat punch indenter*

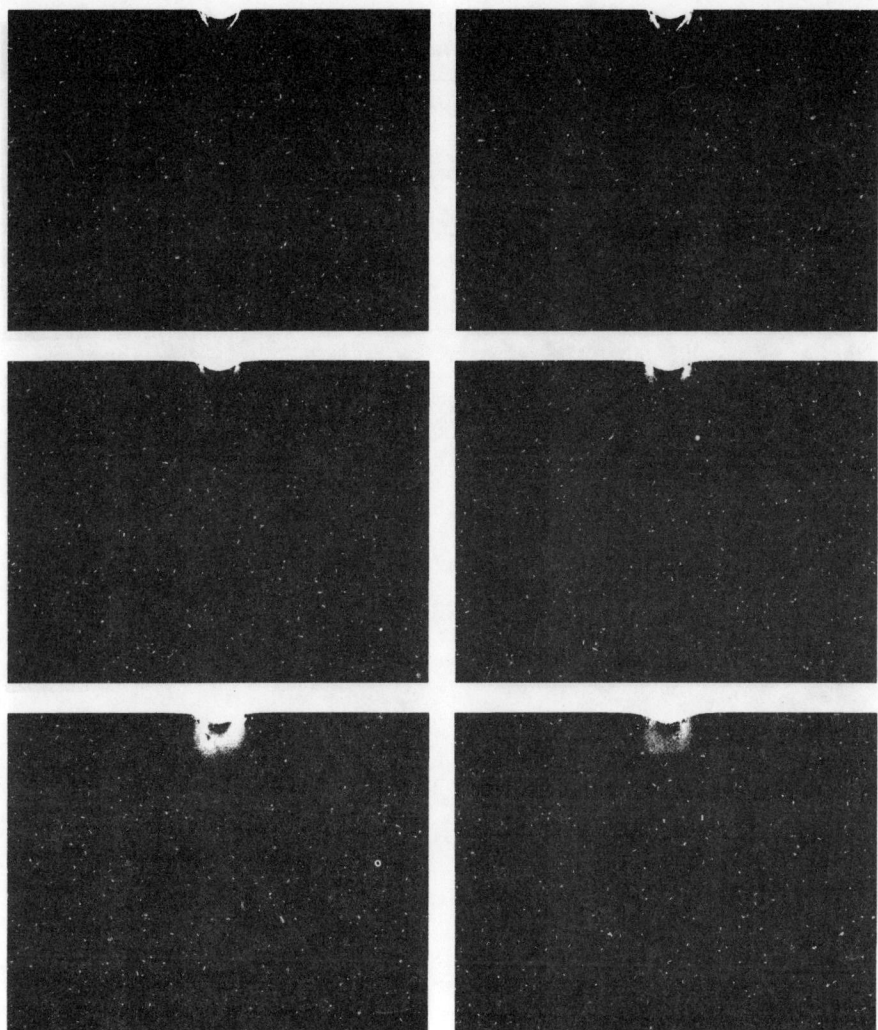

Figure 8. *Distribution of incremental plastic strain (at final load step) for local theory and for gradient theory with l/r = 1/8, 1/4, 1/2, 1 and 2. Brinell type indenter*

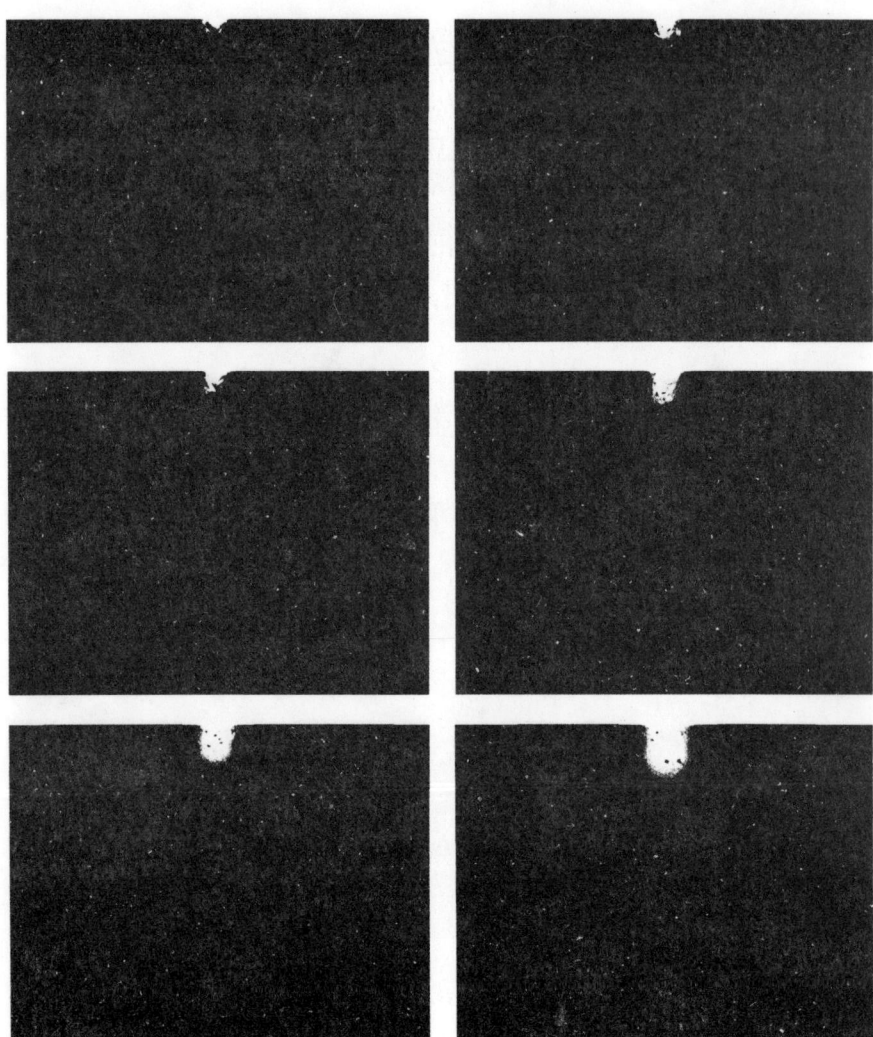

Figure 9. *Distribution of incremental plastic strain (at final load step) for local theory and for gradient theory with l/r = 1/8, 1/4, 1/2, 1 and 2. Vickers type indenter*

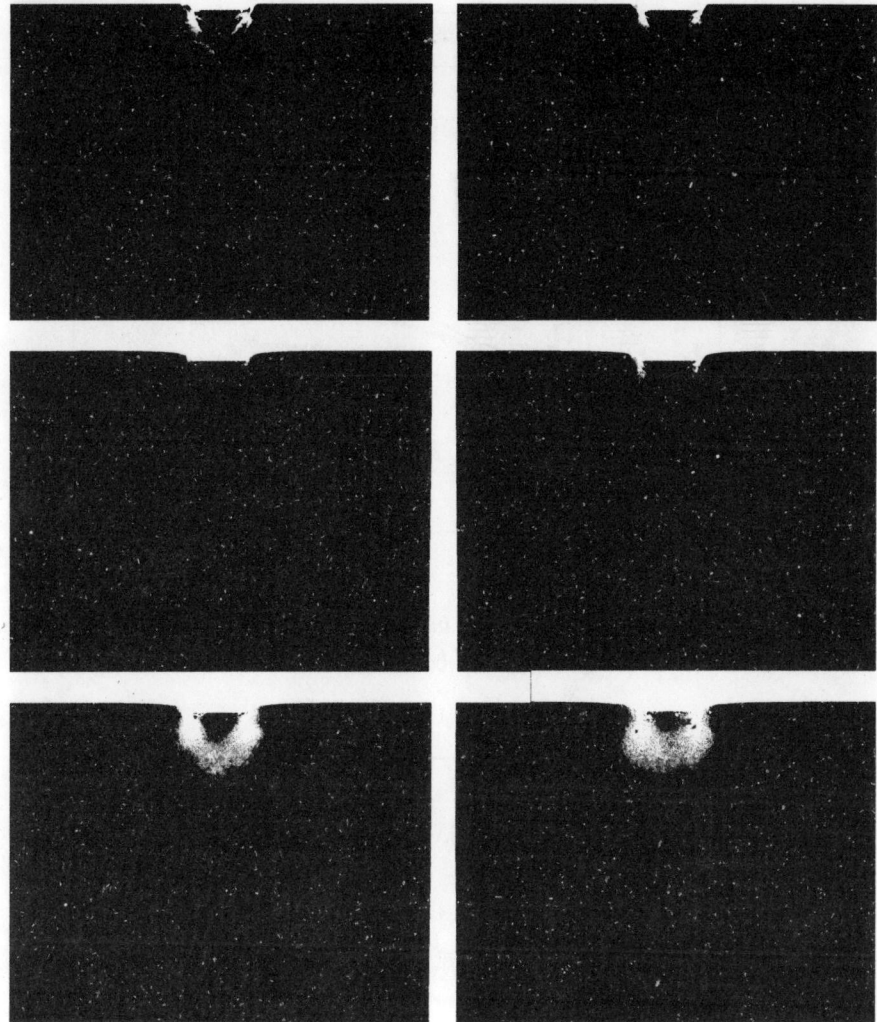

Figure 10. *Distribution of incremental plastic strain (at final load step) for local theory and for gradient theory with l/r = 1/8, 1/4, 1/2, 1 and 2. Flat punch indenter*

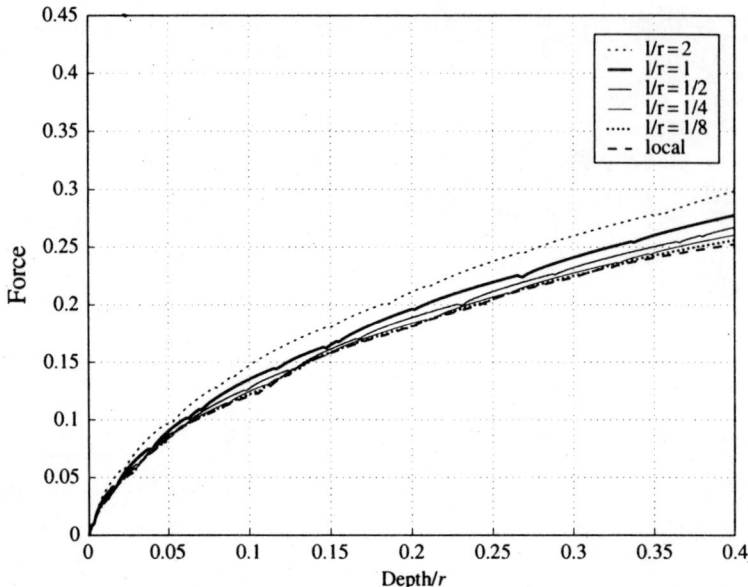

Figure 11. *Total applied force versus indentation depth for local theory and for gradient theory with different l/r. Brinell type indenter*

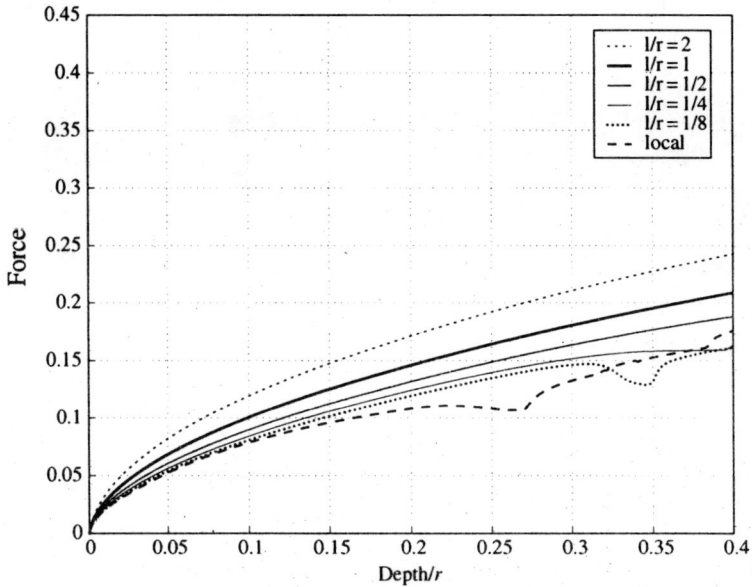

Figure 12. *Total applied force versus indentation depth for local theory and for gradient theory with different l/r. Vickers type indenter*

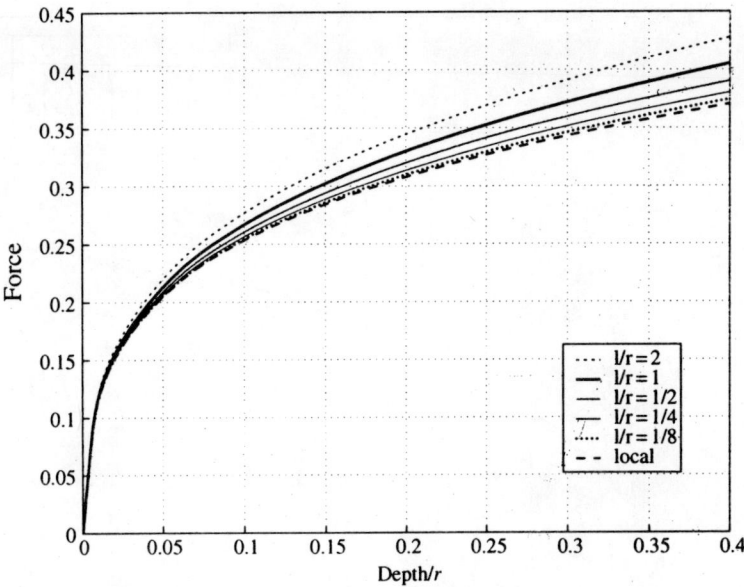

Figure 13. *Total applied force versus indentation depth for local theory and for gradient theory with different l/r. Flat type indenter*

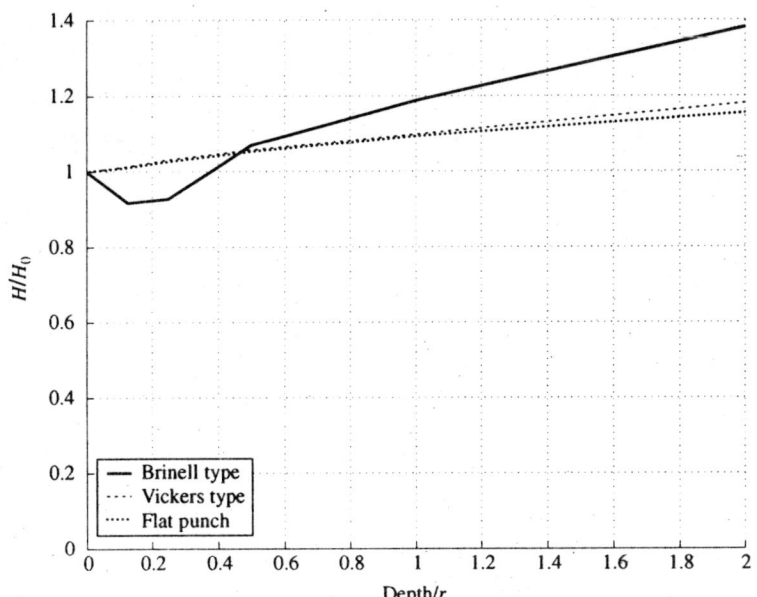

Figure 14. *Relative hardness versus indenter size (l/r) for the different indenter shapes at the final indentation depth*

Chapter 15

A Simplified Model Of Thermal Contact Resistance Adapted to Forging Simulation

Brahim Bourouga and Jean-Pierre Bardon
Laboratoire de Thermocinetique, Polytechnic University of Nantes, France

Vincent Goizet
ARDEM, FORTECH, Pamiers, France

1. Introduction

To conclude a simulation of hot forging, a good knowledge of the thermal conditions at the boundaries is essential in particular at the workpiece–die interface. In general, more than 90% of the thermal losses of the workpiece occur in the die. The quality of the numerical resolution of the simulated problem depends on the accuracy on the Thermal Contact Resistance (TCR). TCR is the reverse of contact heat transfer coefficient. Many works were listed on this subject. All implicitly consider a hypothesis of TCR whatever the conditions and the forging mode (Semiatin *et al.*, 1987; Burte *et al.*, 1990; Jain *et al.*, 1991; Pietrzyk *et al.*, 1991; Nshama *et al.* , 1994; Malinowski et al. , 1994; Li *et al.* 1995, 1996; Hu *et al.*; 1998; Goizet *et al.* , 1998, 1999; Goizet, 1999). To simplify, a majority of authors consider also that the TCR is uniform and remains constant during all the forging process. In other words, they suppose that the contact interface remains static and does not undergo any deformation in time. The large diversity of studied configurations (standard of the press, material and size of the workpiece, type of the interface...) allied with the great disparity of the approaches (instrumentation, data processing models, various assumptions...) gave place to a large dispersion of results (Li *et al.*, 1995; Goizet, 1999).

In a recent study, we showed that the contact workpiece-die in forging is not static (Goizet *et al.*, 1998, 1999; Goizet, 1999; Bourougat *et al.*, 2000). It is a dynamic contact where the structure of the interface varies according to the normal stress field at the interface and to the flow stress of the workpiece. The temporal evolution and the local final value of the TCR are related to the forging force and to the constitutive law of the workpiece under the influence of the thermal conditions which prevail during the hot forging process. The idea to study this evolution to analyse the influence of the principal forging parameters supposes that the assumption of TCR remains valid. By accurate experiment, we have validated this assumption in the case of hot forging on hydraulic press (low forging velocity) (Bourougat *et al.*, 2000).

In practice, the knowledge of the TCR requires either a representative measurement of the interface thermal resistance or its estimation from a model. It would be unrealistic to consider the first solution at each operation of simulation. It is thus advisable to develop models which may be easily exploitable.

The many models of TCR developed in the case of the static contact cannot be applied to the interfaces of dynamic contact such as the workpiece-die contact for forging. In fact, these models concern the contact after loading when the interface structure does not evolve any more. Moreover, the authors refer to a contact pressure supposed constant and uniform through all the interface extent. So the prediction always gives a constant and uniform value of the TCR on all the interface. It is obvious that these models cannot describe the dynamic contact which strongly differs from the static contact. From the mechanical point of view, the dynamic contact characterizes the transient stage of loading. That means that the structure of the interface changes as much as the effort of contact varies. Then the TCR is not still constant but becomes a variable space-time parameter describing the thermal effect of space non-uniformity and time evolution of the thermal contact.

The main objective of this contribution is to propose a simplified model of dynamic TCR in the case of a dry or a mixed lubrication contact and another in the case of a coated workpiece by means of a lubricating or insulating film. But before exposing the proposed model, we present some measurement results which give an order of magnitude of the variations in space and time of the TCR at a plane circular workpiece-die interface during a hot forging operation. That will help to justify the need of such a model and the relevance of its development.

2. Space-time distribution of the TCR at the workpiece-die interface

2.1. *Summary presentation of the measurement principle of the TCR space-time distribution*

To study this distribution, we carried out experiments of hot forging of cylindrical parts on a hydraulic press of 250 tons. The experimental methodology developed for this purpose is exposed in detail in the reference (Goizet *et al.*, 1998; Goizet, 1998). We evoke here summarily the measurement principle and the results obtained from a hot forging experiment carried out on a part made of aluminium alloy 2214. This part is of cylindrical form. It has a diameter of 80 mm and a 160 mm height. The die is made of Waspalloy. It has also a cylindrical form with a diameter of 280 mm and a 60 mm thickness.

We retained the idea of a multiple radial instrumentation on both sides of the interface as represented in Figure 1. Four different radial measurement positions were chosen on a same radius of the interface. They are located respectively in $r_1 = 0$ mm, $r_2 = 11$ mm, $r_3 = 22$ mm and $r_4 = 33$ mm. On both sides of each location, we have introduced a thermocouple in the workpiece at 0.7 mm from the interface and two thermocouples in the die at 0.5 and 1 mm respectively from the interface. The three thermocouples are aligned according to the normal in the contact plan. Thus, the two thermocouples in the die constitute a sensor for the determination of the parietal heat flux. They can be considered as a heat fluxmeter.

Because the small thickness of the instrumented zone, compared with the radial distance separating two measurement sites, what occurs in a measurement point remains independent of the thermal events which proceed at the other measurement points. In other words we suppose that the local heat transfer remains one-dimensional. The recorded temperature fields are used to estimate the thermal contact resistance at the interface. The principle of the thermal contact resistance estimation is based on the identification of the temperature jump and heat flux density across the workpiece-die interface. Identification of the die side temperature and the heat flux density crossing the interface is obtained from the temperature field measurement in the die at the close vicinity of the interface by using a nonlinear inverse technique of heat conduction. Knowing the estimated heat flux density at the interface and one temperature measurement at the bottom of the workpiece, the workpiece side temperature can be calculated by the solution of the direct problem.

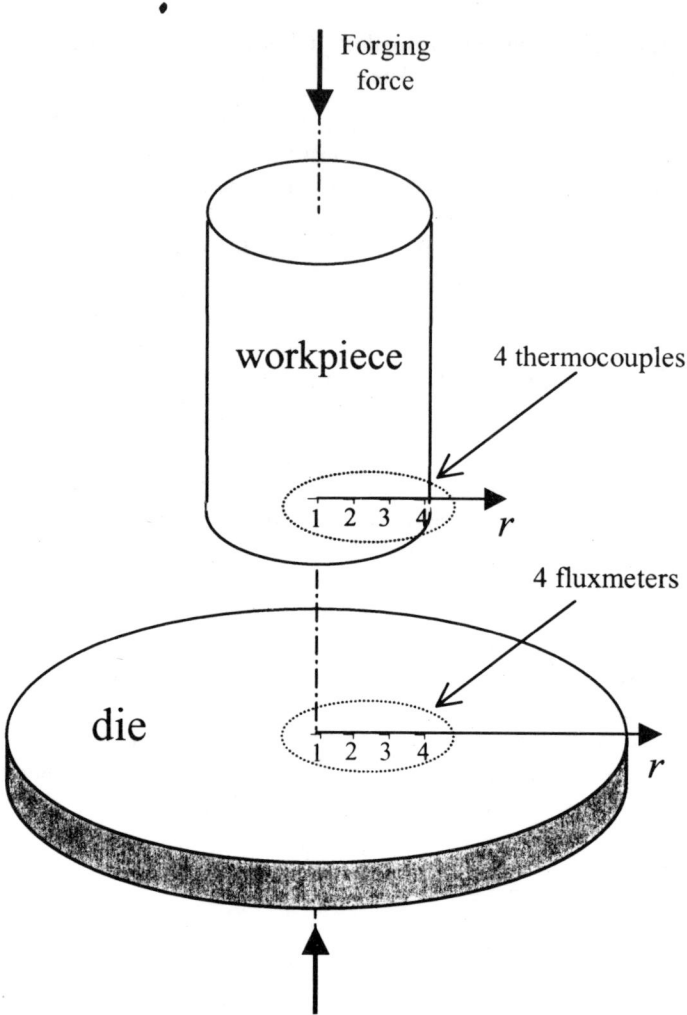

Figure 1. *Schema of measurement principle*

2.2. Space-time distribution of the TCR

The initial value $R_c(0,0)$, considered just before the beginning of the forging at the point $r = 0$, is used as a reference value to reduce all the curves and to return them without dimension. In Figure 2, we present the temporal evolution of TCR experimental different radial measurement positions at the interface.

During the free resting time of the workpiece on the die before deformation, one notes that the initial values of the TCR are constant, with a maximum value at the $r = 0$ position. The three other points of measurement present practically the same value. Theoretically, the four initial values corresponding to four measurement sites should be all equal to the thermal resistance of the equivalent air blade since the contact pressure due to the weight only of the part is very weak and uniform through all the interface. The higher value displayed at the $r = 0$ point is probably related to the machining operation of the instrumented face of the workpiece. The fact that the three other values are equal can be explained by a good uniformity of the two surfaces states in contact (absence of wavyness and/or of contaminant).

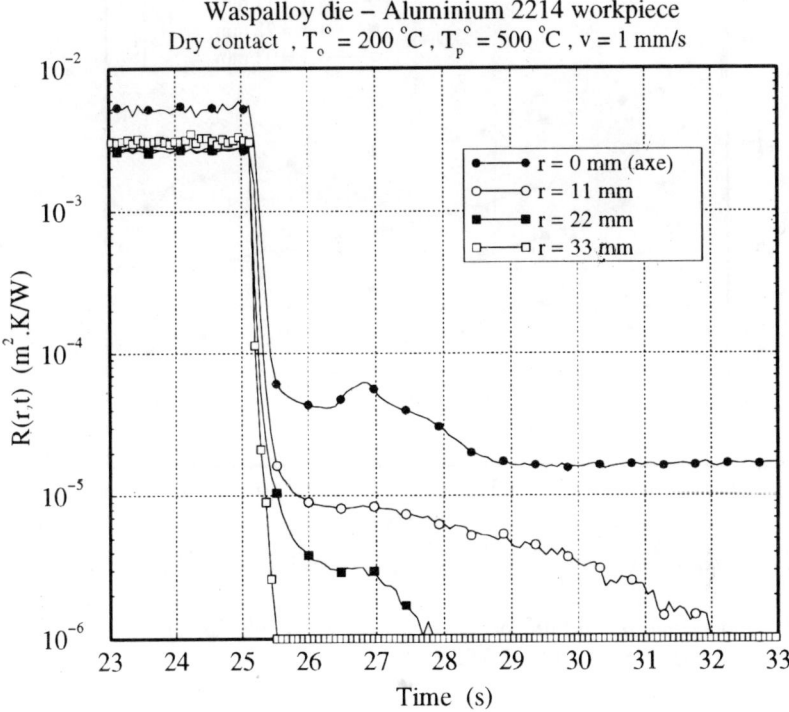

Figure 2. *Space-time distribution of the TCR: part of aluminium alloy 2214*

During the period of deformation, the evolutions of $R_c(r, t)$ at the measurement points are decreasing and very different. At each time, it is noted that the TCR remains maximum in the interface axis ($r = 0$ mm) and decreases according to the radial coordinate. One explains this result by the non-uniform distribution of the normal stress field in the contact plan during the forging operation. Indeed, the simulation of the experiment on the numerical code FORGE II watches that the compression normal stress at the workpiece-die interface remains monotonous, increasing according to the radius r and time t during all the forging process. Illustrated in Figure 3, this result is also confirmed by the bibliography (Baillet, 1994).

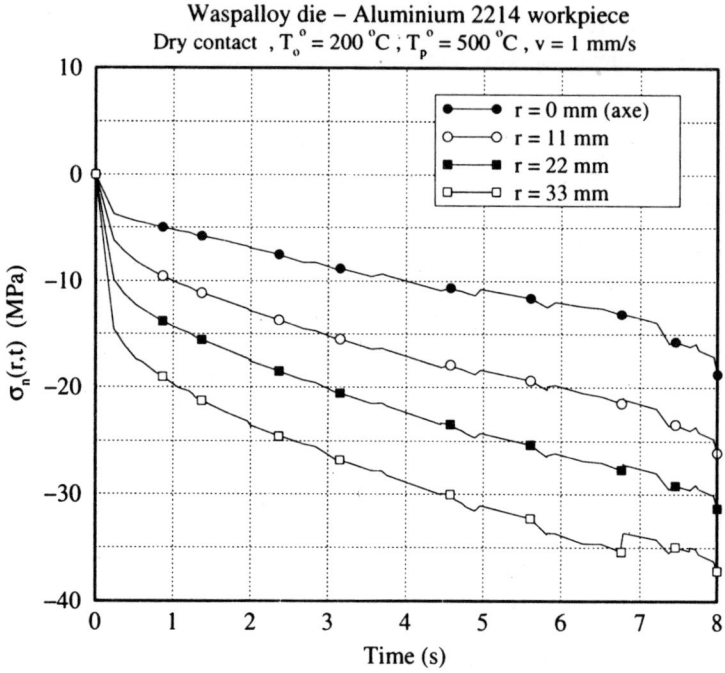

Figure 3. *Space-time evolution of normal stress at workpiece-die interface. Simulation of the experiment on FORGE2 code*

In Figure 4, we present the spatial distribution of the TCR, on the contact plane, at various moments. Before deformation ($t < 25.12$ S), one checks partly that TCR is constant and spatial distribution uniform. During forging, the TCR distribution $R_c(r,t)$ becomes a decreasing monotonous function according to r and t. The difference of TCR between the two locations r_1 and r_4 is large. Whatever the moment considered during forging operation, an order of magnitude separates the two values. Thus, at the end of forging process and in the case of Aluminium 2214, the TCR value is divided by 10^2 approximately on the axis and more than 10^3 on the periphery. One checks also that the TCR locally presents an opposite direction of evolution to that of the normal stress at the interface.

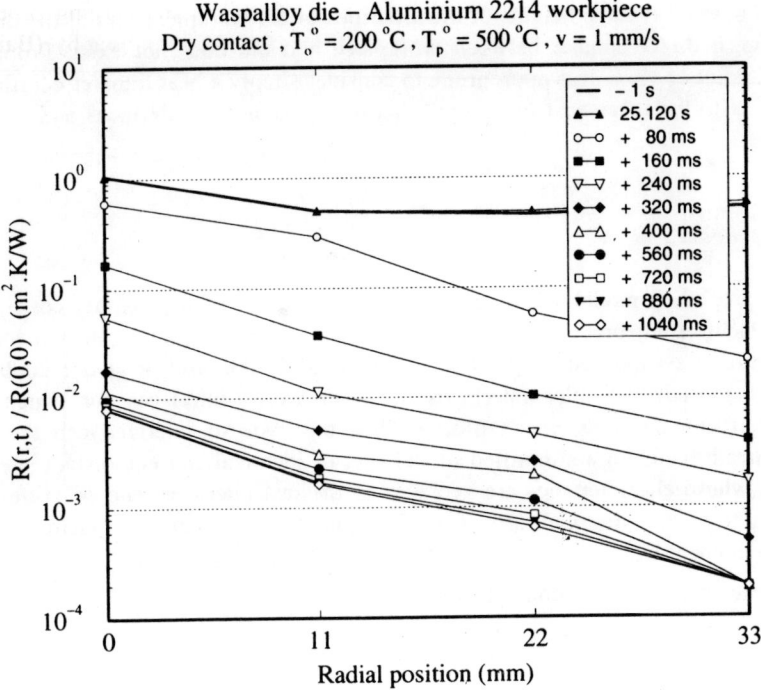

Waspalloy die – Aluminium 2214 workpiece
Dry contact , $T_o^o = 200\ ^oC$, $T_p^o = 500\ ^oC$, v = 1 mm/s

Figure 4. *Space-time distribution of the TCR: another presentation form*

This type of experiment reflects well the reality of the dynamic thermal contact but it implies a heavy and expensive experimental implementation; it justifies the need to develop models to predict directly the TCR in this configuration.

3. Model of dynamic TCR

The model distinguishes two current situations in hot forging. The first corresponds to the case where initial thickness of the lubricant is of the same order of magnitude as the sum of the initial average heights of the asperities of the workpiece and of the die. It brings together the cases of the dry contact and that of mixed lubrication. In both cases, there is direct contact between the asperities of the workpiece and those of the die at the interface. The imperfect thermal contact is characterized by a TCR that one notes as R_c or by its reverse, the contact heat transfer coefficient $h = 1 / R_c$.

The second situation, extremely different, relates to a workpiece coated by means of a solid lubricant or an insulator the thickness of which is large compared with the sum of average heights of the asperities of the workpiece and of the die. In this case, a direct contact between workpiece and die does not exist. From the thermal point of view, it is appropriate to consider simply a heat transfer coefficient equivalent to the reverse of the overall thermal resistance of interfaces and coating film.

3.1. *The model*

We propose a model which describes the evolution of the contact conditions during the forging process. It is a dynamic model in which the phenomena of constriction are supposed established instantaneously (Bourougat *et al.*, 2000). It supposes moreover regular interfaces, in other words interfaces for which the density of contact spots n is uniform. Then this type of interface can then be schematized thanks to a simplified model, say of identical and equidistant sites of contact, whose characteristics are averages of the real interface. The asperities are supposed to be of cylindrical form. Thus the geometry of the contact is characterized at every moment by:

– the average radius of contact spots: $\overline{a}$

– the rate of actual contact: $s*(t)$,

– the density of contact spots: $n(t)$,

– the average heights of asperities of the workpiece and of the die:
$\overline{\delta}_1(t)$ and $\overline{\delta}_2(t)$.

The model allowing the calculation of the TCR is an extension of the model of equidistant asperities of Bardon (Bardon, 1972). In this model, the TCR still results from the combination of two thermal resistances in parallel R_s and R_f characterising respectively the passage of heat by direct contact (solid-solid) and through the interstitial fluid (fluid way) such that:

$$\frac{1}{R_c} = \frac{1}{R_s} + \frac{1}{R_f}$$ [1]

In this extended model, the parameters describing both ways of passage become temporal functions and the thermal constriction term covers a broader field of values of s^* (0–0,8 instead of 0–0,1 in usual elastic contacts). In the forging of aluminium, we have observed that s^* could exceed 0,7 (Goizet, 1999). Thus the function of constriction is given by the Roess series limited to four terms (Roess,). And the R_s term takes the form [2]:

$$R_s = \frac{1}{2\lambda n\bar{a}}\left(1 - 1{,}40925\sqrt{s^*} + 0{,}29591\sqrt{s^*}^3\right.$$
$$\left. + 0{,}05254\sqrt{s^*}^{-5}\right) + \frac{2\delta_{eq}}{\lambda s^*}\left(1 - s^*\right) \qquad [2]$$

with:

$$\frac{2}{\lambda} = \frac{1}{\lambda_1} + \frac{1}{\lambda_2} \quad \text{and} \quad \delta_{\acute{e}q} = \frac{1}{2}\left(\frac{\lambda}{\lambda_1}\bar{\delta}_1 + \frac{\lambda}{\lambda_2}\bar{\delta}_2\right) \qquad [3]$$

where $\bar{a}$ is the average contact radius. λ_1 and λ_2 are the thermal conductivities of the workpiece and of the die. Equation [2] shows that the R_s resistance is composed of two terms: the first accounts for the phenomenum of thermal constriction and the second is the contribution of the internal resistance of the asperities in contact.

The resistance R_f of the fluid gap which has an average thickness $\bar{\delta} = \bar{\delta}_1 + \bar{\delta}_2$ is given by:

$$R_f = \frac{\bar{\delta}}{\lambda_f(1 - s^*)} \qquad [4]$$

λ_f represents the equivalent conductivity of interstitial fluid (conductive and radiative heat transfer).

3.2. Estimation of interfacial parameter: simplified approach

The rate of actual contact s^* which is the ratio between the actual surface of contact and the nominal surface of contact is given by:

$$s^*(t) = \frac{s(t)}{S} = \frac{\sigma_n}{\sigma_e(\theta_i)} \qquad [5]$$

where σ_n and $\sigma_e(\theta_i)$ are respectively the normal stress and the flow stress at the contact temperature θ_i which is metallurgical data. They are given by the computed solution of the thermomechanical problem. The numerical code gives the local normal stress at each time step. The computed temperature fields give the superficial temperatures T_1^s and T_2^s of the workpiece and of the die at the interface from which the contact temperature is deduced:

$$\theta_i = \frac{b_1 T_1^s + b_2 T_2^s}{b_1 + b_2} \tag{6}$$

in this expression, b_1 and b_2 ($b_i = \sqrt{\lambda_i \rho_i C_i}$) are the thermal effusivities. We have validated this relation [6] in many experiments (Goizet, 1999). Other parameters $\overline{a}$, n, $\overline{\delta}_1$, $\overline{\delta}_2$ are deduced from the surface analysis of workpiece and die, carried out before forging and from the knowledge of s^* at each time step.

With an aim of simplification one supposes that the density of contact point n keeps a constant value. Indeed, in the first moments of the forging process, s^* is weak and the contact coefficient of transfer h is governed primarily by the heat transfer through the fluid. $1/R_s$ is negligible in front of $1/R_f$. So n has a very weak influence on the value of the coefficient of transfer given by: $1/h = R_c \approx R_f$.

When s^* increases, it is primarily by a very fast multiplication of contact sites. And n reaches very quickly a maximum value. It can be estimated starting from the formula:

$$n = \left(\frac{1}{Sa_1} + \frac{1}{Sa_2} \right)^2 \tag{7}$$

Where Sa_1 and Sa_2 are the average space periods of the peaks given by the profilometric statements of both surfaces, carried out before forging. Moreover, in this approach, we neglect the phenomenon of coalescence which somewhat decreases n beyond this maximum.

The roughness gives the height of average asperity which is presumed constant ($\overline{\delta}_1$ and $\overline{\delta}_2$). This assumption is justified by the very weak slope of the asperity (at the most few degrees for the angle at the base).

In the denominator of the first term of R_s given by [2], the product $n\overline{a}$ is estimated starting from the relation:

$$\frac{1}{n\overline{a}} = \sqrt{\frac{\pi}{ns^*}} \sqrt{1 + \frac{\sigma_a^2}{\overline{a}^2}} \tag{8}$$

where σ_a is the standard deviation of the contact radii. If we retain the assumption that $\sigma_a \ll \overline{a}$, the knowledge of $s^{\bullet}$ allows the calculus of the average contact radius $\overline{a}$. In the opposite case, it is appropriate to estimate σ_a.

4. The thermal resistance of interfaces and coating film

In the case of a lubricating or insulating film, the contact heat transfer coefficient is equal to the reverse of the overall resistance of both interfaces and film. This coefficient is given by:

$$h = \frac{1}{R_{c1} + R_l + R_{c2}} = \frac{1}{R_l} = \frac{\lambda_c\left(\overline{T_i^s}\right)}{\delta_c\left(T_1^s\right)}$$ [9]

where R_{c1} and R_{c2} are thermal contact resistances at the interfaces workpiece-coating and coating-die and R_l is the internal thermal resistance of the coating. Generally, $R_{c1} \ll R_l$ and $R_{c2} \ll R_l$ during the hot forging process; so that the heat transfer coefficient can be regarded as being simply the reverse of the thermal resistance of the coating such as is given by [9]. The coating on its thickness is the seat of a large temperature gradient and one considers its conductivity at the average temperature of interface given by:

$$T_i^s = 0,5 \cdot \left(T_1^s + T_2^s\right)$$ [10]

and corresponding to the previous time step. The thickness of the coating film δ_c is supposed to be variable decreasing and depending on the workpiece temperature at the interface T_1^s. If $T_1^s < T_f$, T_f being the melting point of the coating, then: $\delta_c = \delta_c^0$, where δ_c^0 is the initial thickness of the coating. On the other hand, if $T_1^s \geq T_f$ then: $\delta_c = \delta_c^f$, δ_c^f being the final thickness known by experiment.

It is important to underline how easy the exploitation of this model is. It can be introduced into the numerical code quickly in the form of a subroutine. It however requires the profilometric statements of surfaces before forging. That can be carried out on mechanical profilometers often available in the Research and Development services of forging companies.

5. Conclusion

We propose a simplified model which allows estimation of the contact heat transfer coefficient at the workpiece-die interface during a hot forging process. The model considers two situations. In the first which considers the dry contact or the mixed lubrication, we propose a formula giving the thermal resistance of dynamic contact to each time step according to the local normal stress and the flow stress of the workpiece corresponding to the contact temperature. This formula requires the knowledge of the profilometric statements of both surfaces before the forging process.

In the second situation, the heat transfer coefficient at the workpiece-die is represented by the reverse of the thermal resistance of the coating which can be variable. But the variation domain is limited to two values: the initial and final values of the thermal resistance of the coating.

The model is simple and presents an easy exploitation. It requires only the knowledge of simple profilometric statements of surfaces.

References

Semiatin S.L., Collings E.W., Wood V.V. and Altan T., "Determination of the interface heat transfer coefficient for non-isothermal bulk-forming processes", *Journal of Engineering for Industry*, Vol. 109, February 1987, 49–57.

Burte P.R., Im Y.T., Altan T. and Semiatin S.L., "Measurement and analysis of heat transfer and friction during hot forging", *Journal of Engineering for Industry*, Vol. 112, 332–339, November 1990.

Jain V.K. and Goetz R.L., "Determination of contact heat transfer coefficient for forging of high temperature materials", ASME, *National Heat Transfer Conference Minneapolis*, July, 28–31, 1991.

Pietrzyk M., Kusiak H., Lenard J.G. and Malinowski Z., "Heat exchange between the workpiece and the tool in the metal forming processes", *Proc. Conf. Formability*, Ostrava, 329–338, 1994.

Nshama W., Jeswiet J. and Oosthuizen, "Evaluation of temperature and heat transfer conditions at the metal PH", *Forming Interface, Journal of Materials Processing Technology*, Vol. 45, 637–642, 1994.

Malinowski Z., Lenard J.G. and Davies M.E., "A study of the heat transfer coefficient as a function of temperature and pressure", *Journal of Materials Processing Technology*, Vol. 41, 125–142, 1994.

Li Y.H. and Sellars C.M., *Literature review: interface heat transfer during hot forging and rolling* (for DTI-NPL Project PMP9), March 1995.

Hu M.Z., Brooks J.W. and Dean T.A., "The interfaciale heat transfer in hot die forging of titanium alloy", *Proc. Instn Mech. Engrs*, Vol. 212, Part C, 485–496, 1998.

Li Y.H. and Sellars C.M., "Evaluation of interfacial heat transfer and its effects on hot forming processes", *Iron Making and Steel Making*, Vol. 23, 1996.

Goizet V., Bourouga B., Bardon J.P., " Etude expérimentale du transfert thermique à l'interface pièce-outil lors d'une opération de forgeage à chaud ", *Rev. de Métallurgie*, pp. 601–608, May 1999.

Goizet V., Bourouga B., Bardon J.P., "Influence of the hot forging parameters on the workpiece-die thermal contact", *Int. J. Forming Processes,* Vol. 1, No. 4, 485–500, 1998.

Goizet V., *Etude experimentale des mécanismes de transfert thermique à l' interface pièce-outil lors d'une opération de forgeage à chaud*, PhD Thesis Nantes University, N° ED 82–373, ISITEM, 1999.

Bourouga B., Goizet V., Bardon J.P., "Thermal contact at a non-static workpiece-die interface during hot forging process", *Proc. 3rd E.C.T.S.*, Heidelberg 2000, pp. 127–132, Ed ETS, 2000.

Baillet L., *Modélisation des frottement pour les opérations de matriçage*, Thèse de doctorat de l'université de Lyion, Lyon, 1994.

Bardon J.P., *Introduction à l'étude des résistance thermiques de contact*, RGT n° 125, 1972.

Roess M., Theory of spreading conductance, Appendix to: *Thermal measurements of joints formed between stationary metal surfaces*, Trans. ASME, Vol. 71–3, Weills & Ryder.

Chapter 16

Validation of the Cockcroft and Latham Fracture Criterion for Cold Heading of Steel Fasteners Using Drop Weight Compression Testing and Finite Element Modeling

Nicolas Nickoletopoulos and Michel Hone
Ivaco Rolling Mills, Ontario, Canada

Yves Verreman
Ecole Polytechnique, Department of Mechanical Engineering, Quebec, Canada

James A. Nemes
Department of Mechanical Engineering, McGill University, Quebec, Canada

1. Introduction

Cold heading is used to produce fasteners such as bolts, rivets and nuts. It is a high strain rate forging operation performed without external heat. A force applied to the end of a metal workpiece contained between a die and a punch displaces metal to a pre-determined contour in one or several blows of the punch [DAV 88].

The ability to assess the formability of cold heading materials and to model the cold heading process is of importance since failures can result in equipment downtime, material waste and product recalls [NIC 01].

There are two sets of requirements for cold heading quality materials. One set speaks to the properties necessary for cold forming operations [MAH 78]. The other set addresses the properties relating to product specifications [MAT 80].

Material properties that influence cold headability include rod surface quality, cleanliness, chemical composition in terms of residual elements and nitrogen content, and microstructure. Process parameters, in particular, strain rate, deformation temperature, pre-draw, and lubrication must be accounted for in the development of a practical cold heading model [NIC 01].

Cold headability is often assessed in industry using tensile and/or compression tests. These tests have certain shortcomings that impede efforts to obtain reliable ductility parameters for cold upsetting. In tensile testing, fracture does not initiate at the surface as it does in cold forging; and the stress and strain states at the crack initiation site are different [OLS 86]. Furthermore, fracture initiates at much smaller nominal strains than those observed during compression testing. The compression test involves compressing a cylindrical workpiece between parallel dies, and circumvents the necking problem associated with tensile testing [NIC 01]. However, the test is usually performed at significantly lower strain rates than those observed in cold heading, and the use of flat dies limits specimen barreling, hoop stress and the propensity for fracture.

In order to have conditions representative of industrial operations, a Drop Weight Compression Test (DWCT) was first developed to compress end-clamped cylindrical specimens to failure at high strain rate. The DWCT was then modeled with a FEM program previously used to simulate the upset stages of a cold heading process [VER 00] [VER 01]. Inputs to the model include the geometrical and kinematical parameters of the test, the stress-strain behavior of the material, and the friction conditions. Here the Cam Plastometer Test (CPT) was employed to determine with accuracy the flow stress evolution of these materials, and the Friction Ring Test (FRT) to determine the friction factor [NIC 01].

Finally, by combining data from the DWCT and the FEM model, the Cockcroft and Latham fracture criterion was calibrated, and the validity of the criterion in assessing materials for cold heading was investigated.

2. Drop weight compression testing

2.1. *Drop Weight Test Machine*

A DWCT machine enables a falling weight to compress a workpiece at strain rates in the range 100–2000 s^{-1}, rates that cannot be obtained by servo-hydraulic machines [VER 00, NIC 01, FOL 89].

The DWCT machine (Figure 1) used in this work was constructed at McGill University in Montreal, Canada [VER 00, NIC 01]. It consists of a tower enabling an assembly with changeable weight plates to be dropped from a height of up to 3 meters, a compression fixture, and an anvil. The compression fixture transfers the impact load from the crosshead to the workpiece through a shaft. Pneumatic shock absorbers avoid secondary loading due to rebounding.

Figure 1. *Drop weight test machine – (a) weight plates, (b) die-set, and (c) pneumatic shock absorbers*

The design of a die-set for the DWCT machine that does not fracture during testing [NIC 01, SAR 99, SHI 88] is a challenge that was overcome by the guided cylindrical-pocket die-set shown in Figure 2. A sleeve acts to guide the dies during testing. Air vents in the sleeve prevent pressure build-up.

Figure 2. *Photo of guided cylindrical pocket die-set for DWCT - (a) lower die, (b) die sleeve, and (c) air vents*

Figure 3 presents a cylindrical test specimen having ends clamped between the upper and lower dies. This assembly rests atop the load cell, a 222 kN PCB quartz force sensor, located between the die-set assembly and the anvil. A 100 MHz oscilloscope was set to obtain a total of 10,000 data points in a time frame of 4 ms. The voltage-time history was converted into load-time history using a linear voltage-load calibration. Double integration of these data yielded displacement versus time information that was plotted in terms of load versus displacement. The energy absorbed by the test specimen is represented by the area under this curve.

DWCT testing can be performed by varying the drop weight or the drop height. In this work, the drop height was kept constant at 1.5 m to maintain a constant initial strain rate. For each material, the drop weight was progressively increased in a series of specimens to determine the deformation to fracture, i.e., the height reduction $\Delta H/H_o$ [VER 00, NIC 01].

2.2. *Fracture evaluation*

Figure 3. *Photo of test specimen between the upper and lower dies. The assembly rests atop the load cell*

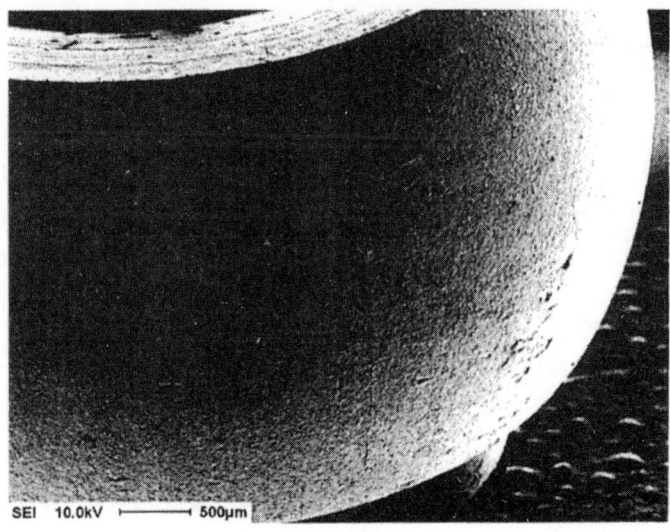

Figure 4. *Photograph of typical crack at fracture onset*

There is no industry standard available to evaluate the surface quality of cold headed parts. Some manufacturers use 100% visual product inspection, while others employ eddy current or laser equipment to detect the presence of surface discontinuities. In this work, the presence of a crack visible under a stereo microscope at x25 signifies specimen failure. Figure 4 illustrates a typical crack at the onset of fracture. Attention was paid to regions of undeformed material in the die pockets above and below a crack to ensure that no prior surface irregularities such as seams, wire drawing scratches or mechanical damage were present to invalidate the results.

2.3. Test materials

Grade 1038 steels were selected for testing. The materials matrix was designed to evaluate the impact of material parameters, *i.e.*, residual element content, nitrogen content, and microstructure, on fracture behavior. Table 1 lists the chemical composition of four materials selected for testing.

Heats A and B are continuously cast steel produced from scrap steel. These were selected to test the effect of low and high nitrogen content, i.e., 107 versus 60 ppm. Heats C and D which are continuously cast steel produced from pig iron were also used to test the effect of nitrogen content, i.e., 101 versus 44 ppm. The latter materials are generally used in more demanding cold heading applications.

Table 1. *Chemical analysis of test materials (wt%)*

Element	Heat A	Heat B	Heat C	Heat D
C	0.38	0.38	0.41	0.41
Mn	1.04	0.99	0.78	0.80
Si	0.24	0.22	0.23	0.22
Cu	0.16	0.15	0.05	0.04
Ni	0.06	0.07	0.07	0.07
Cr	0.25	0.26	0.04	0.04
Mo	0.026	0.020	0.006	0:006
N	0.0107	0.0060	0.0101	0.0044

The test materials were fast cooled following hot rolling on a Stelmor conveyor to maximize the level of solute nitrogen [THA 95a, THA 95b] and thereby exaggerate the negative impact on cold heading.

2.4. *Specimen preparation*

The test materials were hot rolled to a diameter of 5.5 mm and controlled fast cooled to yield a microstructure consisting of pearlite in a ferrite matrix. Half of these test specimens were then spheroidize annealed in an atmosphere-controlled industrial batch furnace. The rod specimens were then drawn from an initial diameter of 5.50 ± 0.10 mm to a final diameter $D_0 = 5.21 \pm 0.01$ mm, a 10% reduction. Typical industrial pre-draw reductions range from 8 to 14%. Materials were then stored at $-6°C$ to minimize aging after cold working.

Specimen aspect ratio (H/D) influences the fracture behavior of the material, as evidenced by the workability diagrams ($\varepsilon_{zz}-\varepsilon_{\theta\theta}$) of Lee and Kuhn [LEE 73, KUH 78]. These researchers found that lower friction values and higher aspect ratios decrease bulge curvature, thereby delaying fracture. In this study, wire specimens were cut to obtain DWCT specimens with three aspect ratios, H/D = 1.0, 1.3, and 1.6, on a micro-lathe.

2.5. *Selected DWCT results*

Figure 5 illustrates the DWCT fracture ranges for the four test materials with an aspect ratio of 1.3. The fracture range has an upper limit defined by the height reduction at which fracture occurred. The lower limit is the highest height achieved without fracture. The fracture strain rests between these extremes. Testing was performed in coarse and fine stages that involved testing through a range of mass increments in multiples of 2.3 kg and 0.3 to 0.6 kg respectively. Repeat experiments (3 for each material) were performed to validate the reproducibility of the results.

It is evident from Figure 5 that for the spheroidized materials, the average global axial strains to fracture are roughly 20% larger than those of the as-rolled materials. The lower flow stress spheroidized microstructures have a better ductility. Another critical parameter is the chemical composition. Heats A and B exhibit significantly lower axial strain to fracture results than Heats C and D. This can be attributed to the significantly higher residual element content. Heats A and B have virtually the same chemical composition (see Table 1) except for nitrogen. The lower nitrogen content material (Heat B) exhibited notably larger axial strains to fracture. Heats C and D also have virtually the same chemical composition except for nitrogen. Again, the lower nitrogen content material (D) exhibited larger axial strains to fracture than the high nitrogen content material (C).

These results show that the DWCT is sensitive to microstructure, and to residual element and nitrogen content.

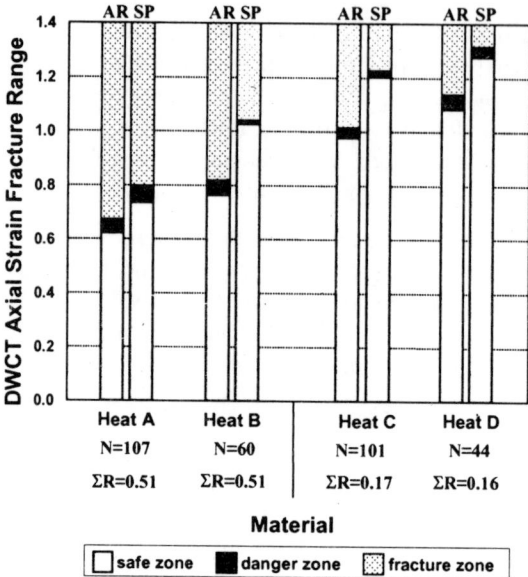

Figure 5. *DWCT fracture ranges (ln(H₀/H))*
AR = as-rolled; SP = spheroidized; N = nitrogen content in ppm
ΣR = sum of residual elements including Cu, Ni, Cr, Mo, and Sn (wt%)

3. Finite element modeling

A commercial FEM software package, Forge2, was employed to simulate the upsetting of Heat C DWCT specimens in the as-rolled and the spheroidized conditions [NIC 01]. The objective of this work was to model the stress and strain evolution in the specimens upset using the pocket die-set configuration, and ultimately to investigate the validity of the Cockcroft and Latham ductile fracture criterion. Simulations were performed at the two extreme aspect ratios, *i.e.*, H/D = 1.0 and 1.6, to determine if this criterion is independent of specimen geometry.

Thermo-mechanical analyses were performed to account for the effect of adiabatic heating. A Von Mises rigid plastic behavior with the following relation for the flow stress evolution has been selected:

$$\sigma_{eq} = K_0 \sqrt{3} (\sqrt{3} \, \dot{\varepsilon}_{eq})^m \, (\varepsilon_{eq} + \varepsilon_0)^n \, \exp(\beta / T) \qquad [1]$$

where $\dot{\varepsilon}_{eq}$, ε_{eq} and T designate the equivalent strain rate, the equivalent strain and the temperature respectively. The constitutive parameters include the strength coefficient (K$_o$), the strain rate sensitivity (m), the work hardening coefficient (n), and the temperature coefficient (β). These parameters were derived from Cam Plastometer Tests performed on the actual materials at high strain rates (20–150 s^{-1}) and at T = 22°C and 500°C [NIC 01]. The friction parameter utilized in the simulations was derived from Friction Ring Tests performed on the DWCT using representative strain rates, temperature, and lubrication conditions. A shear friction factor of 0.13 was determined using Friction Ring calibration curves available in the literature and FEM simulations of the Friction Ring Test [NIC 01].

In all simulations, the prescribed tool velocity is continuously decreasing in order to satisfy the conservation of the initial energy, i.e., the kinetic energy of the actual DWCT mass required to fracture a given specimen. Six-noded triangular mesh elements were used with a maximum mesh size of 0.25 mm was used. Automatic remeshing occurred when the code was no longer able to perform additional computations due to mesh distortion or mesh penetration into the dies [VER 01].

The theoretical initial velocity was calculated to be 5.42 ms^{-1} by equating the kinetic and potential energies. The DWCT crosshead is not actually in a free fall due to friction and air resistance, and had to be determined experimentally for the FEM simulations. The actual initial velocity, 5.29 ms^{-1}, corresponds to initial strain rates of 1040 s^{-1} and 550 s^{-1} for aspect ratios 1.0 and 1.6, respectively. These strain rates are between 3 and 5 times the strain rates observed in industry, i.e., ~200 s^{-1}; nevertheless, they are more characteristic of cold heading than the strain rates associated with servo-hydraulic machines, i.e., 0.1–1 s^{-1}. The higher strain rates were selected to exaggerate the strain rate effect.

3.1. Description of mesh

The initial mesh was generated using the automatic mesh generation function. Quadratic six-noded triangular elements were used. Various mesh refinement rates were tested for solution convergence before final selection of the initial mesh parameters. In these preliminary simulations homogeneous upsetting was performed using frictionless flat dies. Figure 6 shows initial meshes and the pocket dies. Table 2 presents the initial mesh parameters employed for the FEM modeling. All simulations were carried at a constant drop height of 1.5 m, and employed the actual DWCT drop mass required to fracture a given specimen.

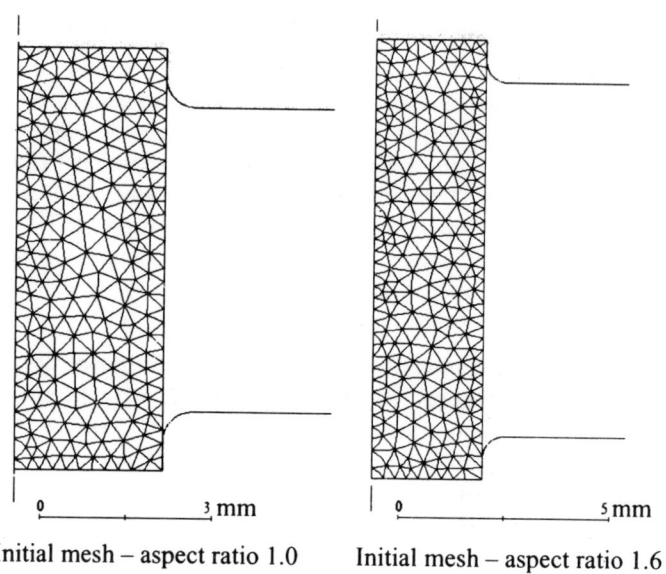

Initial mesh – aspect ratio 1.0   Initial mesh – aspect ratio 1.6

Figure 6. *Initial mesh and die geometry for aspect ratios of 1.0 and 1.6*

Table 2. *Mesh parameters for FEM models*

Mesh Parameter	Aspect Ratio 1.0	Aspect Ratio 1.6
plane area [mm^2]	18.8	26.8
axisymmetric volume [mm^3]	153.5	219.6
number of corner nodes	272	290
number of elements	458	486
nodes of quadratic triangular element	1001	1065
initial maximum element size [mm]	0.43	0.52
remesh maximum element size [mm]	0.25	0.25

3.2. Selected finite element modeling results

The FEM simulations predicted final specimen heights that were lower than those observed in the actual DWCT experiments. The height discrepancies were roughly 1.1 mm and are attributed to compliance of the DWCT machine.

Nevertheless, only the evolutions of stress and strain components until the height reduction to fracture were considered.

Figures 7 and 8 are plots of hoop ($\sigma_{\theta\theta}$) and axial (σ_{zz}) stress components versus equivalent strain until the fracture point. The plots represent the history of the material point on the equatorial surface of the barreling specimens. The radial component of the stress at this point remains at zero throughout the deformation.

From Figure 7, it is evident that for both as-rolled and spheroidized materials, the hoop stress rises faster with smaller aspect ratio. This is in accordance with the $\varepsilon_{\theta\theta}$-$\varepsilon_{zz}$ workability diagrams [LEE 73, KUH 78], where a quicker rise in the hoop strain is observed for smaller aspect ratios. In each case the hoop stress rises to a maximum then it decreases to a local minimum. The decrease in hoop stress occurs as the material comes into sliding contact with the die faces.

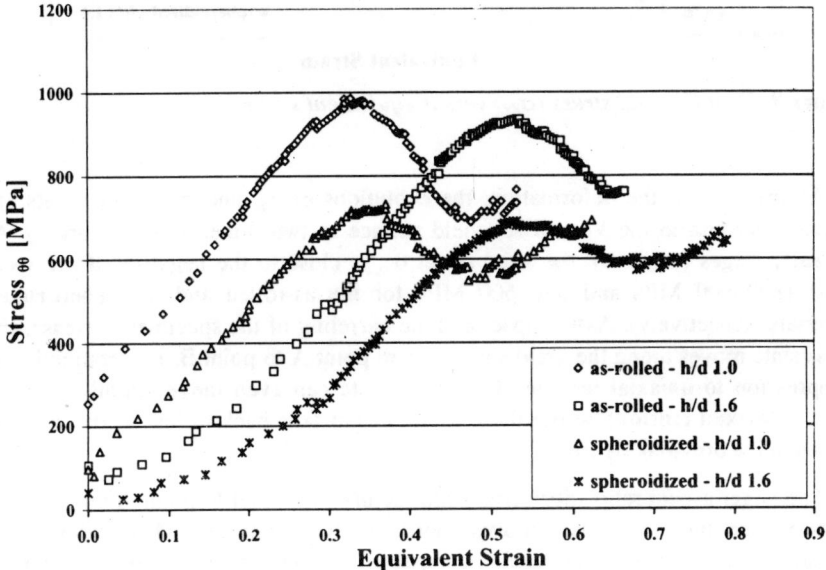

Figure 7. *Heat C: hoop stress ($\sigma_{\theta\theta}$) versus equivalent strain*

Figure 8 illustrates that similar increases were obtained with the axial stress plot (decreasing compression). The axial stress also rises faster for the two materials with the smaller aspect ratio. Contrary to the hoop stress, the axial stress for the spheroidized materials rises more slowly than the axial stress for as-rolled materials at a given aspect ratio.

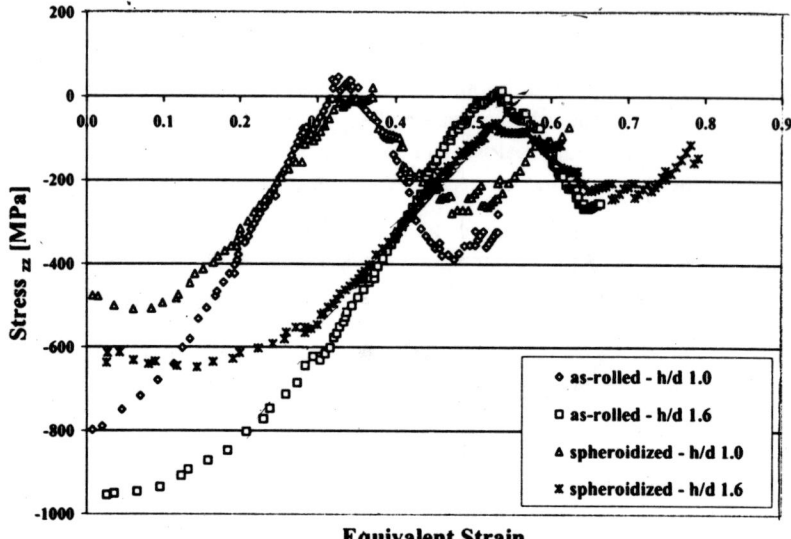

Figure 8. *Heat C: axial stress (σ_{ZZ}) versus equivalent strain*

At any time of the deformation, the evolutions of σ_{zz} and $\sigma_{\theta\theta}$ components are linked according to the Von Mises yield surface in two dimensions (Figure 9). At the early stages ($\sigma_{\theta\theta} \cong 0$), the axial stress σ_{zz} is close to the negative of the yield stress (800–900 MPa and 550–600 MPa for the as-rolled and the spheroidized materials respectively). As the upset and the barreling of the specimen increase, the stress state moves along the yield surface from point A to point B, i.e. from uniaxial compression to uniaxial tension. The stress state can even move slightly towards point C (biaxial tension) before the σ_{zz} component falls back to just below zero, as shown in the previous figure.

The larger aspect ratio specimens begin deforming closer to point A on the yield surface, then the stress state moves more slowly to point B when deformation increases. A similar behavior would be expected for lower friction conditions. In the limit case of homogeneous compression, the stress state would always remain at point A during all the deformation (Figure 9).

Figure 10 is a plot of the stress triaxiality (hydrostatic stress over equivalent stress) versus equivalent strain until the fracture point. As the deformation increases, the triaxiality increases from level A (–0.33 for uniaxial compression) to level B (+0.33 for uniaxial tension), then it decreases to lower values as the material comes into sliding contact with the die faces. On this plot, the curves almost coincide for a given aspect ratio. The stress triaxiality as well as the hoop stress increase faster for the smaller aspect ratio. This is the reason why fracture occurs at lower equivalent strains.

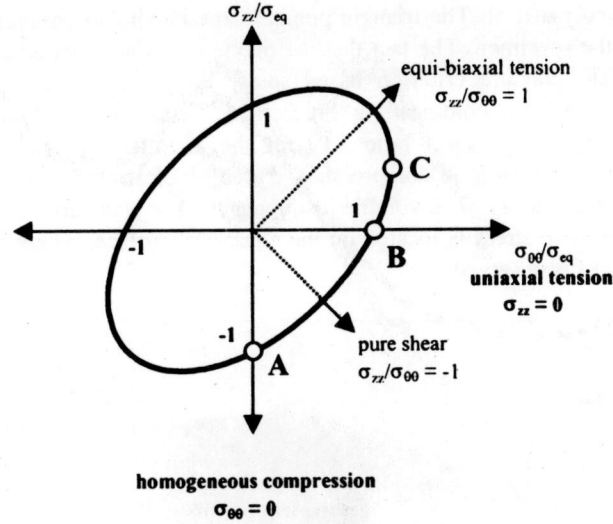

Figure 9. *Stress history along the yield surface*

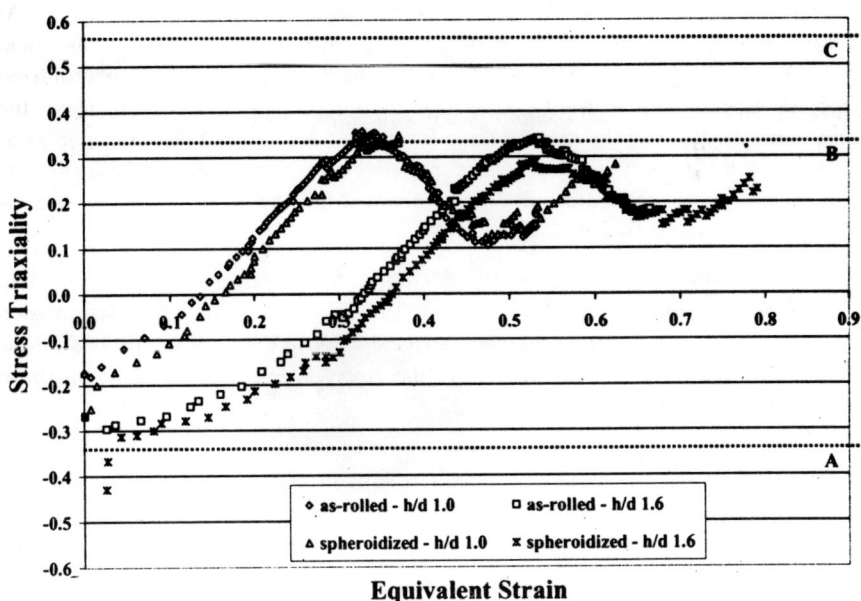

Figure 10. *Heat C: stress triaxiality versus equivalent strain*

Figure 11 is a contour plot of the equivalent strain at fracture (aspect ratio 1.0) for the as-rolled material. The triangle points to the maximum equivalent strain at the center of the specimen. The fact that the maximum is not located at the surface implies that any fracture criterion based solely on the equivalent strain is not adequate for upsetting in cold heading. Figure 12 is a contour plot of the hoop stress component at fracture (aspect ratio 1.0) for the as-rolled material. The central volume of the specimen is in compression, thereby indicating that it is capable of withstanding large strains. The volume just under the barreled surface is in tension. The maximum hoop stress is located on the equatorial surface, where the initiation of vertical cracks is observed.

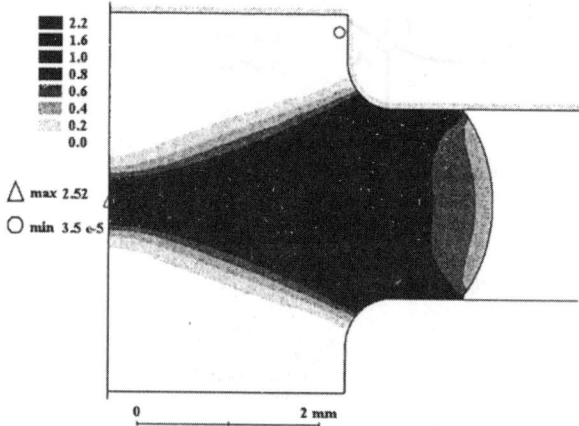

Figure 11. *Heat C as-rolled: contour plot of equivalent strain*

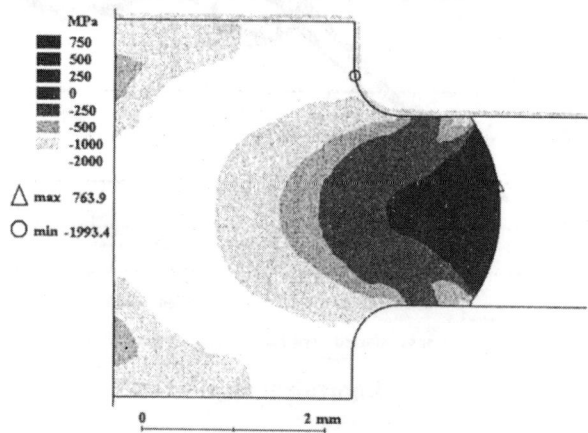

Figure 12. *Heat C as-rolled: contour plot of hoop stress ($\sigma_{\theta\theta}$)*

3.3. *Cockcroft and Latham Criterion*

The Cockcroft and Latham criterion states that fracture occurs when the work done by the maximum tensile stress attains a critical energy density value 'C' (equation 2), a material property [SHI 88, COC 68]. According to this criterion, fracture occurs when:

$$\int_0^{\varepsilon_{eq.f.}} \sigma_1 \, d\varepsilon_{eq} = C \qquad\qquad [2]$$

where   C = fracture criterion constant

ε_{eq} = equivalent strain = $\displaystyle\int \dot{\varepsilon}_{eq} \, dt$

$\varepsilon_{eq.f.}$ = equivalent strain to fracture

σ_1 = maximum principal tensile stress.

In the case of upsetting, the maximum principal tensile stress in the region of fracture, i.e., at the equator, is the hoop stress, as indicated in the previous section. This criterion recognizes the coupled roles of tensile stress and plastic strain in promoting ductile fracture [SHI 88, COC 68]. When only compressive stresses are operating, *e.g.,* in hydrostatic compression, σ_1 is equal to zero and no fracture can occur.

Figure 13 illustrates the contour plot of the Cockcroft and Latham constant at fracture (aspect ratio 1.0) for the as-rolled material. The constant is calculated by time integration of the maximum hoop stress and the equivalent strain rate product. Here, the maximum is located at the equatorial surface, the location at which fracture occurred during DWCT testing.

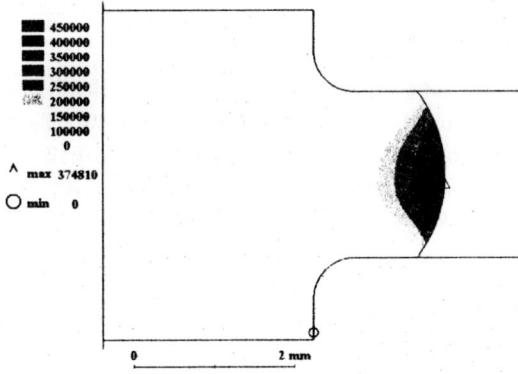

Figure 13. *Heat C as rolled: contour plot of Latham and Cockcroft criterion*

3.4. *Validation of the Cockcroft and Latham criterion*

The Cockcroft and Latham constants were evaluated at the height reductions measured at fracture of the four Heat C materials (as-rolled and spheroidized, and initial aspect ratio H/D = 1.0 and 1.6). These are presented in Tables 3 and 4.

For the as-rolled material, between the two aspect ratios, there is a 6% difference in Cockcroft and Latham constants for a 25% difference in equivalent strains to fracture. For the spheroidized material, there is only a 1% difference in the constants for a 29% difference in equivalent strains to fracture. These results suggest that the Cockcroft and Latham criterion is valid for upsetting in cold heading.

Table 3. *Cockcroft and Latham constants for Heat C as-rolled*

Aspect Ratio	Cockcroft and Latham constant [MPa]	Equivalent Strain at Fracture
1.0	373	0.53
1.6	352	0.66
difference	−6%	25%

Table 4. *Cockcroft and Latham constants for Heat C spheroidized*

Aspect Ratio	Cockcroft and Latham constant [MPa]	Equivalent Strain at Fracture
1.0	320	0.62
1.6	324	0.81
difference	−1%	29%

The Cockcroft and Latham constant for the as-rolled material is 11% larger than that for the spheroidized annealed material, while the axial strain to fracture is lower for the as-rolled materials. To enable comparison of the Cockcroft and Latham constants between materials, it is proposed that the constants be normalized, i.e., be divided by the respective yield stresses of the materials [NIC 01].

The Cockcroft and Latham constants obtained in this work on grade 1038 steel are comparable to those found in the literature. For example, Frater and Petrus [FRA 90] determined a Cockcroft and Latham fracture constant of 395 MPa for an as-rolled grade 1045 steel in tension.

4. Conclusion

The DWCT generates fracture conditions representative of industrial cold heading operations. The test is sensitive to chemical composition and to microstructural differences between as-rolled and spheroidize annealed materials. Drop weight compression testing can thus be used to evaluate the formability of materials for upsetting. FEM modeling of the DWCT allows calibration of the Cockcroft and Latham fracture constants. This criterion appears to be valid in evaluating for materials designated for upsetting in cold heading operations. Numerical modeling could then be used to predict the risks of fracture in an industrial process, i.e. to predict critical geometries for a given material, or to predict the minimum material ductility required for a given application and to select a cost-effective material.

Acknowledgements

The authors thank Ivaco Rolling Mills for supporting this scientific approach to the study of cold heading.

References

[COC 68] COCKCROFT M.G. and D.J. LATHAM, "Ductility and the workability of metals", *Journal of the Institute of Metals*, Vol. 96, 1968, pp. 33–39.

[DAV 88] DAVIS J. [SENIOR EDITOR], "Cold heading", ASM Metals Handbook, *Forming and Forging,* 9TH Edition, Metals Park, Ohio, Vol. 14: 1988, pp. 291–297.

[FRA 90] FRATER J.L. and G.J. PETRUS, "Combining finite element methods and the Cockcroft and Latham criterion to predict free surface workability of cold forgings", *Transactions of NAMRI/SME*, 1990, pp. 97–102.

[FOL 89] FOLLANSBEE P.S., "High strain rate compression testing", *ASM Metals Handbook*, Vol. 8: Mechanical Testing, 9th Edition, Metals Park, Ohio, 1989, pp. 190–207.

[KUH 78] KUHN H.A., "Workability testing and analysis for bulk forming processes", *Formability Topics – Metallic Materials*, ASTM STP 647, 1978, pp. 206–219.

[LEE 73] LEE P.W. and KUHN H.A., "Fracture in cold upset forging – a criterion and model", *Metallurgical Transactions*, Vol. 4, 1973, pp. 969–973.

[MAH 78] MAHESHWARI, M.D., DUTTA B., and MUKHERJEE T., "Quality requirements for cold heading grades of steel", *Tool & Alloy Steels*, 1978, pp. 247–251.

[MAT 80] MATSUNAGA, T. and SHIWAKU, K, "Manufacturing of cold heading quality wire rods and wires", *SEAISI Quarterly Journal*, Vol. 9, No. 1, 1980, pp. 45–55.

[NIC 01] NICKOLETOPOULOS N., *Physical and numerical modeling of steel wire rod fracture during upsetting for cold heading operations*, Ph.D. Thesis, McGill University, 2001.

[OLS 86] OLSSON, K., KARLSSON S., and MELANDER A., "The influence of notches, testing geometry, friction conditions, and microstructure on the cold forgeability of low carbon steels", *Scandinavian Journal of Metallurgy*, Vol. 15, No. 5, 1986, pp. 238–256.

[SAR 99] SARRUF Y., CAO B., JONAS J.J., and NICKOLETOPOULOS N., "Criteria and tests for cold headability", *Wire Journal International*, Vol. 32, No. 2, 1999, pp. 98–105.

[SHI 88] SHIVPURI R., TIETMANN A., KROPP E., and LILLY B., "Investigation of a new upset test to study the formability of steels for cold forging", *NAMRAC Conference, XVI*, Illinois, 1988.

[THA 95a] TAHERI, A.K., MACCAGNO T.M., and JONAS J.J., "Effect of cooling rate after hot rolling and of multistage strain aging on the drawability of low-carbon-steel wire rod", *Metallurgical and Materials Transactions A*, Vol. 26A, 1995, pp. 1183–1193.

[THA 95b] TAHERI, A.K., MACCAGNO T.M., and JONAS J.J., "Effect of quench aging on drawability in low carbon steels", *Material Science and Technology*, Vol. 11, 1995, pp. 1139–1146.

[VER 00] VERREMAN Y., NICKOLETOPOULOS N., and HONE M., "Prediction of fracture in cold-headed steel fasteners by numerical simulation", *Proceedings (CD-ROM) of Int. Symposium "Mathematical Modeling in Metal Processing and Manufacturing"*, The Metallurgical Society of CIM, 2000.

[VER 01] VERREMAN Y. and HONE M., "Simulation par éléments finis du forgeage des métaux: application à la frappe à froid de boulons d'acier", *Matériaux et Techniques*, No. 1–2, 2001, pp. 37–43.

Chapter 17

Industrial Forging Design Using an Inverse Technique

Luísa C. Sousa, Catarina F. Castro, Carlos C. António and
Abel D. Santos
Faculty of Engineering, University of Porto, Portugal

1. Introduction

Since early research work on simulation of forming processes, the finite element method (FEM) has the capability of producing detailed deformation information for hot and cold forging processes, and solves a large variety of problems by changing the model and process parameters. However, design modifications by process simulation using a trial and error procedure may easily become an endless and tedious task.

Nowadays, new procedures became available, which are succeeding as efficient alternative predicting "tools" in the design stage of a component, namely the backward tracing method [PAR 83], [HAN 93], [ZHA 95] and the forward tracing method [GRA 93] and [BER 95]. Other authors developed so-called inverse techniques. The goal of these inverse problems is to determine one or more of the direct problem input data, leading to a prescribed result. Some of the first research work in this area has been concerned with 2D shape inverse problems, which were successfully solved and presented by Fourment *et al.* [FOU 94, 96, 01], Zabaras *et al.* [ZAB 93], [BAD 95], Zhao *et al.* [ZHA 97], and Chung *et al.* [CHU 98]. More recently, research work on die shape optimal design in three dimensional shape metal extrusion [JOU 98] and 2D microstructure optimization in design of forging [GAO 00] has been presented. Also Rodic *et al.* [ROD 99] has developed a computer system for automatic optimization of pre-form shapes, tool geometries, tribological conditions and prestressing.

Industrial demand for optimization of material forming is increasing. The primary object of a metal forming process design is to produce a desired shape and to reduce production costs while maintaining material property levels. This paper presents new research work on 3D forging inverse problems. Its goal is to minimize the gap between the FEM final geometry and the prescribed one.

The design algorithm is implemented around an existing rigid viscoplastic finite element program [CES 96]. The solution obtained from the mechanical direct problem is combined with the sensitivity analysis and a modified sequential unconstrained minimization technique [CAS 00] to achieve the optimal design of the initial workpiece.

2. Direct problem

The constitutive model assumes a rigid, isotropic, strain hardening viscoplastic incompressible deformation; at any instant of time the viscoplastic strain rates for given directions and shear planes are proportional to the instantaneous deviatoric stresses and to the shear stresses. The flow rule is based on the Von Mises criterion and the Perzyna model. Considering the strain rate tensor $\dot{\boldsymbol{\varepsilon}}$, the deviatoric stress tensor s_{ij} is defined by:

$$s_{ij} = \frac{2}{3}\frac{\overline{\sigma}}{\dot{\overline{\varepsilon}}}\dot{\varepsilon}_{ij} \qquad or \qquad \mathbf{s} = \mathbf{D}\dot{\boldsymbol{\varepsilon}} \qquad [1]$$

where $\dot{\overline{\varepsilon}}$ is the effective strain rate and $\overline{\sigma}$ the effective stress, which is a function of strain, strain rate and temperature.

In forming processes, the deformation is imposed by tools and the friction between workpiece and tools influences the forces and the required energy. As the tools are of a much harder and stronger material than the workpiece, if the tangential or interfacial shear stress equals the workpiece flow strength in shear, material will not slide; otherwise if the interfaces are lubricated, the workpiece slides under the tools.

Friction has been characterized by a modified sticking friction law, where a friction factor m is defined varying from zero (no friction) to one (sticking friction). The shear stress is assumed constant and equal to the workpiece material $\sigma / \sqrt{3}$, and the frictional shear stress τ_f is expressed as:

$$\tau_f = m\frac{\overline{\sigma}}{\sqrt{3}} \qquad [2]$$

Friction and contact are modelled by interface elements of zero thickness, formulated on the basis of local normal and tangential relative displacements. To prevent penetration of the 3D element nodes of the workpiece mesh into the spatial triangular elements used for the discretization of the tools, a penalty formulation is used considering a high normal stiffness. The tangential stress τ is evaluated from the relative sliding velocity at the interface $\Delta\dot{\mathbf{u}}_r$:

$$\tau = \frac{\tau f}{\left\| \Delta\dot{u}_r \right\|} \Delta\dot{u}_r.$$ [3]

The finite element solution is built over a mixed formulation having as variables the velocity field $\dot{u}$ and the pressure p which is used as a Lagrange multiplier function to impose the incompressible deformation constraint on a variational formulation associated with the minimization of the total energy system. The resulting equations are [CES 96]:

$$\int_{\Omega} \delta\dot{\boldsymbol{\varepsilon}}^T s\, d\Omega - \int_{\Omega} p\, tr(\delta\dot{\boldsymbol{\varepsilon}})\ d\Omega - \int_{\Omega} \delta\dot{u}^T b_v\, d\Omega - \int_{d\Omega} \delta\dot{u}^T t\, d\Omega +$$

$$\int_{\delta d\Omega} \delta(\Delta\dot{u}_r^{\,T})\tau\, ds = 0$$ [4]

$$\int_{\Omega} \delta p\, tr(\dot{\boldsymbol{\varepsilon}})\, d\Omega = 0$$

considering Ω the workpiece part, $\delta\Omega$ the workpiece boundary and $\delta d\Omega$ the workpiece tool contact zone; b_v and t are the body forces and surface traction. Approximating the independent variables in the usual manner in the FEM a non-linear system equation is obtained:

$$\begin{bmatrix} K_d + K_f & Q \\ Q^T & 0 \end{bmatrix} \begin{bmatrix} \dot{u} \\ p \end{bmatrix} = \begin{bmatrix} f \\ 0 \end{bmatrix}$$ [5]

in which: $\dot{\boldsymbol{\varepsilon}} = B\dot{u}$ $K_d = \int_{\Omega} B^T DB\ d\Omega$, $K_f = \int_{\delta\Omega} B_f^T D_f B_f\ d\Omega$,

$Q = -\int_{\Omega} B^T mN_p\ d\Omega$ and $f = \int_{\Omega} N_{\dot{u}}^T b_v\, d\Omega + \int_{\delta\Omega} N_{\dot{u}}^T t\, d\Omega$.

The system of equations can be written as:

$$\begin{bmatrix} K_d + K_f & Q \\ Q^T & -\alpha^{-1}I \end{bmatrix} \begin{bmatrix} \dot{u} \\ p \end{bmatrix} = \begin{bmatrix} f \\ -\alpha^{-1}p \end{bmatrix}$$ [6]

where α^{-1} is a very small value. The system can be solved by the following iterative scheme [TOY 85]:

$$(K_d + K_f + K_\alpha)\dot{u}^{(i)} = f^{(i-1)} - Qp^{(i-1)}$$

$$p^{(i)} = p^{(i-1)} + \alpha\, Q^T\dot{u}^{(i)}$$ [7]

where: $\mathbf{K}_\alpha = \alpha \mathbf{Q}^T \mathbf{Q}$.

If a penalty function is used to impose the incompressibility the terms of $\mathbf{K}_\alpha$ matrix must be evaluated with reduced integration technique to avoid possible locking effects.

Due to the possibility of large deformations an updated Lagrangian approach is adopted which makes easier to follow the material deformation. For two-dimensional analysis the workpiece is discretized in 4 node linear elements and for three-dimensional analysis in solid 8 node linear elements.

3. Inverse problem

3.1. *Optimization problem formulation*

Forging is a complex problem due to its non-steady nature involving the evolution of boundary conditions. Let us consider the one step forging of a workpiece with a prescribed final shape. The only information known beforehand is the final product shape and the material. The following inverse problem is considered: given a forging tool corresponding to a prescribed final shape find the optimal geometry of the initial workpiece. This inverse problem is an optimization problem with an objective function measuring several parameters of the process. The optimization goal is the minimization of a functional $\varphi(\mathbf{b})$ that quantifies the distance between the current shape at the end of the process and the prescribed shape. The optimization problem can be formulated as:

$$\text{Minimize} \qquad \varphi(\mathbf{b}) = \int_{\delta\Omega_{end}} \|\pi(\mathbf{X}) - \mathbf{X}\|^2 \, ds$$

subject to the equilibrium equation [7]

[8]

where $\mathbf{b}$ is the design vector defining the workpiece initial shape and $\pi(\mathbf{X})$ the projection of a material point $\mathbf{X}$ of the workpiece boundary $\delta\Omega_{end}$ onto the surface of the prescribed shape at the end of the simulation process.

The optimization algorithm is based on a modified sequential unconstrained minimization technique and a gradient method. Numerical or pseudo-analytical sensitivities would be quite expensive and cannot be considered for actual industrial problems. Hence analytical sensitivities of $\varphi(\mathbf{b})$ are required and we adopt the calculation of discrete derivatives based on the differentiation of the equations of the discrete problem.

3.2. Sensitivity analysis

Considering $\varphi^*_d(b)$ the discretization of the objective function, the total derivative with respect to the shape design vector b is:

$$\frac{d\varphi^*_d}{db} = \frac{\partial\varphi^*_d(X^{(t_{end})})}{\partial X} \frac{dX^{(t_{end})}}{db} \qquad [9]$$

where $X^{(t)}$ is the nodal co-ordinate vector at instant t. Assuming the velocity field $\dot{u}^{(t)}$ constant during the time increment Δt, the domain $\Omega^{(t+\Delta t)}$ at time $t + \Delta t$ is calculated:

$$X^{(t+\Delta t)} = X^{(t)} + \dot{u}^{(t)} \Delta t \qquad [10]$$

The derivatives of equation [10] with respect to the design variables b can be obtained considering:

$$\frac{dX^{(t+\Delta t)}}{db} = \frac{dX^{(t)}}{db} + \frac{d\dot{u}^{(t)}}{db} \Delta t \qquad [11]$$

The gradients of the functional $\varphi^*_d(b)$ can be calculated if the nodal velocity sensitivities with respect to the design variables are known. At every time t, the nodal velocity sensitivities $d\dot{u}^{(t)}/db$ can be calculated using the direct differentiation method. The equilibrium equation [7] can be written as:,

$$K\bar{q} = \bar{f} \qquad [12]$$

Differentiating the equilibrium equation [12] with respect to the b variables:

$$\left(\frac{\partial K}{\partial q}\bar{q} + K\right)\frac{d\bar{q}}{db} = \frac{\partial\bar{f}}{\partial X}\frac{dX}{db} - \frac{\partial K}{\partial X}\frac{dX}{db}\bar{q} \qquad [13]$$

where X represents the nodal velocity co-ordinates and $\bar{q} = [\dot{u}\ p]^T$ is the velocity and pressure vector at time t. Equation [13] can be simply written as:

$$K^{(T)} \frac{d\bar{q}}{db} = \frac{\partial\bar{F}}{\partial b} \qquad [14]$$

representing a linear system of algebraic equations where $K^{(T)}$ is the tangent of matrix $K = K_d + K_f + K_\alpha$, the material and process-dependent nonlinear stiffness matrix obtained after convergence of the iterative process at every time t. The right end side of equation [14] is updated at the end of every time increment. The solution of this system, $d\bar{q}/db$, is now introduced in equation [11] and the sensitivities of the nodal co-ordinates with respect to the design variables are obtained, allowing the calculation of the sensitivities of the functional $\varphi_d^*(b)$ using equation [9].

The proposed algorithm to solve the inverse forging problem is described multifold:

Given an initial guess of the workpiece shape, described by a design parameter vector b, the finite element analysis of the forging process is performed.

1/ In every incremental time step and after convergence of the finite element solution of the direct problem, the sensitivities of the nodal velocities with respect to the design variables are obtained using the direct differentiation method. The sensitivities of nodal co-ordinates with respect to the design variables are dependent on the deformation history and they have to be updated iteratively.

2/ When the forging simulation is finished, the resulting sensitivities of the nodal co-ordinates with respect to the design variables are used to calculate the gradients of the objective function. The objective function is calculated according to the actual final forging shape and the prescribed one.

3/ After each complete simulation a new design parameter vector b will be provided. At this point, the optimization algorithm also checks to see if the stopping criteria are satisfied.

4/ If the convergence conditions are not satisfied, the die shape is updated and the next optimization iteration is performed. If the convergence conditions are satisfied with the current shape, the design objectives are met.

Therefore, the optimization is based on the simulation final results.

For a one-stage forming process the objective of the optimization problem is to design the workpiece preform shape so that, after forging, the required final part without defect is obtained. The shape of tool is a prescribed datum and the initial workpiece shape becomes the unknown of the design. The geometric design of the initial workpiece is different according to each forging application example.

4. Applications

Two examples of forging are presented to validate the proposed inverse technique. The first example optimizes the compression of a cylindrical part, which is used to test the accuracy of the sensitivities considering the direct differentiation method. The second example corresponds to a forged industrial component enhancing the proposed optimization approach.

4.1. *Validation of the direct differentiation method*

The direct differentiation method is first validated by studying the forging of an academic axisymmetric upsetting process considering a 2D analysis. The geometry of the free surface of the billet has been defined using a Bspline curve. Control points determine the shape of a Bspline curve and the Bspline curve is bounded by its control polygon. For 2D problems each control point P_i, $i = 0,...,N$ has two co-ordinates $(P_i(x), P_i(y))$. In this example the Bspline curve considers 6 optimization points uniformly distributed (see Figure 1a). The x displacements of selected active control points of the Bspline function become the design parameters. Five active control points were considered defining a design parameter vector **b** of 5 components. Starting from a straight cylinder and considering bilateral contact conditions with a constant shear friction factor $m = 0.9$, the deformed shape obtained after 10 time increments corresponding to a 20% deformation is shown in Figure 1b).

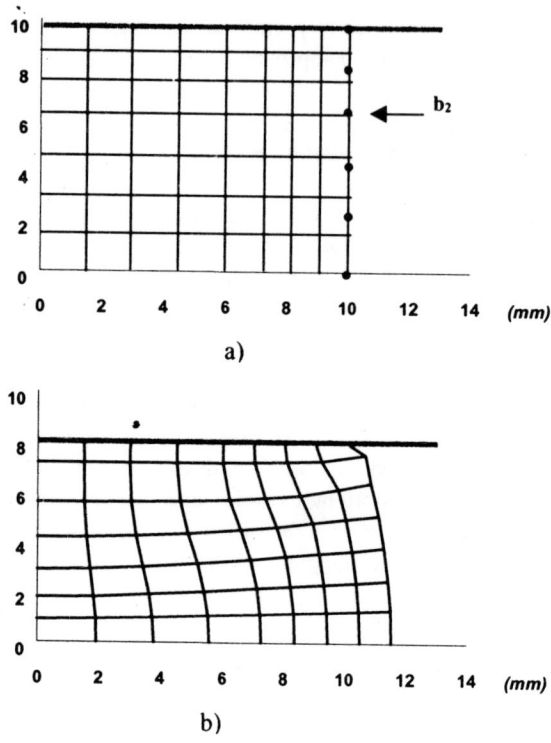

a)

b)

Figure 1. *Upsetting process. a) Initial billet shape. b) Deformed final shape*

The accuracy of the analytical derivatives is estimated by comparing the derivative values $\left.\dfrac{d\mathbf{X}^{(t)}}{db_j}\right|_{DDM}$ given by equation [11], obtained by the direct differentiation method, with the numerical values obtained by a finite difference scheme:

$$\left.\frac{d\mathbf{X}^{(t)}}{db_j}\right|_{FDM} = \frac{d\mathbf{X}^{(t)}(b_j + \zeta) - d\mathbf{X}^{(t)}(b_j)}{\zeta} \tag{15}$$

where ζ is a small numerical perturbation of the parameter value. As an illustrative example, Figure 2 shows the comparison of the sensitivities obtained by the two methods with respect to the second design parameter b_2, considering $\zeta \approx 10^{-1}\,mm$ and point $\mathbf{X}^{(t=0)} = (x, y)_{initial} = (10.0, 4.5)\,mm$ for the referenced point on the free surface. The discrepancy between the analytical and the numerical values is around 2 per cent. This discrepancy is probably due to accumulation of numerical errors, the method of computing analytical derivatives and to poor accuracy of the numerical derivatives themselves. Globally the discrepancy between the derivatives is considered satisfactory and sufficient for the optimization.

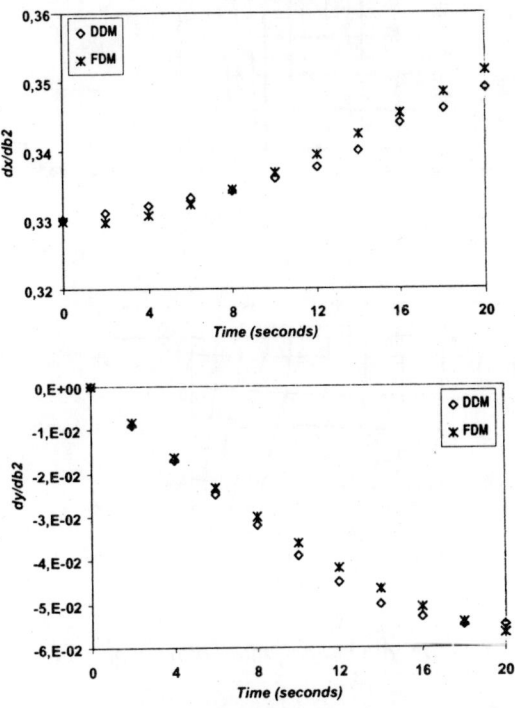

Figure 2. *Sensitivities of a free boundary node with respect to a design variable*

These derivatives are applied to the academic problem of finding the preform shape allowing to obtain a straight cylinder after upsetting. Convergence is obtained after 12 iterations. Figure 3 shows the optimal initial shape obtained and the final shape forged with this optimal billet.

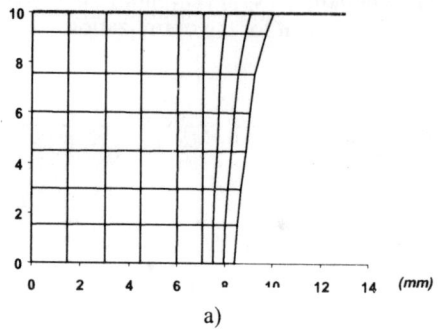

a)

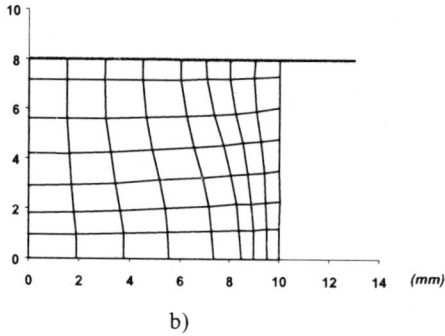

b)

Figure 3. *Optimal solution. a) Optimal initial billet shape. b) Deformed final shape*

4.2. Industrial design

The shape optimization technique is also applied to design the initial workpiece of a bicycle headset produced by forging an aluminium 6000 series alloy. It is assumed an isothermal analysis and the aluminium workpiece is previously heated to 440 °C.

Figure 4 shows the part intended to be produced. Bilateral contact conditions are considered and a constant shear friction factor $m = 0.6$ is assumed between the workpiece and the die; the material's behaviour is modelled by the constitutive relation [SAN 99, 01]

$$\bar{\sigma} = 100 * (0.003 + \bar{\varepsilon})^{0.25} \quad kPa \quad (T = 440^{\circ}C)$$

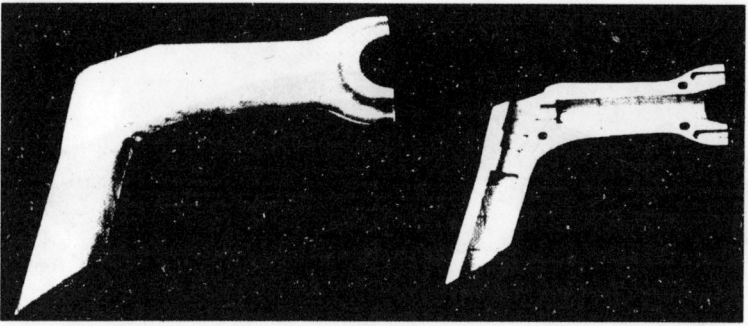

Figure 4. *Headset geometry (solid model obtained by Laminated Object Modelling)*

The 3D optimization problem consists in determining the initial shape of the workpiece in order to provide the desired final geometry after the forming process. A previous study using a constant 30mm diameter cylinder workpiece showed that the material deformation created a great flash along most of the part and did not fill the die at one end of the component [REI 01]. Therefore, a head for the initial workpiece cylinder was predicted to be needed to form the part. For the optimization problem, the considered shape design vector has three components: R_{max} – the largest radius of the head at one end of the cylinder, R_{min} – the minimum radius at the other end and H_{length} – the head length of the initial workpiece. So, the design parameter vector **b** has 3 components, $\mathbf{b} = [R_{max}, R_{min}, H_{length}]$. The workpiece discretization depends on a polynomial fit of the three components. Figure 5 shows the tool discretized in spatial triangular elements and Figure 6 the initial workpiece solid eight node linear element mesh.

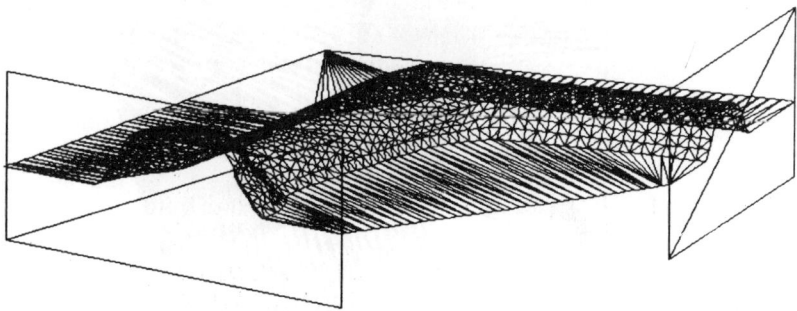

Figure 5. *Tool finite element mesh*

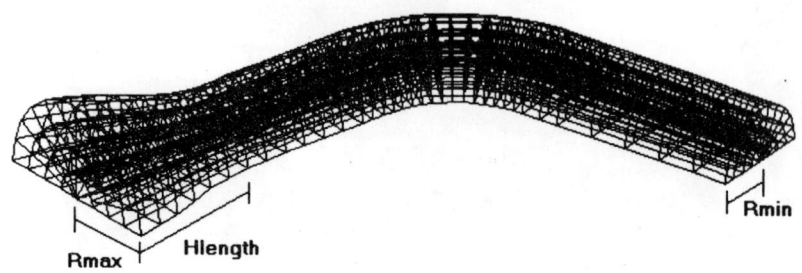

Figure 6. *Initial workpiece finite element mesh*

Figure 7 shows the simulation result of the first iteration of the optimization process using the initial guess [R_{max} = 0.025m, R_{min}= 0.012m, H_{length}= 0.040m]. The thicker lines represent the workpiece mesh and the thinner ones correspond to the triangular elements of the tool. So as Figure 7 shows, at the end of the forging the die is not filled and it creates a flash at the head of the part.

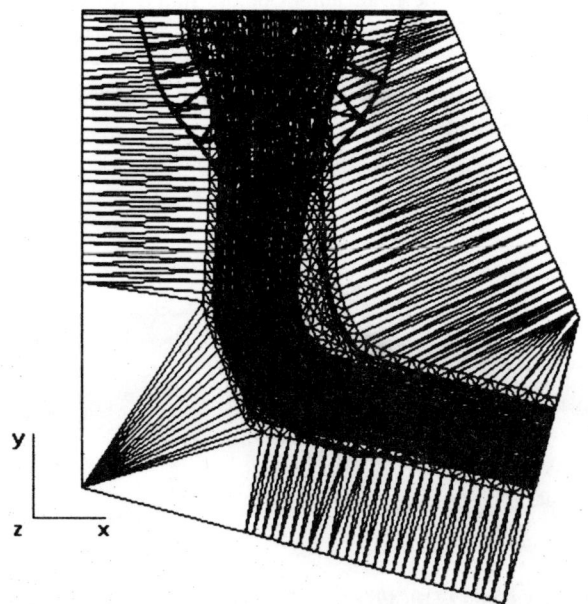

Figure 7. *Simulated forged part using the initial guess*

After six iterations the optimization process converged to an optimal solution. Figure 8 shows the sequence of shapes for the head workpiece profile that have been tested.

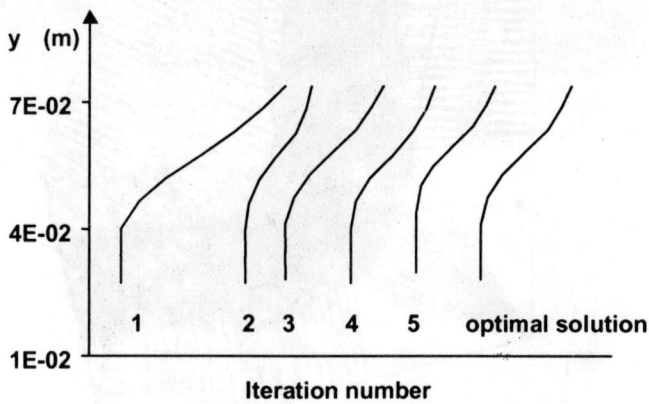

Figure 8. *Optimization process history for the head of the part*

Finally, Figure 9 shows the forged part using the optimal geometry for the workpiece [R_{max}= 0.021m, R_{min}= 0.015m, H_{length}= 0.035m]. A perfect filling of the die can be noticed. The obtained thin flash sugests it can be easily eliminated with a simple machining operation.

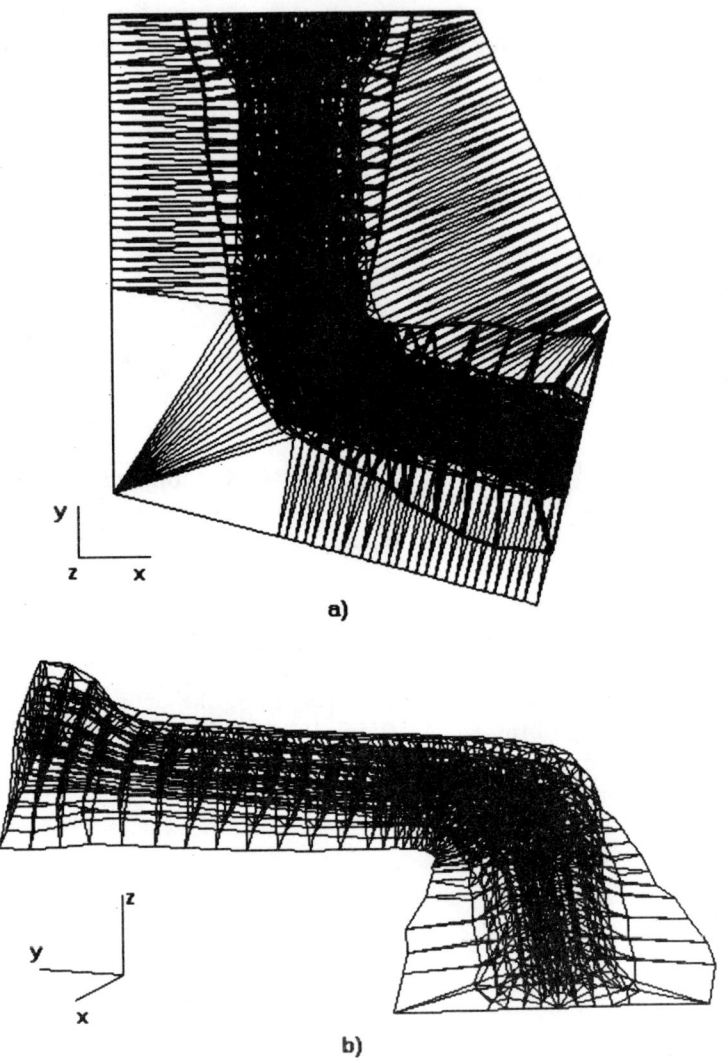

Figure 9. *Final configuration of the optimal solution. a) Deformed workpiece and die. b) Deformed workpiece mesh*

The industrial design results here presented correspond to a fine mesh of 1470 eight node linear elements for the workpiece and 3811 spatial triangular elements for the tool; each optimization iteration, corresponding to one complete forging simulation, took around one hour to run on a Pentium III 860 MHz. Similar results where presented by the authors at the 2001 ESAFORM Conference [Sou 01] where a coarse mesh of 234 elements for the workpiece was considered. Each iteration of this last optimization process took only twenty five minutes to run on the same computer. Figure 10 shows the deformed optimal solution using this coarse mesh. Comparing Figures 9 and 10 we observe that the filling of the die is almost the same but with the finer mesh optimal solution we get a smaller flash. The good agreement of the optimization results considering coarse and fine meshes validates the optimization scheme and the sensitivity analysis.

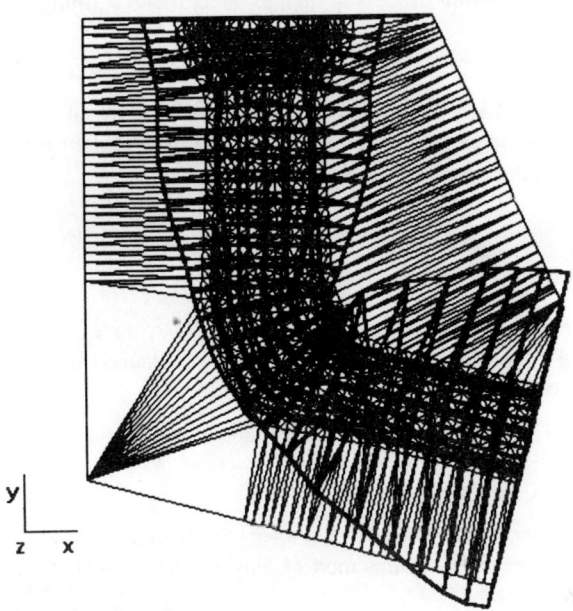

Figure 10. *Final configuration of the optimal solution using the coarse mesh (R_{max} = 0.022m, R_{min} = 0.015m, H_{length} = 0.038m)*

5. Conclusion

An algorithm for optimal design of 3D initial workpiece shapes in metal forging is presented. The objective function sensitivity is calculated by the accumulated sensitivities of the nodal co-ordinates throughout the entire simulation of the process.

Two examples are presented: an axisymmetric upsetting process to validate the sensitivity analysis and an industrial example to show the capability of the method. Both validate the computed sensitivity fields since the final shapes are very close to the intended ones.

The 3D analysis is applied to design an initial workpiece shape in the forging of a bicycle headset. The obtained final shape of the workpiece and the intended geometry are very close and so the potential manufacturing costs are reduced as there is no need of extensive trial-and-error procedure either experimental or direct FEM simulation. For the present industrial example the optimal initial workpiece obtained by the optimization method is also easy to manufacture.

Each optimization iteration needs just one complete forging simulation. It takes only a few iterations to obtain an acceptable design result and the procedure is very efficient. Our near future efforts will be dedicated toward multistage 3D forging problems and other forming processes such as sheet metal forming.

Acknowledgements

The authors would like to thank Prof. J. Ferreira Duarte and Prof. Ana Reis for their cooperation in the experimental work and also the Portuguese Company Ibérica for its help concerning parts to model.

References

[BAD 95] BADRINARAYANAN S., ZABARAS N. and CONSTANTINESCU A. "Preform design in metal forming", *Proceedings of 5th Int. Conference on Numerical Methods in Industrial Forming Processes,* Ithaca, New York, 1995, pp. 533–538.

[BER 95] BERG J.M., and MALAS J., "Open-loop control of a hot forming process", *Proceedings of 5th Int. Conference on Numerical Methods in Industrial Forming Processes,* Ithaca, New York, 1995, pp. 539–544.

[CAS 00] CASTRO C.F., COSTA SOUSA L., CONCEIÇÃO ANTÓNIO C.A. and CÉSAR DE SÁ J.M., "A multilevel approach to optimisation of bulk forming processes", *Proc. ECCOMAS 2000,* Barcelona, Spain, 2000.

[CÉS 96]. CÉSAR DE SÁ J.M, COSTA SOUSA L. and MADUREIRA M.L., "Simulation model for hot and cold forging by mixed methods including adaptive mesh refinement", *Engineering Computations,* 13 (2–4), 1996, pp. 339–360.

[CHU 98] CHUNG S.H. and HWANG S.M., "Optimal process design in non-isothermal, non-steady metal forming by the finite element method", *International Journal for Numerical Methods in Engineering,* 42, 1998, pp. 51–65.

[FOU 94] FOURMENT L. and CHENOT J.L., "The inverse problem of design in forging", *Inverse Problems in Engineering Mechanics,* Bui, Tanaka *et al.* (eds) A. A. Balkema, Rotterdam, Holland, 1994, pp. 21–28.

[FOU 96] FOURMENT L., BALAN T. and CHENOT J.L., "Optimal design for non-steady-state metal forming processes - II. Application of shape optimization in forging", *International Journal for Numerical Methods in Engineering,* 39, 1996, pp.51–65.

[FOU 01] FOURMENT L., CHUNG S.H., VIELLEDENT D. and CHENOT J.L., "Computer aided forging design using shape optimization techniques", *Proc. 4th ESAFORM Conference on Material Forming,* Liège, Belgium, 23–25, April, 2001, pp. 545–548.

[GAO 00] GAO Z., GRANDHI R.V., "Microstructure optimization in design of forging processes", *International Journal of Machine Tools & Manufacture*, 40, 2000, pp. 691–711.

[GRA 93] GRANDHI R.V., KUMAR A. ,CHAUDHARY A. and MALAS J., "State-space representation of optimal control of a non-linear material deformation using the finite element method", *International Journal for Numerical Methods in Engineering*, 36, 1993, pp. 1967–1986.

[HAN 93] HAN C.S., GRANDHI R.V., SRINIVASAN R., "Optimum design of forging die shapes using nonlinear finite element analysis", *AIAA Journal, American Institute of Aeronautics and Astronautics*, Inc.(ed), 31, 1993, pp. 774–781.

[JOU 98] JOUN M. S. and HWANG S.M., "Die shape optimal design in three-dimensional shape metal extrusion by the finite element method, *International Journal for Numerical Methods in Engineering*, 41, 1998, pp. 311–335.

[PAR] PARK J.J., REBELO N. and KOBAYASHI S., "A new approach to perform design in metal forming with the finite element method", *International Journal of Machine Design Research*, 23, 1983, pp. 71–79.

[REI 01] REIS A., *Finite element modelling of metal forming processes*, MSc thesis (in Portuguese), University of Porto, Portugal, 2001.

[ROD 99] RODIC T., GREŠOVNIK I., JOLOVŠEK D. and KORELC J., "Optimisation of prestressing of cold forging die by using symbolic templates", *Proceedings of the European Conference on Computation Mechanics*, ECCM, Munchen, Germany, 1999.

[SAN 99] SANTOS A. D., FERREIRA DUARTE J., REIS A., BARATA DA ROCHA A., NETO R., PAIVA R., "Finite Element Simulation of closed die forging for prediction of material behaviour and optimization of process, *Proc. 2nd ESAFORM Conference on Material Forming*, Guimarães, Portugal, 13–17 April, 1999, pp. 295–298.

[SAN 01] SANTOS A. D., FERREIRA DUARTE J., REIS A., BARATA DA ROCHA A., NETO R., PAIVA R., "The use of finite element simulation of metal forming and tool design", to be published in *Int. J. of Forming Processes*.

[SOU 01] SOUSA L. C., CASTRO C.F., ANTÓNIO C.A. C. and SANTOS A.D., "Inverse methods applied to industrial forging processes", *Proc. 4th ESAFORM Conference on Material Forming*, Liège, Belgium, 23–25, April, 2001, pp. 541–544.

[TOY 85] TOYOSHIMA S., *Iterative mixed methods and their application to analysis of metal forming processes*, PhD Thesis, University of Swansea, UK, 1985.

[ZAB 93] ZABARAS N., BADRINARAYANAN S., "Inverse problems and techniques in metal forming processes", *Inverse Problems in Engineering: Theory and Practice*, edts N. Zabaras *et al.*, ASME, 1993, pp. 65–76.

[ZHA 95] ZHAO G., WRIGHT E. and GRANDHI R.V., "Forging preform design with complexity control in simulating backward deformation", *International Journal of Machine Tools and Manufacturing*, 35, 1995, pp. 1225–1239.

[ZHA 97] ZHAO G., WRIGHT E. and GRANDHI R.V., "Preform die shape design in metal forming using an optimization method", *International Journal for Numerical Methods in Engineering*, 40, 1997, pp. 1213–1230.

Chapter 18

Simulation of Face Milling and Turning with the Finite Element Method

Luc Masset

LTAS – Méthodes de Fabrication, Institut de Mécanique et de Génie Civil, Liège, Belgium

1. Introduction

There is a strong need for machining simulation in manufacturing industries. CAD software such as Catia or Pro-Engineer, for instance, only include a limited panel of machining features, process visualization and NC language interface. Today, important research efforts are made to develop knowledge and tools that meet industrial needs. The two main research directions in machining simulation are the simulation of chip formation focusing on the physics aspect of cutting and the simulation of the process itself. In this category many published works deal with part quality and particularly with the form error of the part surfaces. Among them, few consider workpiece flexibility although this can be the main cause of form error. This is the case for workpieces such as cylinder blocks or transmission casings presenting low rigidity due to thin walls or a small thickness.

This paper presents a means to compute the form error due to workpiece deformations occurring in face milling and turning operations. Forces applied by the fixing devices and cutting forces applied by the tool are taken into account. The part geometry is modeled using a finite element (FE) mesh and its deformations are computed with the FE method in Samcef [SAM 01].

What can be the contribution of such a tool in an industrial environment? The pre-production phase can be significantly reduced by replacing a number of experimental tests by simulations. Innovative technical solutions can come out of the simulations. For example new fixture designs can be imagined and tested easily without making expensive actual tests. And finally when problems occur in production, simulation tools can speed up the solution.

2. Method description

2.1. *Literature review*

Few works dealing with form error due to the workpiece flexibility can be found in the literature. The main reason seems to be the high computation effort required for this kind of simulation. The principle of all the existing methods is to compute a large number of part deformations. This can lead to an excessive computational cost if the adopted method is not well suited to this particular type of finite element analysis.

The first work [SCH 93, SCH 94] presents a very complex method. It consists in taking *snapshots* of the workpiece at different moments of the process, each snapshot corresponding to a part deformation. At each step the exact workpiece geometry is modeled using a CAD representation and a corresponding finite element mesh is produced. The forces applied by the tool are computed thanks to the Kienzle cutting force model [KON 73]. Then the part deformation is computed using finite element analysis. The form error is finally obtained by applying an interpolation scheme between the successive deformations.

The main drawback of this method is obviously the time needed to obtain a single result. According to the authors it took several days to perform the simulation on an industrial application, even with a small number of computation steps. Although computer power has increased, this method remains unattractive from a computation time point of view.

Another proposed method [GU 97] is a much more similar to the one presented in this paper. It is limited to face milling. The deformations are computed on a single finite element model. The effect of the spindle-cutter deflection is also taken into account. The most important drawback here is that the defect is only computed on a few lines of the part surface. This method gives only a partial result – the straightness of a few lines – and not the flatness of the whole surface.

2.2. *Hypotheses*

In this research the workpiece deformation is assumed to be the main cause of the form error. So we only consider face milling and turning processes for which we can assume that tool and machine-tool are perfectly rigid. In contrast to boring or end milling, the tool is usually weaker and it cannot be considered as rigid. However fixing devices such as back center, screws or supports can be considered as flexible.

The quasi static component of the deformation is supposed to be greater than the dynamic response. The workpiece material is assumed to have a linear behavior. The thermal effects and the eventual residual stress are neglected. The part deformations are thus computed with the finite element method in linear static.

The exact workpiece geometry is not taken into account here. Only one finite element model of the part is adopted. It usually corresponds to the workpiece geometry after the given process for which the stiffness is the lowest. Consequently the computed deformations are greater and we should obtain a theoretical upper bound of the form error. Neglecting the effect of the decreasing stiffness due to the material removal is valid if the height of removed material is small compared with the workpiece thickness. This assumption is usually true for finishing operations.

A last hypothesis is that the mill inserts are identical and equally spaced. In other words the cases of mills with a scraper and mills with variable spacing, designed to reduce dynamic problems, are not considered.

2.3. *Modeling scheme [MAS 99, MAS 01]*

The form error of a machined surface comes from the cutting height variations occurring throughout the cutting process. These variations are due to the part deformations. At a point on the machined surface, the amount of cut material depends on the point displacement when the tool is cutting it. If it is positive (directed towards the workpiece), the actual depth of cut is smaller than the prescribed one. If it is negative, the actual depth of cut is greater. We have respectively a positive and a negative defect at this point. One can notice that only the displacement component perpendicular to the machined surface induces a depth of cut modification. Figure 1a shows the effect of the clamping forces on the surface obtained.

The proposed method is based on the assumption that the finite element mesh of the surface is sufficient to describe the form error. The principle is to compute the defect of the n surface nodes. So there is no need to use complex interpolation method or time description of the process. The first step is to determine the n tool positions corresponding to the cutting of each node (Figure 1b). Then the forces acting on the part for each tool position are computed using the Kienzle cutting force model [KON 73]. Finite element analysis with n load cases is carried out. The surface geometry is obtained by picking up the displacement of each node out of the corresponding deformed structure. Finally the form error of the machined surface is computed with the *p-norm method* [DEB 98, DEB 99].

In milling the axis tilt effect is also taken into account. In addition, the back cutting of the mill inserts can be detected.

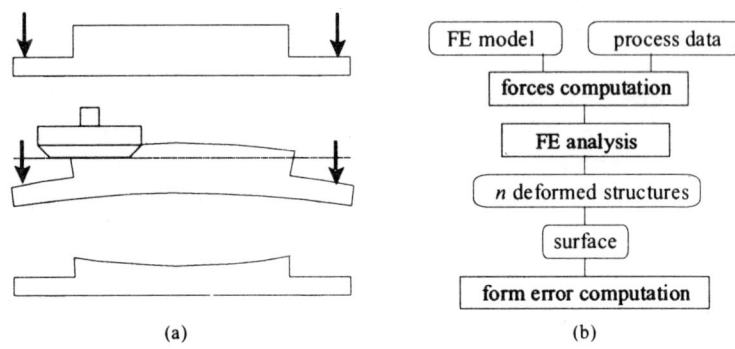

Figure 1. *Effect of the clamping forces a) and method description b)*

3. Force computation

3.1. *Cutting force model*

Kienzle's model is used to compute the cutting forces. The three force components, respectively the main cutting force, the feed force and the passive force are given by

$$F_i = bhk_i \quad \{ i = c, f, p \} \tag{1}$$

where b is the width of cut, h is the thickness of cut and k_i is the cutting pressure, given by a power law of the thickness

$$k_i = k_{i1.1} \, h^{-m_i} \tag{2}$$

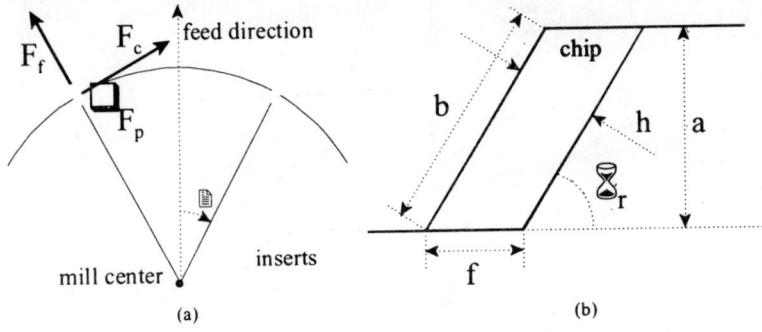

Figure 2. *Cutting forces in milling (a) and chip geometry (b)*

The nominal cutting pressure $k_{i1.1}$ is given by:

$$k_{i1.1} = k^*_{i1.1} C_{i\gamma} C_{i\lambda} C_{i\kappa} C_v C_m C_w \qquad [3]$$

where: $-$ $k^*_{i1.1}$ is the reference cutting pressure,

$-$ $C_{i\gamma}$, $C_{i\lambda}$ and $C_{i\kappa}$ are the adjustment factors for the cutting angles,

$-$ C_v is the adjustment factor for the cutting speed v,

$-$ C_m is the adjustment factor for the cutting material,

$-$ C_w is the adjustment factor for the tool wear,

The six constant values $k^*_{i1.1}$ and m_i for common steels and cast irons were measured by König and Essel [KON 73]. For any material, these constants can be computed using cutting forces measurements. Figure 3 shows the values of the adjustment factors $C_{i\gamma}$ (normal rake angle) and C_v. For a detailed description of the adjustment factors one can refer to [KON 73].

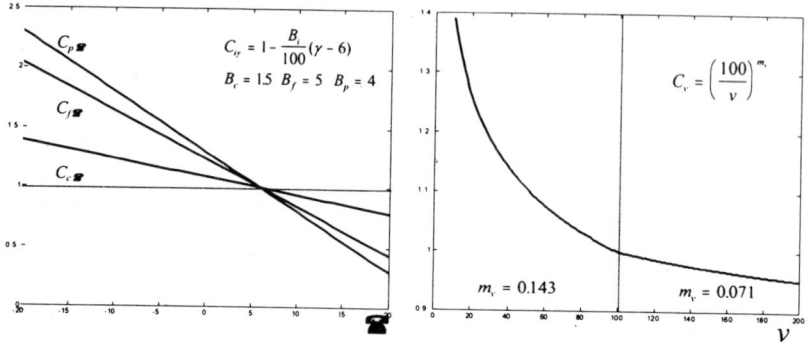

Figure 3. *Correction for normal rake angle* γ *and for cutting speed* v *[KON 73]*

3.2. Load cases computation

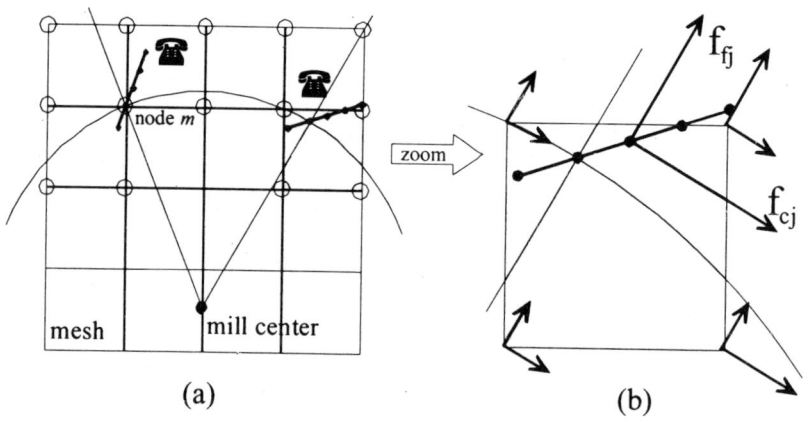

Figure 4. *Inserts position and cutting edges projections (a) and forces distribution on the surface element nodes (b)*

For each surface node, we have to compute the forces applied to the part when that node is cut by the tool. These forces are, on one hand, the forces applied by the fixing devices and, on the other hand, the cutting forces applied by the tool. The clamping forces are part of the data. In order to compute the cutting forces we first need to determine the tool position.

In milling, the mill center is computed thanks to the node coordinates, the tool trajectory and the tool radius. For this particular position (Figure 4a) there is an insert located on node m. Since we have made the hypothesis that all the inserts were identical, it is possible to determine where the other inserts are located. The cutting forces are computed for each insert engaged in the part. Actually these forces act on the cutting edges but here we only have the machined surface and not the actual contact zone between the part and the tool. The cutting forces F_i generated by one insert engaged in the workpiece are thus applied at a given number l of points defined on the cutting edges projections γ on the machined surface so that

$$f_{ij} = \frac{1}{l} F_i \quad j = 1,...,l \tag{4}$$

Finally each force f_{ij} is distributed on the nodes of the surface element where it acts and projected on the structure axes. The distribution is performed using the shape functions of the surface element (Figure 4b). In turning, the tool is simply located on node m so the method is the same but with a single cutting edge. The procedure is applied for the n nodes of the machined surface so that a load matrix g is obtained, each column corresponding to one of the n load cases. For the tool position shown in Figure 4a, the loaded nodes are marked with a circle. Only the degrees of freedom (d.o.f.) corresponding to these nodes and eventually those associated with boundary conditions (clamping forces) are loaded for this load case.

4. Finite element analysis

4.1. Direct method

We call *direct method* the standard way of performing the finite element analysis, *i.e.* the finite element model on which the n load cases are applied. The equilibrium equation to solve is

$$K q = g \tag{5}$$

where the size of the stiffness matrix K is $n_T \times n_T$ and the size of the load matrix g is $n_T \times n$, n_T being the number of d.o.f. of the whole FE model.

The direct method exhibits three severe drawbacks:
- significant amount of stored data,
- high computation cost,
- huge amount of memory required,

so that industrial examples are even impossible to achieve. Moreover, for each data modification (tool trajectory, cutting conditions, etc.), a new complete analysis is required. For these reasons the direct method is inefficient. One can notice that there is only a relatively small number of nodes involved, actually the loaded nodes (machined surface and fixations) and the nodes for which we compute displacements (machined surface). Consequently, there is a large number of nodes unloaded and *"uninteresting"* in terms of FE results. A better suited method is the *superelement* one.

4.2. *Superelement method*

The superelement method (SEM) consists in building a reduced system (the superelement) by condensing a set of the structure d.o.f.. If we denote q_R the n_R retained d.o.f. and q_C the n_C condensed ones, the equilibrium equation [5] can be written in the following way

$$\begin{bmatrix} K_{RR} & K_{RC} \\ K_{CR} & K_{CC} \end{bmatrix} \begin{bmatrix} q_R \\ q_C \end{bmatrix} = \begin{bmatrix} g_R \\ g_C \end{bmatrix} \qquad [6]$$

In this application all the loads g_C are equal to zero so we obtain the reduced system

$$[K_{RR} - K_{RC} K_{CC}^{-1} K_{CR}] q_R = g_R \iff K_{RR}^{*} q_R = g_R \qquad [7]$$

The retained d.o.f. correspond to those of the machined surface and those associated with boundary conditions (fixations). The number of retained d.o.f is small compared with the total number of d.o.f.; otherwise the superelement method would not be so attractive. Since the obtained reduced system is reasonably small it is not too expensive to invert it. The first part of the resolution is to compute the reduced matrix (SE creation step) and the second one to invert it in order to obtain the flexibility matrix S (SE use step).

Thanks to the superelement technique, the simulation tool becomes really powerful. The matrix inversion is only performed once. A simulation result is then obtained almost directly since it requires only a matrix multiplication (Figure 5b). To illustrate the benefit of the SEM, Figure 5a shows the extrapolated values of the CPU time and the disk space that would be required with the direct method for the camshaft cover and the suspension arm cases. These values are obtained with a linear interpolation of the values observed with 50, 100, 200 and 400 load cases (the direct method with the total number of load cases is not feasible on our machine mainly because of the memory required). It can be seen that the benefit of the superelement method increases as the size of the FE model grows.

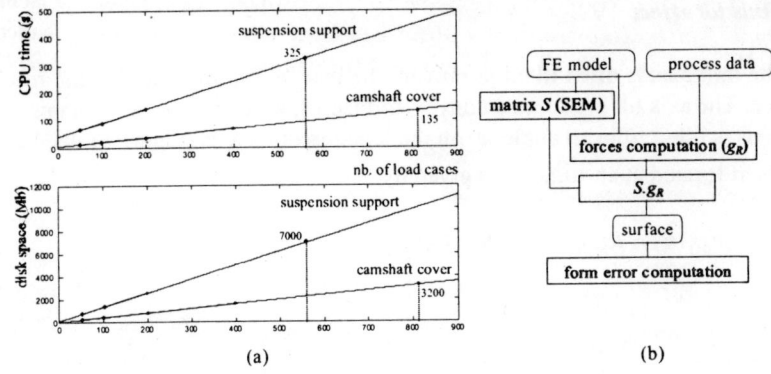

(a)   (b)

Figure 5. *Interpolation of the direct method characteristics (a) and method description with the superelement technique (b)*

4.3. Superelement method performance

Analyses were performed with Samcef V8 [SAM 01] on a standard computer running Windows NT (Pentium III 770 MHz with 512 Mb of memory). Table 1 shows the CPU time and disk space required to obtain the flexibility matrix S on the industrial applications. Even for large models, a very low computation time is achieved thanks to the new sparse solver of Samcef.

Table 1. *Superelement performances on industrial applications*

	d.o.f. (total)	d.o.f. (retained)	CPU (s) (step 1)	CPU (s) (step 2)	disk (Mb) (steps 1,2)
Camshaft cover	36639	2436	21	132	750
Suspension support	121749	1851	77	56	683
Exhaust manifold	184944	3087	147	250	1578
Gear box cover	112365	5328	211	1806	4362

5. Other tool features

5.1. *Axis tilt effect*

The mill axis is often tilted to prevent the rear inserts from cutting the machined surface. The axis tilt produces a convex surface whose form is easily obtained. If the mill axis is tilted from an angle φ on the feed direction axis (Figure 6), the height of the insert located at an angle θ is given by

$$z_{ins}(\theta) = R_{mill}(1 - \cos\theta)\sin\varphi \qquad\qquad [8]$$

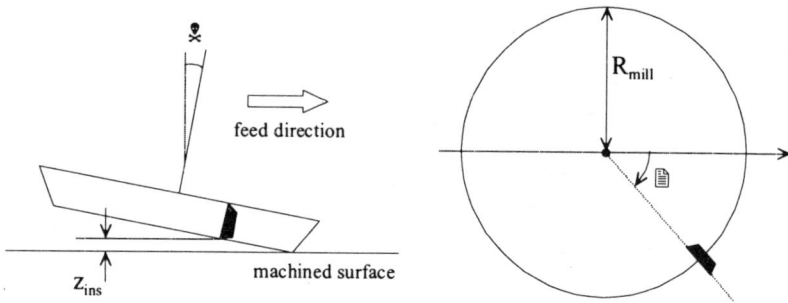

Figure 6. *Mill axis tilt and insert height*

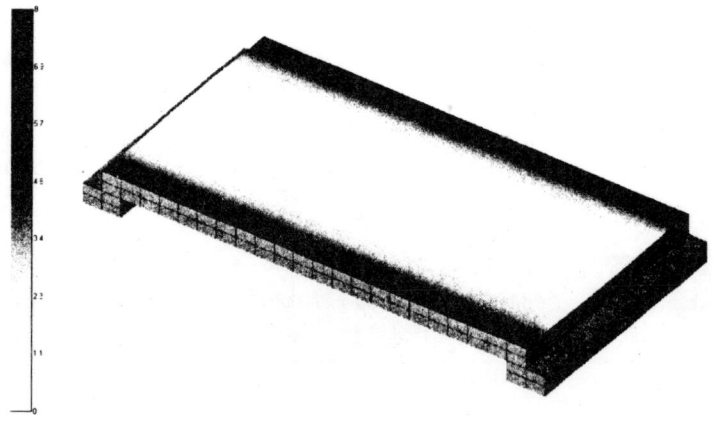

Figure 7. *Surface produced by the mill axis tilt (* $\varphi = 1.146 \cdot 10^{-2}$ deg, $R_{mill} = 65$, *part wide = 120mm)*

The axis tilt effect is included in the method as follows. For each of the n tool positions the height $z_{ins,m}$ of the insert located on node m (Figure 4a) is computed. The defect component of node m due to the axis tilt is then simply equal to $z_{ins,m}$. The surface produced is shown in Figure 7. Axis tilt only applies to single straight line trajectory. When the trajectory is constituted of multiple segments the mill axis has to be perpendicular to the machined surface.

5.2. *Back cutting detection*

A feature that is easy to add to the simulation tool is the detection of the back cutting. We use the principle adopted to compute the nodes defect except that here we look at what happens at the back of the mill. For a given node m, two values are computed:

– the height z_m of the node when the back of the mill passes over,

– the height $z_{ins,m}$ of the insert located over node m.

The node height z_m is the sum of the vertical displacement of node m for this tool position and the defect d_m that has been produced by the front mill inserts. The height $z_{ins,m}$ of the insert is computed with [8]. Back cutting occurs for node m if $\Delta z = z_{ins,m} - z_m$ is negative which means that the insert is located under the surface.

Figure 8 shows the surface geometry obtained, the dark zone indicating that the tool has left material there (positive defect). The back cutting is shown in Figure 9. It can be seen that back cutting occurs in the middle of the part surface where the defect is high.

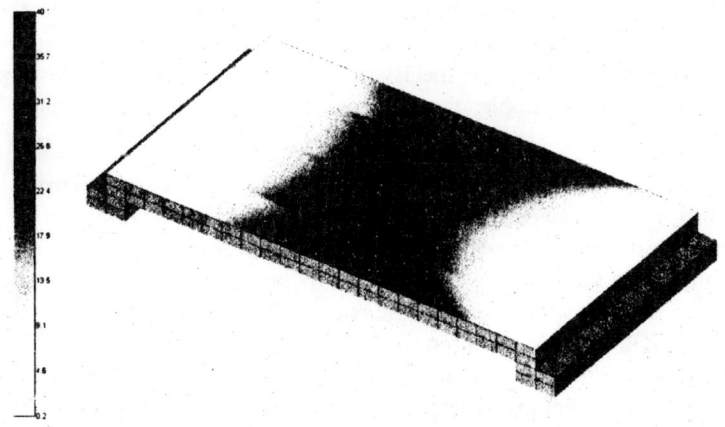

Figure 8. *Surface obtained after a milling process*

This tool feature allows one to correctly set up the axis tilt if it is necessary to avoid back cutting. In the applications presented (see Section 6), the axis tilt is always chosen so as to avoid back cutting. Actually, back cutting is only detected and not modeled. The forces applied by the rear inserts are not taken into account. The reason is that the height of material taken by the back of the mill is usually small so that the cutting force model does not apply any more.

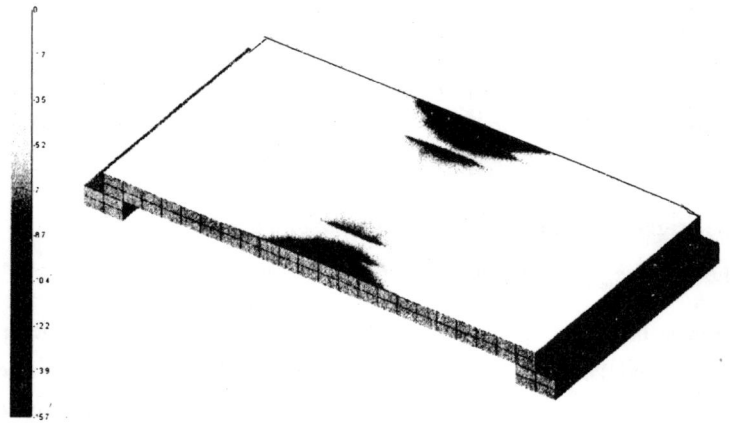

Figure 9. *Back cutting (dark zone)*

5.3. *Form error computation*

Form error computation is a critical problem in metrology. It is certainly one of the fields where the research effort has been the most important for the last ten years. The challenge is to find algorithms able to compute form errors according to the standards (ISO, ANSI, etc.) that specify that the form error is the smallest framing of a set of points. For example, the flatness of a surface is the minimum distance between two parallel planes framing the surface points.

Up to now, the most common method used is the least squares method. It is well known that it always gives an overestimated value of the form error (up to 25%). In production, parts can be rejected even though they actually satisfy the tolerance. This can greatly lower the productivity. That is why numerous research teams have tried to develop algorithms respecting the standards [BAL 91, BOU 88, CAR 95a, CAR 95b, GOC 90]. Most proposed algorithms are based on minimization techniques. The problem is that the convergence towards the global minimum *(i.e.* the smallest framing) is rarely ensured.

A new method called the *p-norm method* has been developed by [DEB 98, DEB 99]. It applies to the most common form errors – straightness, flatness, circularity and cylindricity. Its main advantage is that it has given the correct form error for all the tested cases (actually hundreds of cases with both measured and random sets of points).

The p-norm method is included in the machining simulation tool to compute the form error of the set of points corresponding to the n surface nodes. The form error is visualized as the nodes deviation from a perfect geometric entity (the best fitting plane in milling and transverse turning and the best fitting cylinder in straight turning).

6. Applications

6.1. Boundary conditions modeling

For the following applications, the workpiece fixatings consist of supports and clamps aligned with the supports. The clamping forces are usually given in the manufacturing plan. They are applied to the nodes located at the contact zone between the workpiece and the clamps. For the nodes located on the contact zone between the part and the supports, we use stiffness elements when the support stiffness (axial and transversal) is known. Otherwise these nodes are completely fixed.

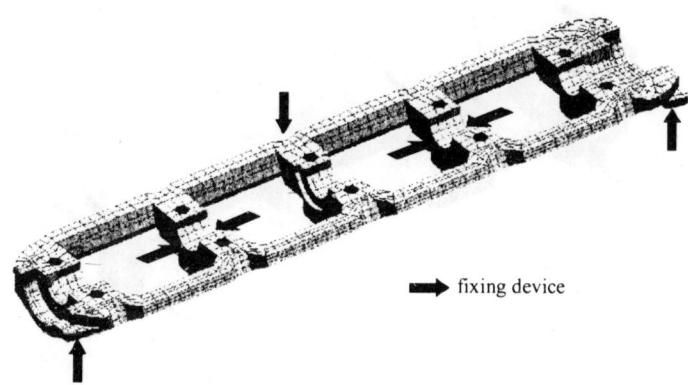

fixing device

Figure 10. *Camshaft cover and fixture design*

6.2. Camshaft cover

The workpiece is made of a silicium aluminum A-S9U3Y40 (Young modulus $E = 75\,000$ N/mm2; Poisson ratio $\upsilon = 0.33$). Figure 10 shows the FE model and the fixing devices. The tool is a 100-mm mill with four carbide inserts. The aim of the simulation is to find the trajectory leading to the smallest form error among the two possible centered trajectories. One of them leads to a much smaller form error than the other. The results obtained are illustrated in Figure 11.

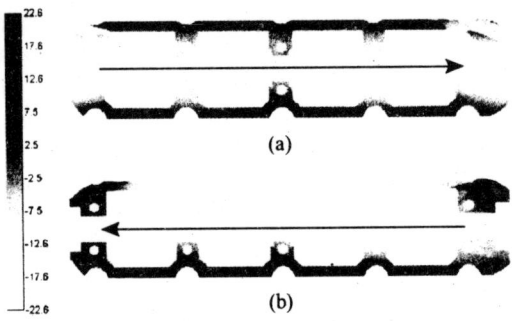

Figure 11. *Flatness errors obtained for trajectory (a) and (b) are equal respectively to 27.4 μm and 45.2 μm*

6.3. Suspension support

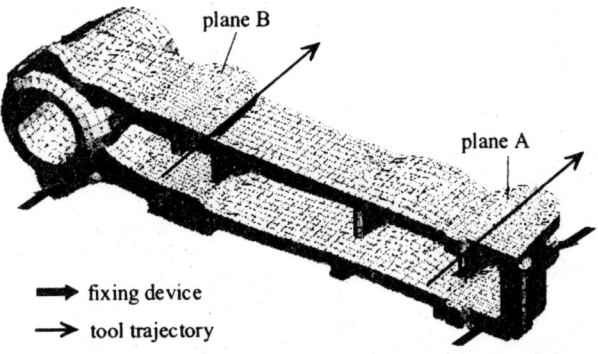

Figure 12. *Suspension support model, fixture design and tool trajectory*

The suspension support is made of cast iron GS52 (E = 180000 N/mm2; υ = 0.29) and face milled with the tool trajectory shown in Figure 12. Two mills are compared in this simulation: a 100-mm mill with 14 inserts and a 140-mm mill with 10 inserts. Figure 13 shows the results obtained for plane A.

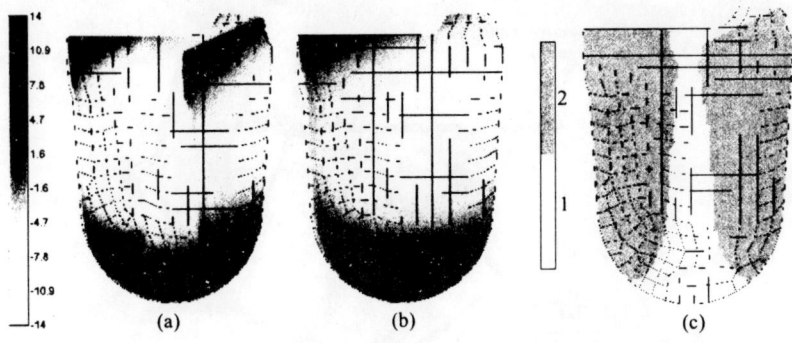

Figure 13. *Flatness errors obtained for 100-mm mill (a) and 140-mm mill (b) are equal respectively to 27.9 μm and 18 μm. Number of inserts engaged with 100-mm mill (c)*

Other interesting information that the simulation tool can give is the number of engaged inserts. For the 100-mm mill there is always at least one insert engaged in the workpiece (Figure 13c). It can be seen that the 140-mm mill gives a smaller form error but then there is only one insert engaged at a time. This can lead to dynamic problems since the part will go up and down.

6.4. *Exhaust manifold*

The material is a cast iron GS53 (E = 165000 N/mm2; υ = 0.3). The tool is a 315-mm mill with 50 inserts. The initial fixture design is shown in Figure 14. The four pipes are clamped with strap-support couples. The simulation aim is to test a lighter part fixation (design b) whose advantage is that the clamping operation is easier and faster. We remove the fixations of the two inner pipes because their lengths are smaller than the outer ones. The flatness tolerance being 200 μm, the new design (Figure 15b) is not satisfactory at all.

A possible solution would be to reduce the feed rate when the tool cuts the two inner pipes so as to limit the cutting forces. This application illustrates well the possible tests to achieve in the phase of pre-production of a part.

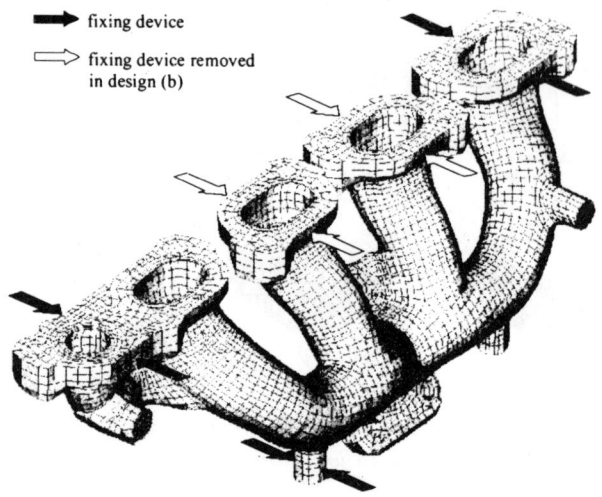

Figure 14. *Exhaust manifold and fixture design (a) and (b)*

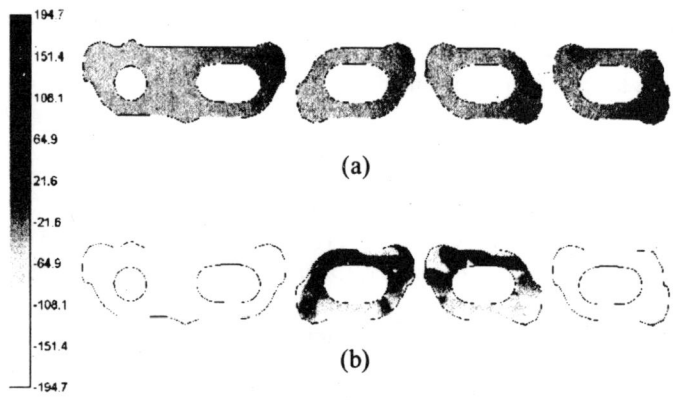

Figure 15. *Flatness errors obtained for initial fixture design (a) and new fixture design (b) are equal respectively to 103.9 μm and 389.3 μm*

6.5. *Gear box cover*

The process is the transverse turning of a gear box cover made of a A-S9U3Y40 aluminum. For the finishing pass, the depth of cut equals 0.5 mm. The fixture is constituted of three strap-support couples (Figure 16). Here, the simulation purpose is simply to check if the machined surface satisfies the imposed tolerance of 30 μm.

The small flatness error obtained (22.8 μ m) is due to the very low cutting forces level in aluminum and the small depth of cut in finishing (the maximum passive force value is equal to 24 N).

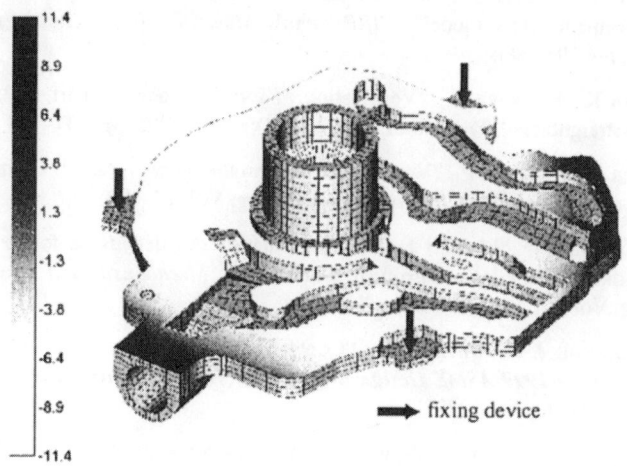

Figure 16. *Gear box cover model and flatness error*

7. Conclusion

The simulation tool proved to be very flexible and cost effective for industrial applications. The influences of the cutting process parameters are easily shown off in order to choose the best settings. In further improvements of the tool, the optimal settings should be found automatically thanks to optimization techniques. Taking the part flexibility into account is probably one of the missing links in process simulation. From now on, the challenge is to develop process simulation tools taking into account all the main effects (tool and part flexibility, dynamics, etc.) in order to meet the industrial needs.

Acknowledgments

The author gratefully acknowledges the financial and technical support of the Renault Department of Powerstrain Structural Analysis, Rueil-Malmaison, France and especially the Renault team members, S. Foreau, T. Dumont, J.-J. Droux, J. Duysens and P. Le Roy. He also wants to thank Professor J.-F. Debongnie and Professor P. Beckers, University of Liège, Belgium, for their help.

References

[BAL 91] BALLU A., BOURDET P., MATHIEU L., "The processing of measured points in coordinates metrology in agreement with the definition of standardized specifications", *CIRP Annals*, Vol. 40, No. 1, 1991, pp. 491–494.

[BOU 88] BOURDET P., CLÉMENT A., "A study of optimal-criteria identification based on small-displacement screw model", *CIRP Annals Manufacturing Technology*, Vol. 37, No. 1, 1988, pp. 503–506.

[CAR 95a] CARR K., FEIRRERA P., "Verification of form tolerances - Part I: Basic issues, flatness and straightness", *Precision Engineering*, Vol. 17, 1995, pp. 131–143.

[CAR 95b] CARR K., FEIRRERA P., "Verification of form tolerances - Part II: Cylindricity and straightness of a median line", *Precision Engineering*, Vol. 17, 1995, pp. 144–156.

[DEB 98] DEBONGNIE J.-F., MASSET L., "Sur l'évaluation des défauts de forme à partir de mesures tridimensionnelles", *European Journal of Mechanical and Environmental Engineering*, Vol. 43, No. 1, 1998, pp. 13–21.

[DEB 99] DEBONGNIE J.-F., MASSET L., "Controlling form errors from 3D measures", *Proceedings of the 1999 ASME Design Engineering Technical Conferences*, Las Vegas, Nevada, 12–15 September 1999.

[GOC 90] GOCH G., "Efficient multi-purpose algorithm for approximation and alignment problems in coordinate measurement techniques", *CIRP Annals*, Vol. 39, No. 1, 1990, pp. 553–556.

[GU 97] GU F., MELKOTE S. N., KAPOOR S. G., DEVOR R. E., "A model for the prediction of surface flatness in face milling", *ASME Journal of Manufacturing Science and Engineering*, Vol. 119, 1997, pp. 476–484.

[KON 73] KÖNIG W., ESSEL K., *Spezifische Schnittkraftwerte für die Zerspanung Metallischer Werkstoffe*, Verlag Stahleisen M. B. H., Düsseldorf, 1973.

[MAS 99] MASSET L., DEBONGNIE J.-F., FOREAU S., DUMONT T., "A model for the prediction of form errors in face milling and turning", *Proceedings of the 1999 ASME Design Engineering Technical Conferences*, Las Vegas Nevada, 12–15 September 1999.

[MAS 01] MASSET L., DEBONGNIE J.-F., BECKERS P., "Face milling and turning simulation with the finite element method", *Proceedings of Fourth International ESAFORM Conference on Metal Forming*, Liège, 23–25 April 2001, pp. 627–630.

[SAM 01] SAMCEF, *Système d'analyse des milieux continus par la méthode des eléments finis*, www.samcef.com.

[SCH 93] SCHULZ H., BIMSCHAS K., "Optimization of precision machining by simulation of the cutting process", *CIRP Annals*, Vol. 42, No. 1, 1993, pp. 55–58.

[SCH 94] SCHULZ H., GLOCKNER C., *Simulation of the deflection of workpieces due to the cutting forces*, Final Report, August 1994, Institut für Produktionstechnik und Spanende Werkzeugmaschinen, Darmstadt.

Chapter 19

Analysis of Material Behaviour at High Strain Rates for Modeling Machining Processes

Paolo Bariani and Stefano Corazza
DIMEG, University of Padova, Italy

Guido Berti
DTG, University of Vicenza, Italy

1. Introduction

Accurate measurement of the material response to high strain-rate deformation is a prerequisite both for accurate analysis and for modeling metalworking processes. In typical machining processes, both strain and strain rate, as well as temperature, vary over a wide range as shown in Table 1 [JAS 99]. In particular in the plastic zone the material can reach very high strain rates, above those that can be investigated using currently available methods based on traditional ordinary mechanical and hydraulic devices. SHPB technique allows rates of straining of $300 \div 10^4$ s^{-1} to be reached.

Table 1. *Machining: typical process parameters range in the workpiece for different materials. Strain and strain rate ranges refer to the primary shear zone [JAS 99]*

Material	Temperature °[C]	Strain	Strain rate [1/s]
Al 6062 T6	20 – 400 °C	1 – 2	$10^3 - 10^5$
AISI 1045	20 – 600 °C	1 – 3	$10^3 - 10^5$

This work aims to investigate material behaviour under conditions similar to those experienced during machining processes (temperature, strain and strain rate).

2. Conditions influencing material response in the machining process

2.1. *Strain*

In machining, operation energy consumption is mainly the result of two phenomena: shearing in the flow separation zone (primary shear zone) and chip formation (secondary shear zone). The most critical conditions, in terms of strain, strain rate and temperature, develop in these zones.

In the primary shear zone common strain values range from 1 to 3 (Table 1) and usually decrease as cutting speed and feed rate increase. In the secondary shear zone, especially where there is sticking and where severe deformations occur, strain is difficult to determine. Some authors have estimated equivalent strains ranging from 10 to 20 [BAN 84; MAT 82].

2.2. *Strain rate*

In order to determine the strain rate, the shear zone thickness in the primary shear zone must first be estimated. Various methods have been proposed for this. For example, Stevenson and Oxley [STE 70] performed orthogonal cutting experiments using a quick-stop explosive device and a printed grid to measure the primary shear zone strain rate. They found a mean strain rate of $15\text{–}20 \cdot 10^3 \text{ s}^{-1}$. This value was considerably reduced when the rake angle was increased, because of the reduction of shear zone thickness.

With the considered SHPB the maximum amount of strain rate achieved was close to 10^4 s^{-1}. This suggests that strain rates obtained in SHPB compression tests could be near those occurring in machining. It is important to note that performing reliable tests at strain rates of 10^5 s^{-1} and over is critical. However, it should be remembered that, in metals, changes in flow stress are significant only when the strain rate is varied by factors of ten [CAR 96].

2.3. *Temperature*

During machining processes maximum workpiece temperature is usually reached in the chip formation zone, not only because of the large amount of plastic work but also because of friction between the chip and the tool face. On the other hand, in the shearing zone, the main cause of increases in the temperature of materials is the conversion of plastic work into heat. Maximum temperatures usually reached when machining an Al 6062-T6 alloy are ~300°C in the shearing zone, and ~400°C in the chip formation zone. Cutting conditions have little impact on shear plane temperature. However, higher feed rate or cutting speed does give a lower chip mean temperature because of the higher rate at which volume is removed.

3. Experimental investigation of material behaviour

3.1. *High strain rate testing device: SHPB*

SHPB systems are based on the measure of pressure waves travelling along bars. The incident wave is generated by the impact of a striker bar on the input bar. A part of this wave is transmitted through the specimen to the output bar and another part is reflected at the input bar-specimen interface. Once measured, these waves permit one to know how material behaves during the test. When used to gather flow stress data during compression tests, SHPB performance is essentially related to (i) the range of deformation conditions (in terms of strain, strain-rate and temperature) that can be reproduced during the tests and (ii) accuracy when evaluating stresses and strains. Both testing conditions and accuracy are directly affected by the design parameters of the SHPB which are related to its physical components and to the data processing unit. Previous work by the authors [BAR 01] described the operational field of SHPB in terms of strain and strain rate and how better accuracy can be achieved with proper test data interpretation and correction. Repeatability of SHPB tests is high, with a difference below 1,5% in the measured stress value.

3.2. *Test temperature conditions*

Figure 1. *The two IR Image-furnaces units of the heating device*

3.3. *Adiabatic softening*

The SHPB compression test could be considered to be adiabatic, as can be demonstrated: thermal diffusion distance (L_T) in the material describes the distance travelled by heat flow in a given period of time (t), depending on the material thermal diffusivity (α):

$$L_T = 2\sqrt{\alpha \cdot t}$$

[1]

Pressure bars are made of maraging steel with thermal diffusivity $\alpha \sim 19 mm^2/s$. Duration t of an SHPB test, with the striker 0.8 m long, is 370μs. L_T is then 0.17 mm, a very short distance if compared to the typical length of a specimen $\sim$ 8 mm. Therefore there is not enough time for heat to flow from the specimen to the bars and the SHPB test could be considered adiabatic. During the test instantaneous specimen temperature can be estimated as initial temperature plus the temperature rise generated by i) plastic deformation and ii) specimen/bar interfacial friction (the latter is less than 5 °C). When calculating the rise in temperature due to deformation work [2], deformation is assumed to be homogeneous along the axis of the bars.

$$\Delta T = \frac{\int_0^{\varepsilon_f} \sigma \, d\varepsilon}{\rho c} \cdot \eta$$

[2]

where η is the fraction of deformation work converted into heat, i.e. the total amount of deformation work obtained in the flow curve less the elastic deformation work, ρ is material density and c is the material specific heat. The value of η depends on material elastic behaviour and on the amount of final deformation. For SHPB tests on metals this value is usually assumed to be 1. To verify this assumption by experiment, a specimen temperature was acquired both before and during the test. A thermocouple was welded onto the specimen, to monitor its temperature. The time constant of the thermocouples used was in the order of 0.1–0.2 s. Just before the test, the specimen is slowly cooling to the target temperature. After the compression test, during specimen cooling, the thermocouple measures specimen temperature with some delay. Another source of delay is due to the configuration of the measuring system: thermocouples are welded onto the specimen side surface where the smaller deformation occurs. Time is then needed in order to achieve thermal equilibrium in the specimen between the hotter core and the cooler external surface. Both delays must be evaluated and taken into account in order to correct the measured temperature (Figure 2).

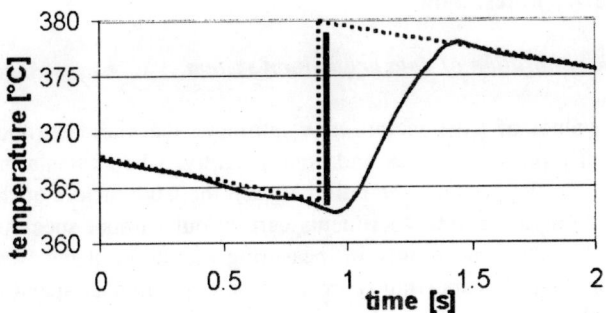

Figure 2. *Experimental acquisition of specimen temperature increase during the test (continuous line). The calculated temperature increase due to deformation work is shown as a thick line. The extrapolation of specimen temperature from temperature acquired in the experiment is represented with dotted line*

Test temperatures shown in Figures 4, 5 and 6 have to be considered to be the initial temperature of the specimen, just before compression begins. The calculated (using $\eta = 1$) temperature increase fits the experimental increase if both the uncertainty of thermocouples ($\pm$ 3 K) and any specimen heating due to interfacial friction (2 ÷ 5 K) are considered.

Thus, temperature increases during SHPB tests influence the response of the material adiabatic softening) and must therefore be taken into account when calculating material behaviour at a defined temperature.

3.4. *Maximum amount of strain*

The maximum amount of strain considered in the experiment was 0.9 both because of the limitations of the SHPB technique and also because of the compression test itself. However, it is clear (Figures 4 and 5) that at strains close to 1, flow stress, as a function of the strain, is almost constant. For this reason it is possible to estimate flow curves at higher strains and to identify a reliable constitutive model of the material even in a strain range larger than that used in the experiment. The amount of strain can be increased especially at medium strain rate by using a longer SHPB striker [BAR 01].

4. Interpretation of test data

4.1. *Dynamic calibration of data acquisition system*

Nominal values of gage factors may introduce inaccuracies (about 5 to 10%) when measuring pressure waves and, subsequently, when calculating the stresses and strains of the flow curve. The ratio between the gage factors of the two bridges is evaluated through impact experiments carried out without specimens. With this testing procedure, the uncertainty in measuring the strain at the two Wheatstone-bridges depends on the uncertainty in measuring the impact speed of the striker, which is less than 1%.

4.2. *Wave synchronization*

Wave attenuation and, above all, wave dispersion (due to sound velocity variation with frequency) cause the change in shape of the pressure waves [GRA 75].

Accurate calculation of the strains and stresses inside a specimen requires (i) reconstruction of the wave shapes at interfaces starting from those measured at the two bridges and (ii) synchronization of the waves. Reconstruction is performed by compensating for attenuation and dispersion [GOR 83; GOR 96].

4.3. *Friction and inertia*

The axial and radial components of inertia forces affect flow stress data by superimposing a three-dimensional stress state within the specimen [BER 75]. Likewise inertia, the friction at bar-specimen interfaces has the effect of increasing material flow resistance. The combined effects of friction and inertia can be corrected using equation [MAL 86]:

$$\sigma_c = \sigma_m \cdot \left(1 - \frac{\mu d}{3l}\right) - \frac{\rho_s}{12}\left(l^2 - \frac{3d^2}{16}\right)\frac{d^2\varepsilon}{dt^2} + \rho_s\left(\frac{3d^2}{64} \cdot \frac{d^2\varepsilon}{dt^2}\right) \qquad [3]$$

where σ_c and σ_m are the corrected and measured values of the flow stress respectively; l, d and ρ are the length, diameter and material density of the specimen and μ is the friction coefficient at interfaces, estimated to be 0.2 with a procedure presented in [BAR 01a]. In that work Rastagaev specimens are compressed under hydrodynamic lubrication conditions in order to have negligible friction tests. Then through comparison with standard specimen testing and inverse analysis, the friction coefficient is identified. A 3D stress state is induced in the specimen when friction increases as highlighted in Figure 3.

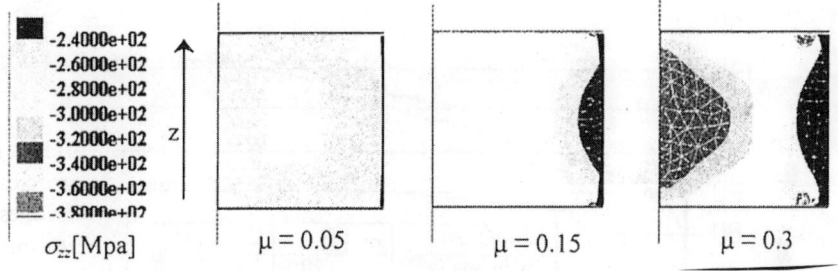

■	-2.4000e+02
	-2.6000e+02
	-2.8000e+02
	-3.0000e+02
	-3.2000e+02
■	-3.4000e+02
	-3.6000e+02
	-3.8000e+02

σ_{zz}[Mpa] $\mu = 0.05$ $\mu = 0.15$ $\mu = 0.3$

Figure 3. *FEM numerical simulation of SHPB compression test at 1050 s⁻¹ with different friction coefficient. The sections show stresses in the axial direction (σ_{zz})*

5. Material flow curves

The Al 6062-T6 flow curves obtained at different strain rates and temperatures were used to identify parameters on the Johnson-Cook constitutive model. Figures 4 and 5 show different rate sensitivity at different material temperatures. Figure 6 shows the temperature sensitivity at 1200 s⁻¹ strain rate. Tables included in the figures indicate the calculated rise in temperature due to plastic work and the corresponding initial specimen temperatures are reported in the figure captions. The quasi static tests were performed using a Gleeble 3800 thermomechanical simulator, the other tests used 16.2 mm bar diameter SHPB.

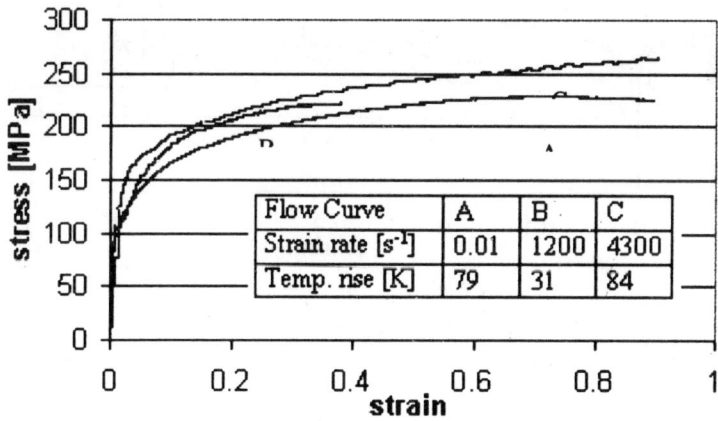

Flow Curve	A	B	C
Strain rate [s⁻¹]	0.01	1200	4300
Temp. rise [K]	79	31	84

Figure 4. *Flow curves for Al6062-T6 at different strain rates and 20°C initial specimen temperature. Cylindrical specimens are used with diameters from 8 to 12 mm and lengths from 5 to 10 mm*

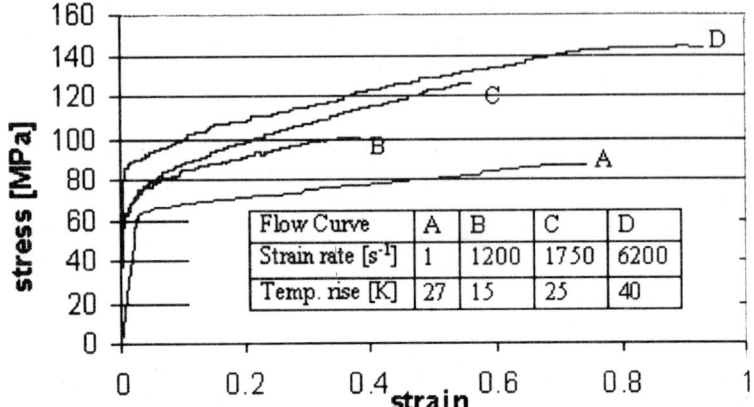

Figure 5. *Flow curves for Al6062-T6 at 360°C initial specimen temperature and different strain rates*

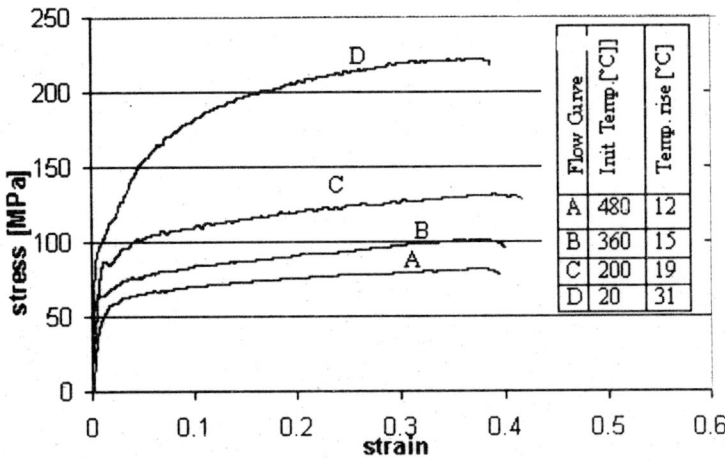

Figure 6. *Flow curves for Al6062-T6 at 1200 s⁻¹ strain rate and different temperatures*

6. Material constitutive model

The Johnson-Cook semi-empirical model was used [4] to describe the constitutive behaviour of the tested material. $\dot{\varepsilon}^*$ was the dimensionless strain rate (strain rate divided by $\dot{\varepsilon} = 1.0\ s^{-1}$), T^* was the homologous temperature, A, B, C, n and m were considered to be material constants. The homologous temperature was defined as $T^* = (T - T_0) / (T_{melt} - sT_0)$ where T_0 was the room temperature. The Johnson-Cook constitutive equation considered was:

$$\sigma = [A + B\varepsilon^n] \cdot [1 + C\ln\dot{\varepsilon}^*] \cdot [1 - T^{*m}]$$

[4]

The relevant parameters for Al 6062-T6 were identified using multivariable regression of the Johnson-Cook constitutive model [JOH 83]. Results are summarized in Table 2.

Table 2. *Johnson-Cook model parameters for Al 6062-T6*

A	B	n	C	M
−1.76 Mpa	187 Mpa	0.165	0.055	1.33

Where the model was identified, strain rate ranged from 1 to 6000 s-1 for maximum strains of 0.5 ÷ 1. Temperature range of material behaviour investigation was from room temperature (20°C) to 360°C of the initial test temperature. Material strain hardening for high strains is rather low, leading to small errors when extrapolating the stress value for ε>1. Figure 7 shows a good fit between experiments and the model.

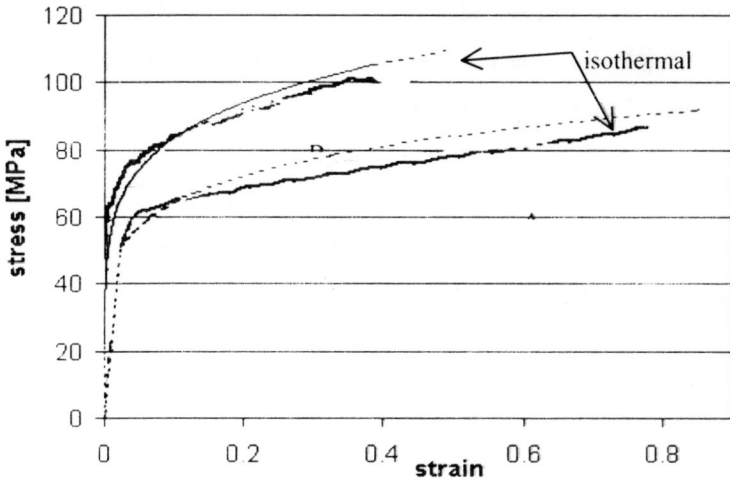

Figure 7. *Comparison between experimental flow curves (black) and Johnson-Cook model curves (grey) for Al6062-T6 at 1 s⁻¹ (A) and 1200 s⁻¹ (B). The initial test temperature is 360°C. The flow curves obtained with the model for an isothermal test are shown by a dotted line*

7. Conclusion

This study describes an investigation into material behaviour over a wide range of strains, strain rates and temperatures, which are typical of machining processes. Particular attention has been paid to increases in temperature during the test: the flow curves in which temperature showed a significant increase during specimen compression. This aspect must be taken into account when attempting to obtain isothermal flow curves and identifying a reliable material model. The advanced SHPB system presented seems to be a good solution for material behaviour investigation in cutting conditions.

References

[BAN 84] BANDYOPADHYAY B.P., 1984, "Mechanism of formation of built-up edge", *Precision Engineering*, Vol. 6(3), pp. 148–151.

[BAR 01] BARIANI P., BERTI G., CORAZZA S., "Enhancing perf. of SHPB for determination of flow curves", *CIRP 2001 Annals*, 2001.

[BAR 01a] BARIANI P., BERTI G., CORAZZA S., "Experimental study for friction effect investigation on SHPB", *AITEM2001*, Bari (I), 2001.

[BER 75] BERTHOLF L.D., KARNES C.H., "Two-dimensional analysis of the SHPB system", *J. Mech. Phys. Solids*, Vol. 23, pp. 1–19, 1975.

[CAR 96] CARSI M. *et al.*, "The strain rate as a factor of influencing the hot forming simulation etc.", *Mat. Sci. Eng.*, A216, pp. 155–60, 1996.

[GRA 75] GRAFF, K.F., *Wave motion in elastic solids*, Clarendon Press, Oxford,1975.

[GOR 83] GORHAM D.A., "A numerical method for the correction of dispersion in pressure bar signals", *J.Phys. E.: Sci. Instrum.*, Vol. 16., 1983.

[GOR 96] GORHAM D.A., WU X.J., "An empirical method for correcting dispersion in pressure bar measurements of impact stress", *Meas. Sci. Technol.*, No. 7, pp. 1227–1232, 1996.

[JAS 99] JASPERS S., 1999, *Metal cutting mechanics and material behaviour*, Technische Universiteit Eindhoven Thesis.

[JOH 83] JOHNSON G.R., COOK W.H., "A constitutive model and data for metals subjected to large strains, high strain rates and high temperatures", 7th *International Symposium of Ballistics*, The Hague, pp. 541–547, 1983.

[MAL 86] MALINOWSKI J.Z., KLEPACZKO J.R., "A unified analytic and numerical approach to specimen behaviour in the SHPB", *Int. J. Mech. Sci.*, No. 6, pp. 381–391, 1986.

[MAT 82] MATHEW P., OXLEY P.L.B., "Predicting the effects of very high cutting speeds on cutting forces etc.", *Annals of CIRP*, Vol. 31(1), pp. 49–52, 1982.

[STE 70] STEVENSON M., OXLEY P.L.B., "An experimental investigation of the influence of speed and scale in a zone of intense plastic def.", *Proc. of Inst. of Mech. Eng.*, Vol. 184, pp. 561–576. 1970.

Chapter 20

A Comparative Study of Crystallization of iPP and PA6 Under Pressure

Vicenzo La Carrubba and S. Piccarolo
Chemical Engineering Department, University of Palermo, Italy

Valerio Brucato
Chemical Engineering Department, University of Salerno, Italy

1. Introduction

In the most widely employed industrial processes, such as extrusion and injection molding, the solidification process is a complex phenomenon where flow fields, high thermal gradients and high pressures determine the final morphology and the resulting properties. Lack of information about the influence of processing conditions on solidification behavior restricts the possibility of modeling the industrial material transformation processes. As a matter of fact the development of a model able to describe polymer behavior under drastic solidification conditions is a goal so far to be achieved in the state of present knowledge. For semicrystalline polymers, the application of a model describing crystallization behavior under processing conditions requires experimental results in order to check the reliability of the theoretical modeling and of the predicted results. Thus a good insight into polymer forming processes demands a detailed knowledge of material behavior under extreme conditions that are very far from the usual conditions to which data normally available in literature refer.

In fact, owing to the experimental difficulties, any investigation of polymeric structure dependence upon pressure and cooling rate has been mainly performed using conventional techniques, such as dilatometry and differential scanning calorimetry under isothermal conditions or non isothermal conditions but at cooling rates several orders of magnitude lower than those experienced in industrial processes.

Among the experimental results concerning dilatometry, data presented in Thermodinamik (Thermodynamik, 1979) represent a very interesting source of information on PVT behavior of polymeric materials (amorphous and semicrystalline) under medium cooling rates (up to 2–3°C/s). Zoller and Fakhreddine (Zoller *et al.*, 1994) and elsewhere He and Zoller (He *et al.*, 1994) employed a "confining fluid dilatometer" by GNOMIX Inc. in order to measure the specific volume of some semicrystalline polymers, such as isotactic polypropylene (iPP), polyamide 66 (PA66), polyethylenetherephthalate (PET). They found a decrease of final solid density with increasing pressure for iPP, while they noticed an increase of density with pressure for PA66 and PET. Chiu and Liu (Chiu *et al.*, 1995) developed a method for measuring PVT relationships of thermoplastic materials using an injection molding machine. Polystyrene (PS), acrylonitrile-butadiene-styrene resin (ABS) and low density polyethylene (LDPE) were investigated by using an empirical correlation based on the Tait equation. Their results display that in the low pressure region the specific volumes determined from this method and from a dilatometric apparatus are close to each other, whereas at higher pressures the results obtained by the two methods do not match. Fleischmann and Koppelmann (Fleischmann *et al.*, 1990) modified the PVT diagram of iPP by accounting for crystallization induced by the flow field, in which the material is subjected to in an injection molding process. They also take into account the influence of cooling rate. This investigation shows that a modification of PVT diagrams is necessary to obtain a better description of the melt behavior under injection molding conditions. Dee *et al.* (Dee *et al.*, 1994) obtained PVT data for an extensive series of polytetrafluoroethylenes (PTFEs) and PTFE oligomers, determining also crystallinities by ambient densities of the solids and also by DSC heat of fusion. Bhatt and Mc Carthy (Bhatt *et al.*, 1990) studied the PVT behavior of LDPE, PS and ABS filled and unfilled, in order to calculate thermal diffusivity values in the melt phase. A standard capillary rheometer was used for their experiments. Rodriguez and Filisko (Rodriguez *et al.*, 1987) applied a rapid hydrostatic pressure field on HDPE and LDPE samples under adiabatic conditions. The temperature changes were reported as a function of pressure and temperature by using a curve fitting analysis based on an empirical equation. Then data were analyzed by determining the so called "thermoelastic coefficient" derived from the Thomson equation.

Some other studies have concentrated on the dependence of the mechanical properties of materials on pressure during transformation processes. Shlykova *et al.* (Shlykova *et al.*, 1993) have studied the effect of injection molding at high pressures (up to 500 MPa) on some relevant properties for HDPE. Mears *et al.* (Mears *et al.*, 1969) have shown that under hydrostatic pressures up to 210 MPa the ultimate strains are large, but they decrease gradually with an increase in pressure. At higher pressures, the ultimate strains decrease abruptly. Silano *et al.* (Silano *et al.*, 1975) and Zihlif (Zihlif, 1975) investigated the dependence of stress on deformation and pressure for iPP. Their results show that the dependence of shear modulus on pressure breaks for pressures above 200 MPa. The dependence of yield strength on pressure may be considered to be linear. Baltà Calleja *et al.* (Baltà Calleja *et al.*, 1986) have carried out microhardness (MH) measurements on HDPE crystallized under different pressures and crystallization rates. Their results show a large increase of microhardness when the material is crystallized under high pressure.

As for the effect of cooling rate, Choi and White (Choi *et al.*, 2000) described structure development of thin melt spun iPP filaments, obtaining conditions under which different crystalline forms of iPP were obtained as a function of cooling rate and spinline stress. On the basis of their experimental results the authors constructed first a diagram, which indicates the crystalline states that form at different cooling rate in isotropic quiescent conditions. Continuous cooling transformation curves (CCT) were also reported on the diagram. Then the authors constructed a structural map reporting the cooling rate as a function of spinline stress (Choi *et al.*, 2000), where phase stability regions are clearly shown. According to the authors' interpretation, at low cooling rates and high stresses monoclinic   structure was formed, whereas at high cooling rates and low stresses a large pseudo-hexagonal/smectic (the "mesomorphic phase") region was evident. Finally the effect of uniaxial stresses was shown to be a shift of the nose of the curve towards lower values of time (*i.e.* higher crystallization rates). In other words, an increase of uniaxial stresses promotes the formation of   -phase with respect to the mesomorphic phase. Piccarolo (Piccarolo, 1992) analyzed a very large set of iPP morphologies obtained by varying the cooling rate at ambient pressure under quiescent conditions. A continuous variation of morphology and crystal structure was obtained with cooling rate. Brucato *et al.* (Brucato *et al.*, 1991; 1998) studied the effect of thermal history on morphology of PA6 samples crystallized from the melt at ambient pressure under quiescent conditions. The results show that cooling history relevant to quenched sample morphology is confined in the interval 110 to 160°C.

The aim of this work is to supply reliable experimental data for semicrystalline polymers crystallized under pressure and controlled cooling conditions, in order to quantify the combined effects of solidification pressure and cooling rate on the final properties of the product obtained. For this purpose a new experimental route based on quenching polymer melts under a hydrostatic pressure field has been designed and widely tested on different classes of semicrystalline polymers (Brucato *et al.*, 2000, La Carrubba *et al.*, 2000, La Carrubba, 2001). A large set of experimental data concerning isotactic polypropylene (iPP) and polyamide6 (PA6) is reported. The experimental techniques adopted tried to assess the dependence of end-product macroscopic properties (density and microhardness) and morphology (by means of wide angle and small angle X-ray scattering, WAXS-SAXS, infrared spectroscopy IR) on processing conditions (cooling rate and solidification pressure). A comparison of the behavior displayed by two different semi-crystalline polymers, iPP and PA6, has been extensively taken out on the basis of the experimental results here achieved.

2. Experimental

For the purpose of our work an injection molding iPP grade named "T30G" kindly supplied by Himont was employed, having the following main features: $M_n = 75100$; $M_w = 483000$; $M_w/M_n = 6.4$. PA6, kindly supplied by DSM, was characterized by $M_w = 25000$, $M_n = 13000$. Materials were quenched under pressure and high cooling rates according to a procedure already reported (Brucato et al., 2000; 2001). The basic idea was to set up a model experiment to couple rapid cooling histories with pressure levels up to 40 MPa under controlled conditions. The experimental apparatus consisted of a modified injection molding machine, used as a source of molten polymer supplied at a pre determinable and maintainable constant pressure (Brucato et al., 2000; La Carrubba et al., 2000). Samples were then solidified under a uniform pressure field and under a non-uniform cooling rate profile, the cooling rate depending upon the distance from the cooled surface. Thus, starting from the cooled surface, a continuous and gradually decreasing cooling rate profile was achieved, which determined a continuous variation of structure with depth (La Carrubba et al., 2000). The relationship between cooling rate and depth was taken into account in the so-called "mapping function" (Brucato et al., 2000; La Carrubba et al., 2000). Another apparatus was used for crystallizing iPP samples at low cooling rates, consisting in a confining fluid pressurized cell where the polymer was confined and maintained under hydrostatic pressure (La Carrubba, 2001).

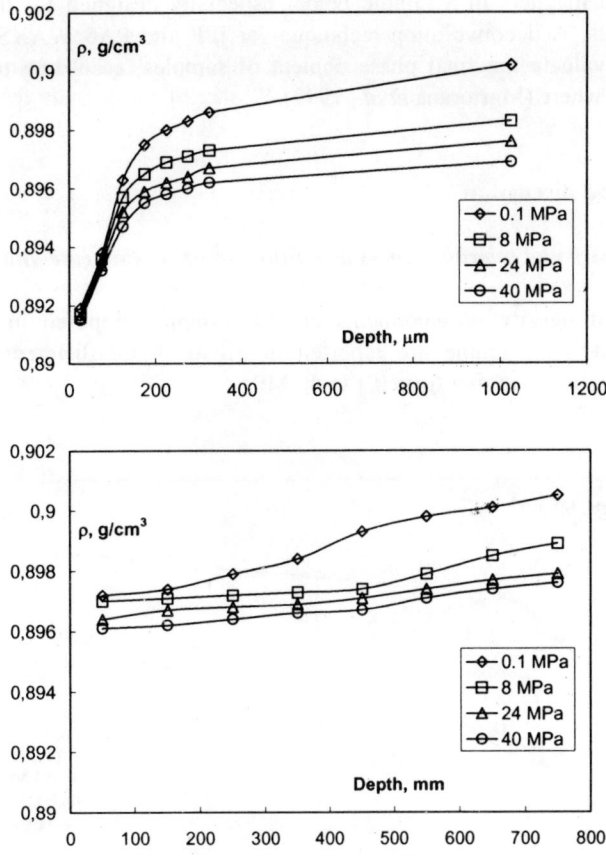

Figure 1. *iPP density depth profile: a) (on the top) surface cooling rate of 100°C/s; b) (on the bottom) surface cooling rate of 20°C/s*

Prepared samples were subjected to the following characterization techniques: density measurements, microhardness measurements, wide-angle X-ray scattering (WAXS), small angle X-ray scattering (SAXS), infrared spectroscopy (IR). Density measurements were performed in density gradient columns filled by mixtures of an appropriate couple of solvents (water and ethyl alcohol for iPP, n-heptane and carbon tetrachloride for PA6). For microhardness measurements a standard Vickers microindenter Anton Paar was employed. WAXS and SAXS patterns were collected at a beam line of two synchrotron light sources: the A2 beam line in Hamburg for iPP samples (radiation wavelength of 0.154 nm) and the SAXS/WAXS beam line of Trieste for PA6 samples (radiation wavelength of 0.143 nm). IR spectra were measured using a Perkin Elmer spectrometer coupled with a microscope with radiation polarized in two directions mutually orthogonal. The slice was fixed between two magnets in a sample holder especially designed for these line scan measurements. A deconvolution technique for iPP and PA6 WAXS patterns was applied to evaluate the final phase content of samples, according to a procedure reported elsewhere (Martorana *et al.*, 1997).

3. Results and discussion

3.1. *iPP density and microhardness as a function of cooling rate and pressure*

Results of density measurements on iPP samples prepared in the modified injection molding machine are reported in Figure 1 for different solidification pressure conditions ranging from 0.1 to 40 MPa.

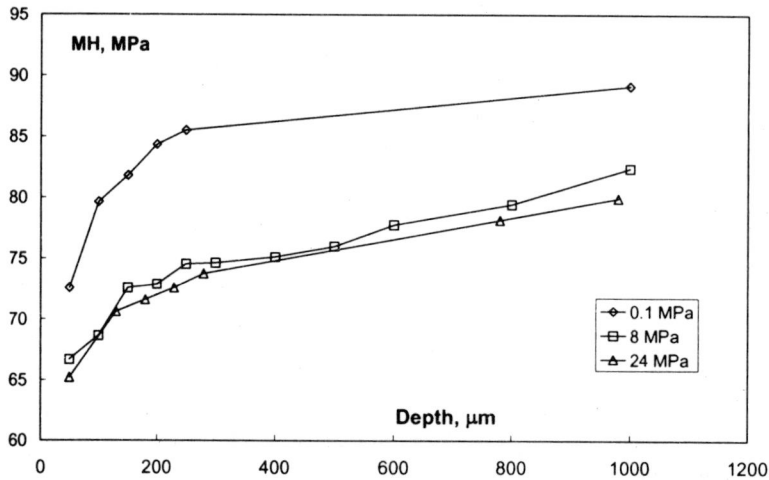

Figure 2. *iPP microhardness depth profile*

Figure 1a reports the density depth profile for samples characterized by a surface cooling rate (at 70°C) of about 100°C/s, whereas Figure 1b reports the density depth profile for samples characterized by a surface cooling rate (at 70°C) of about 20°C/s. Cooling rates have been evaluated at 70°C since it has been shown that the cooling rate calculated at 70°C is a good measure of quench effectiveness for iPP (Piccarolo, 1992; Piccarolo *et al.*, 1992).

Figure 1 shows that for all pressure values, density increases proceeding from the surface to the bulk of the sample, this behavior can be related to an increase of crystallinity since deeper slices are cooled at lower rates. Furthermore, the largest part of the density increase takes place in the neighborhood of the surface and irrespective of the applied pressure. This was qualitatively confirmed by the polarized light micrographs shown elsewhere (Brucato *et al.*, 2000). If we now consider the effect of pressure, Figure 1 shows that density decreases when pressure increases at the same depth. The decrease in density due to the pressure effect is minimum at the sample surface and grows with depth. Furthermore, the majority of the density change is observed by varying the solidification pressure from 0.1 to 8 MPa, this value being quite low especially if compared to the typical pressure values attained in polymer processing. Figure 1 shows also that on increasing pressure density curves tend to level off. Additionally, the greatest density differences are in the internal zones of the samples, which are subjected to the lowest cooling rates. On going from the surface towards the bulk of the sample (i.e. on decreasing the cooling rate) the effect of pressure on the final density tends to disappear.

In Figure 2 microhardness measured along the sample depth is reported. The same dependence on solidification pressure as density is observed, with some minor differences. As a matter of fact it is worth noticing that the curves at pressures above 8 MPa collapse on each other.

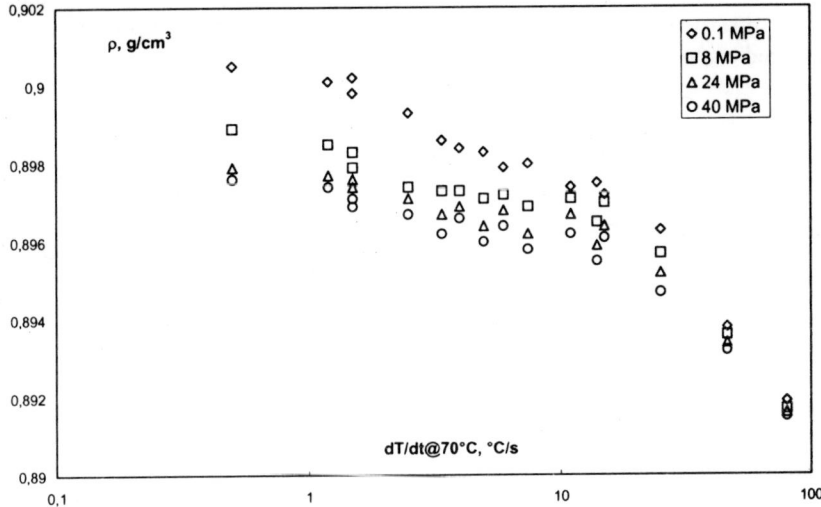

Figure 3. *iPP density versus cooling rate*

Figure 3 is obtained by plotting the density data of Figure 1 versus cooling rate by means of the "mapping function" derived from a simplified transient heat conduction model, as reported elsewhere (Brucato *et al.*, 2000, La Carrubba *et al.*, 2000). Figure 3, like Figure 1, shows that at constant cooling rate final density decreases with solidification pressure. Independently of solidification pressure the material shows the same trend with the cooling rate, with a density drop above 10–

20°C/s. Finally, Figure 3 shows that the decrease of density with pressure vanishes with increasing cooling rate. This implies that the influence of pressure is more pronounced in the bulk of the sample and becoming negligible near the sample surface.

Also microhardness results of Figure 2 were converted into the diagram shown in Figure 4, where microhardness as a function of cooling rate is reported. It is easy to notice that icrohardness decreases on increasing cooling rate at all the adopted solidification pressure values. With respect to density results of Figure 3, microhardness curves do not tend to overlap on each other in the high cooling rate region.

In order to extend the range of cooling rate to be explored below the minimum achievable cooling rate using the modified injection molding machine (around 1°C/s), a confining fluid pressurized cell was used, in order to ensure a constant pressure field during sample crystallization. In both experiments (modified injection molding machine and pressurized cell) the same methodological approach of recording the thermal history experienced by the samples during cooling and then analyzing the resulting sample morphology was adopted. Due to the small size and the low achievable maximum cooling rate (less than 2°C/s), the samples crystallized in the pressurized cell were completely homogeneous and their density was measured in a density gradient column.

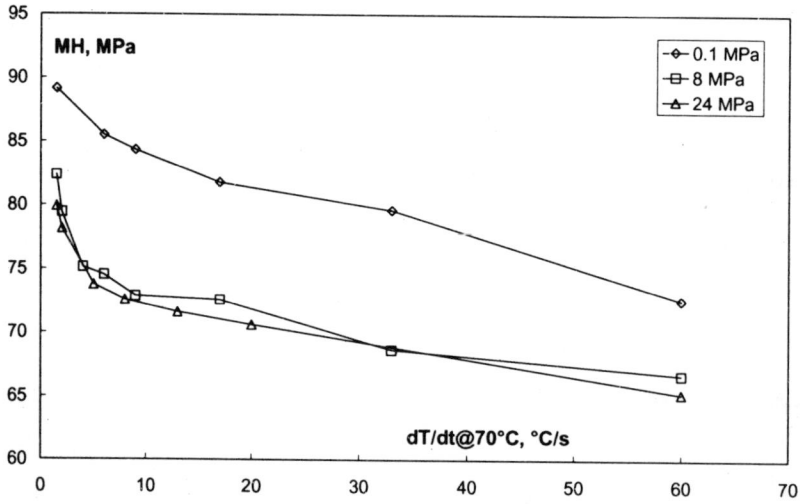

Figure 4. *iPP microhardness versus cooling rate*

Reporting all the data obtained using both apparatus, the density versus cooling rate (evaluated at 70°C) dependence shown in Figure 5 is obtained. It is worth reminding that cooling rate was measured for samples crystallized in the pressurized cell, whilst for samples solidified in the modified injection molding machine cooling rate was calculated by means of a simplified heat transfer model (Brucato *et al.*, 2000). The first result that must be emphasized is that the pressurized cell data obtained by solidification at 2°C/s, indicated with open symbols in Figure 5, exhibit a satisfactory overlapping with the data obtained in the modified injection molding machine, indicated with full symbols in Figure 5. The good matching of data referred to completely different experimental set-ups confirms the reliability of the whole experimental route adopted (La Carrubba *et al.*, 2000).

The other experimental points achieved by means of the pressurized cell were at the cooling rates of 0.1 °C/s and 0.02°C/s (indicated with open symbols in Figure 5). The final density versus cooling rate dependence covers nearly four orders of magnitude in cooling rate (from about 0.01 up to nearly 100°C/s). Three regions can be distinguished within the diagram. A region of high cooling rates (above 20°C/s), characterized by a sharp fall in density, in this zone the role played by the solidification pressure is almost negligible.

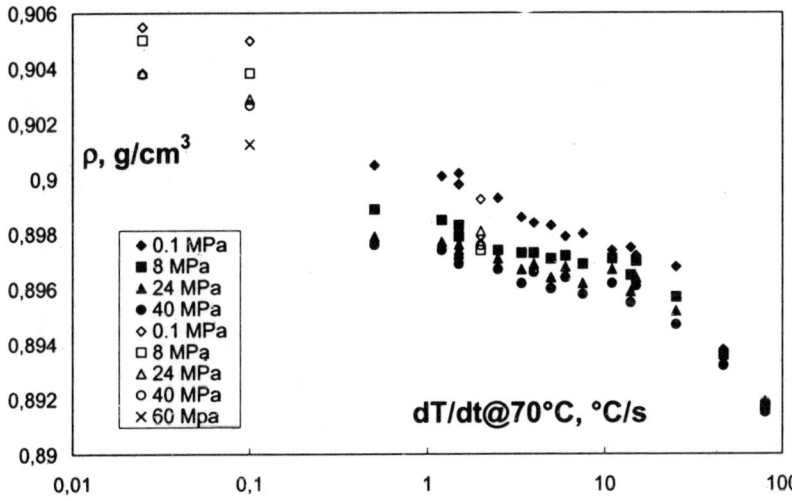

Figure 5. *iPP density versus cooling rate for samples prepared in the modified injection molding machine (full symbols) and in the pressurized cell (open symbols)*

In other words, in the high cooling rate region, the final density value is dominated by cooling rate, and it can be therefore considered to be independent of the pressure applied during the crystallization process. In the second region, spanning from 1 to 20°C/s, density exhibits a gradual fall at all the investigated solidification pressures. Solidification pressure here strongly affects the final density value, the higher the pressure the lower the final density. In this range of cooling rates it may be observed that a cooling rate increase at a constant pressure has the same effect as an increase of solidification pressure at a constant cooling rate (La Carrubba *et al.*, 2000). The third region corresponds to the low cooling rate zone, ranging from 0.01 to 1°C/s. A gradual decrease of density may be noticed with the same influence of solidification pressure on final density, as seen in the region 1–20°C/s. The effect of pressure on the final density may be accounted for by using a different master curve, determined on the basis of the experimental density data of this region. From Figure 5 it is also evident a sharp fall of density from 0.1°C/s to 1°C/s at all the solidification pressures. This result leads to the conclusion that a further process must occur when cooling rate becomes lower than 1°C/s, this process being responsible for the observed increase of density and alpha phase. A "secondary crystallization" process in series with the first one may indeed give a plausible justification of density and WAXS results, as explained elsewhere (La Carrubba *et al.*, 2000).

3.2. *PA6 density and MH dependence upon cooling rate and pressure*

Figure 6 reports density of PA6 samples as a function of depth for all the adopted solidification pressures. In analogy with iPP density results, three different zones can be identified on that diagram.

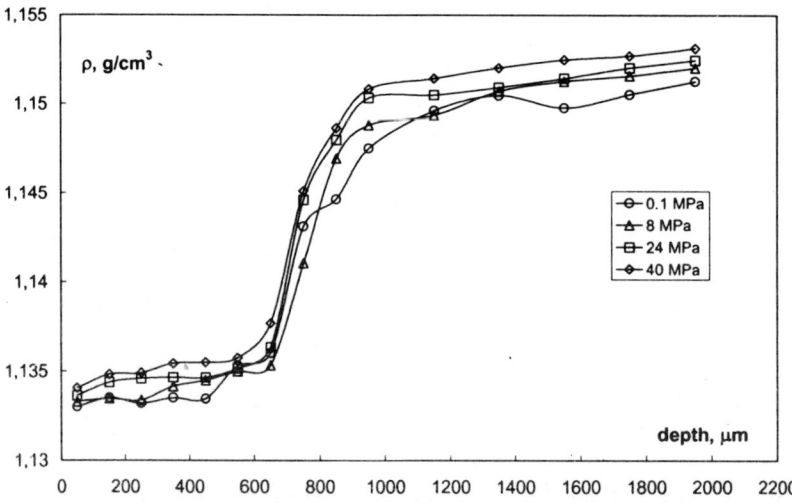

Figure 6. *PA6 density depth profile*

The first zone, near the cooled wall, is characterized by higher cooling rate values, and it can be easily noticed that density, going from the surface up to a depth of around 700 µm, remains nearly constant at a constant pressure. Only a very small gap separates experimental densities of samples crystallized at ambient pressure with respect to the ones crystallized under higher pressure conditions (40 MPa). Anyway, the increase of density with increasing solidification pressure is not negligible, being still detectable.

The second zone corresponds to a sort of "transition zone" where the most significant variations of density versus depth take place. As a matter of fact, in the region where depth ranges from 700 µm up to 1000 µm, a sudden rise of density should be pointed out, independent of the working solidification pressure. In the whole transition zone the effect of pressure on the final sample densities seems to be negligible. Finally, the third region, the "bulk zone", starting from a depth of approximately 1000 µm, is the region where density tends to level off with increasing depth for every solidification pressure. The effect of pressure on the final density across this zone appears to be more pronounced than in the low-density plateau. Specifically, the higher the pressure the higher the resulting final density.

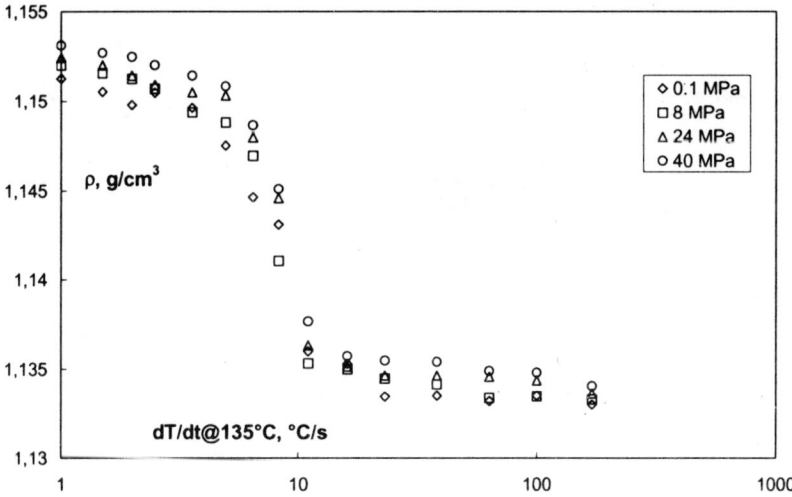

Figure 7. *PA6 density versus cooling rate*

On the whole, the increase of density due to the increase of solidification pressure is of the same order of magnitude as the decrease of density observed in the case of iPP, as already discussed above. This evidence turns out to be in full agreement with the dilatometric results reported by He and Zoller in 1994 (He *et al.*, 1994) who found a decrease of density with solidification pressure of 0.1% in the case of iPP, while they noticed an increase of density of 0.3% in the case of PA66. It should be remembered that the cooling rate adopted during their experiments was very low (1–2°C/min) due to the intrinsic constraints imposed by the apparatus.

If density of PA6 is reported as a function of cooling rate measured at 135°C (Brucato *et al.*, 1991; 1998) the same trend discussed above is observed, as it could be noticed by looking through Figure 7. The low-density plateau corresponds to the high cooling rate region, and its starting point is located above 10°C/s. This region covers more than one order of magnitude in cooling rates, reaching the cooling rate of 250°C/s. The high-density plateau covers nearly one order of magnitude in cooling rate, going from the lowest achieved cooling rate during the experiment (1°C/s) up to 7–8°C/s. As for the transition zone, it should be pointed out that it is located in a very narrow region, centered at a cooling rate of about 8–10°C/s. This cooling rate could be defined as a sort of "critical cooling rate" since small changes of cooling rate around this value could determine large structural and morphological variations that, in turn, determine large density variations. A very similar value of critical cooling rate was already found for a different PA6 some years ago, which is in agreement with the results here presented (Brucato *et al.*, 1991, Brucato *et al.*, 1998).

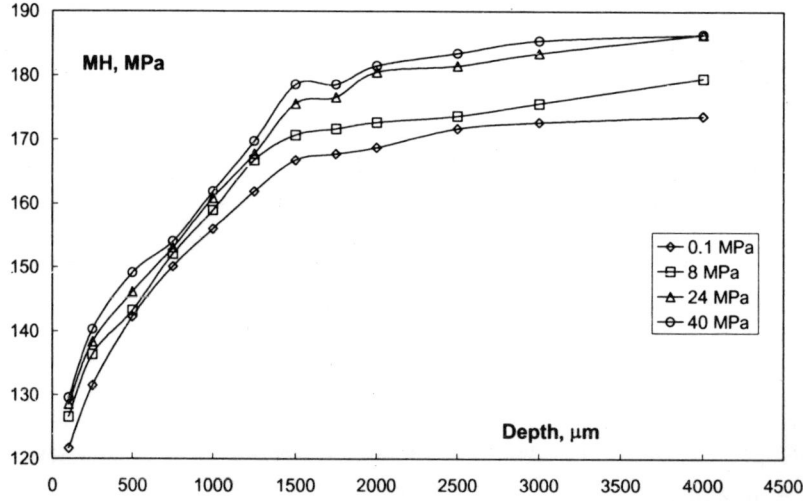

Figure 8. *PA6 microhardness depth profile*

As far as the microhardness behavior of PA6 is concerned, a different sensitivity with respect to the density behavior is to be highlighted, in agreement with the results already presented for iPP. Figure 8 reports microhardness of PA6 samples as a function of depth for all the adopted solidification pressures. The first evidence to be analyzed is a continuous gradual raise of microhardness with depth, starting form the cooled wall. In other words, the low-density plateau, observed in the case of density, does not show up for microhardness. A second issue concerns the plateau corresponding to high microhardness values. As a matter of fact this plateau corresponds only partially to the one observed in density, since its starting point is located at a much larger depth, around 1500 µm. All these differences both for iPP and PA6 could be explained by taking into account that density is a material "bulk" property, while microhardness is a "surface" property and for this reason the latter turns out to be much more sensitive to any structural modification than the former. Generally speaking, these results seem to indicate that the structural and morphological modifications at the cooled wall can be easily highlighted by measuring the microhardness profile (Figures 3 and 8), while in the case of density the variations measured near the surface were much less evident (Figures 1 and 5).

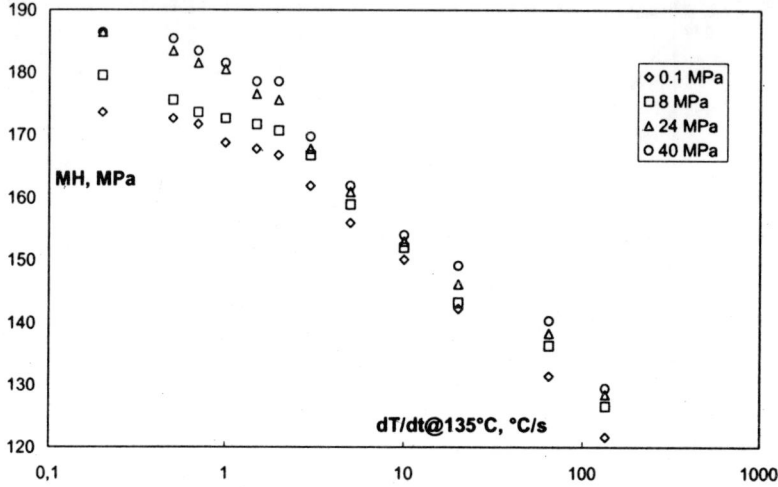

Figure 9. *PA6 microhardness versus cooling rate*

As for the influence of solidification pressure on the final microhardness values, Figure 8 shows that an increase of pressure determines a parallel increase of microhardness in the whole explored depth. Nevertheless, the effect of pressure is more remarkable in the internal region of samples (in the high microhardness plateau zone), where the gap between low pressure and high pressure crystallized samples tends to broaden. When microhardness is reported as a function of cooling rate (Figure 9), the value of cooling rate above which the increase of microhardness due to the increase of pressure turns out to be less pronounced, can be identified. It corresponds to a cooling rate of about 5–6°C/s, *i.e.* lower than the critical value (8–10°C/s) found in the case of density.

3.3. *Phase distribution of iPP and PA6 as a function of cooling rate and pressure*

WAXS experiments were performed on thin slices (50–100 •m) microtomed according to the procedure already reported (Brucato *et al.*, 2000, La Carrubba, 2001). All experiments concerning iPP samples were performed at the synchrotron radiation source of the center DESY in Hamburg, whereas all experiments concerning PA6 samples were performed at the synchrotron radiation source of the center Elettra in Trieste. In both cases a very long accumulation time was applied in order to achieve excellent statistics and a good reproducibility. A deconvolution technique was applied to WAXS patterns in order to evaluate the phase content according to a procedure that has been fully discussed elsewhere (Martorana *et al.*, 1997). Figure 10 reports phase content of iPP samples versus pressure for four different values of cooling rates, ranging from 1.5 to 80°C/s.

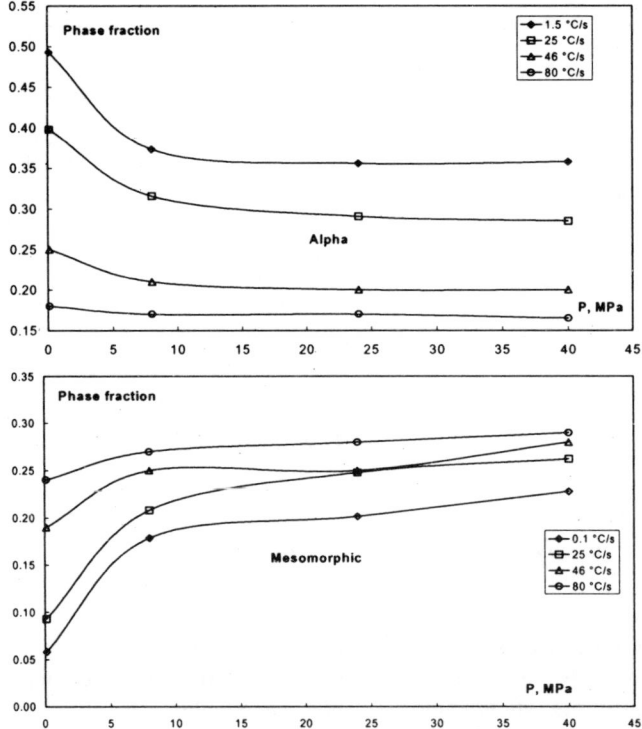

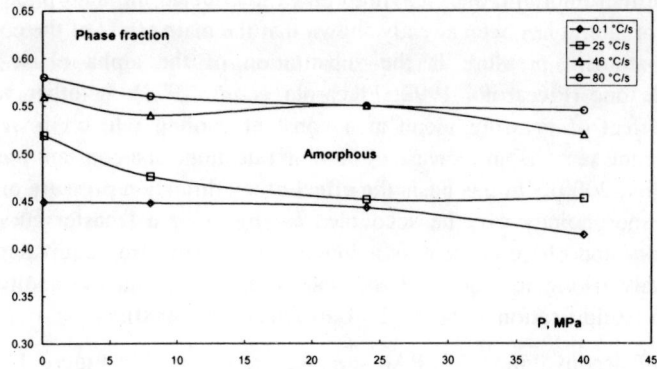

Figure 10. *iPP Phase distribution from WAXS deconvolution*

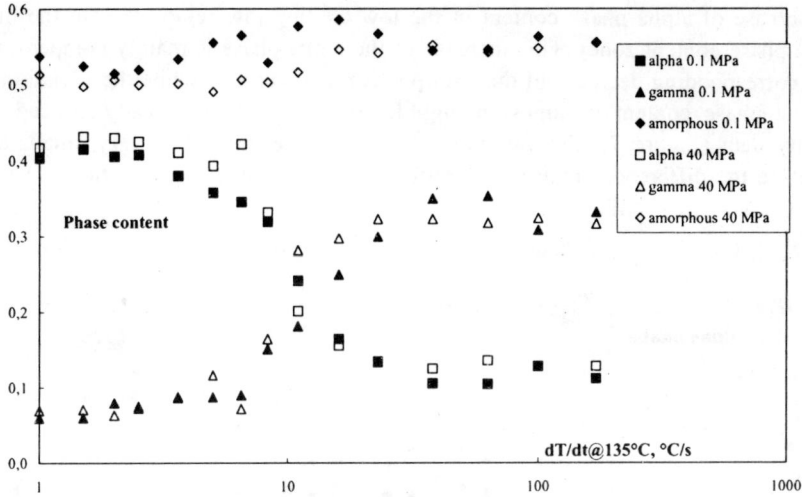

Figure 11. *PA6 Phase distribution from WAXS deconvolution*

It should be noticed a decrease of alpha phase with pressure for cooling rates equal to 1.5 and 25°C/s. In agreement with density and microhardness, the alpha phase trend with pressure shows the largest variations taking place in the range 0.1–10 MPa. By looking at Figure 10 one should also notice that the decrease of the alpha phase with pressure tends to vanish when the cooling rate increases, particularly for cooling rates above 20–30°C/s. Additionally, the decrease of alpha phase is mainly balanced by an increase of the mesomorphic phase content while the amorphous phase seams to be only slightly affected by the pressure increase. This last point is of high relevance, as the main effect of pressure alone is replacing alpha phase by the mesomorphic one, leaving almost unaffected the amorphous content. As a matter of fact, it has been already shown that the main effect of the cooling rate increase at ambient pressure is the substitution of the alpha phase with the mesomorphic one (Piccarolo, 1992; Piccarolo *et al.*, 1992). In other words, the qualitative effect of pressure alone at a constant cooling rate on final structure appears to be the same as an increase of cooling rate alone at a constant pressure (La Carrubba *et al.*, 2000). On this basis the effect of solidification pressure on the final density and morphology may be accounted for by using a "master curve" which reports density and phase content as a function of a so-called "equivalent cooling rate", correctly taking into account the real cooling rate and the additional shift caused by the solidification pressure (La Carrubba *et al.*, 2000).

Results of deconvolutions on PA6 samples are reported in Figure 11, showing phase distribution as a function of cooling rate for samples crystallized at atmospheric pressure and at 40 MPa. It is easy to notice that phase content is in good agreement with density results shown in Figure 7, with the alpha phase drop located around 10°C/s. Moreover, Figure 11 shows that the increase of pressure determines an increase of alpha phase content in the low cooling rate region (*i.e.* in the high alpha phase content zone). This increase of the alpha phase is mainly compensated by a corresponding decrease of the amorphous phase content, while the variation of gamma phase content is almost negligible. In any case, as already noticed for density data (Figure 7), the variation of alpha phase content is very small, and therefore the differences may be quantified only increasing solidification pressure from 0.1 MPa to 40 MPa.

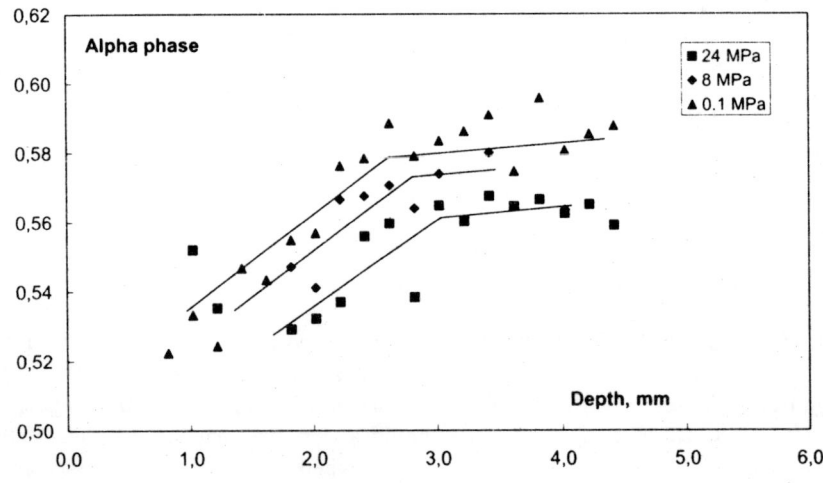

Figure 12. *iPP infrared crystallinity versus depth*

3.4. *Infrared crystallinity measurements on iPP and PA6*

On thin iPP and PA6 slices (10 •m), Infrared transmission spectra were recorded in order to evaluate the crystallinity distribution across the sample depth. In the case of iPP samples, the solidification pressures investigated were 0.1 MPa, 8 MPa and 24 MPa. It is worth remembering that infrared crystallinity, measured for iPP as the ratio between the area under some typical "crystalline peaks" and the total area of typical crystalline and typical "amorphous peaks", does not take into account the presence of mesomorphic phase, since with this technique mesomorphic phase contributes to a sort of "ordered configuration" which cannot be resolved in the contribution of the two phases. Actually the IR crystallinity should be expected to be higher than WAXS crystallinity due to the fact that the contribution to the ordered configuration given by the mesomorphic phase is lumped in the IR crystallinity. Furthermore, the IR crystallinity should also be expected to be higher than the crystallinity calculated from density data, since the latter does not consider the intermediate density phase, whose contribution is relevant for the final IR crystallinity value. Last but not least, it must be emphasized that the results presented below are only qualitative, and a correct evaluation of the crystallinity needs an accurate calibration.

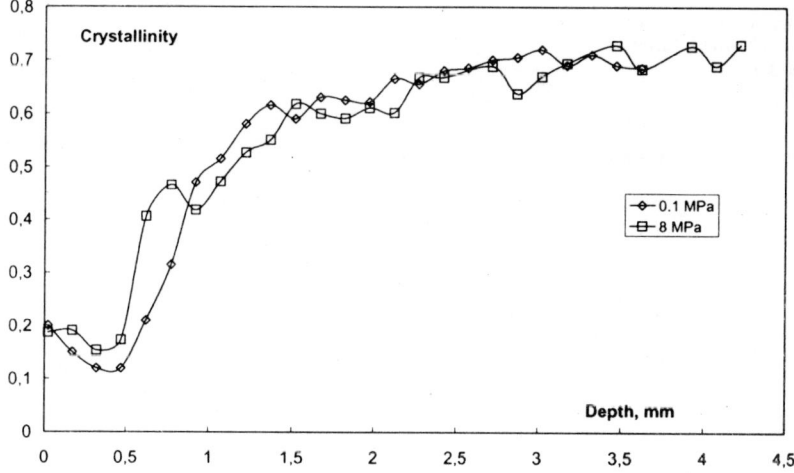

Figure 13. *PA6 infrared crystallinity versus depth*

As for infrared-based crystallinity measurements for PA6, according to literature crystallinity was evaluated by comparing the height of a reference peak which remains always stable and constant in height with the height of a peak characteristic of the crystalline phase of PA6 which is sensitive to structure variation (La Carrubba, 2001).

Crystallinity measurements based on infrared data for iPP samples are reported in Figure 12. Due to the large scattering of data in the zone going from the surface of the sample up to 1000 •m, the results were not reported. For this reason Figure 12 reports IR crystallinity as a function of sample depth from 1000 •m ahead. As it should be easily noticed, crystallinity shows a gradual rise from 1000 •m to 3000 •m, independently of the applied pressure thus confirming the increase of ordered configuration with decreasing cooling rate, independently of whether it is in the alpha or mesomorphic phases. Since at a depth of 1000 •m the calculated cooling rate is around 1°C/s, this change in IR crystallinity could correspond to some structural modifications occurring in iPP samples below this cooling rate. This result will be explained below in more detail when describing the density-cooling rate relationships. Despite the scattering in the initial layers, comparison of the IR crystallinity results for the three solidification pressures adopted and reported in Figure 12 shows that an increase of pressure determines a decrease of crystallinity, confirming the trend already observed by means of many different experimental techniques (density, microhardness, WAXS measurements).

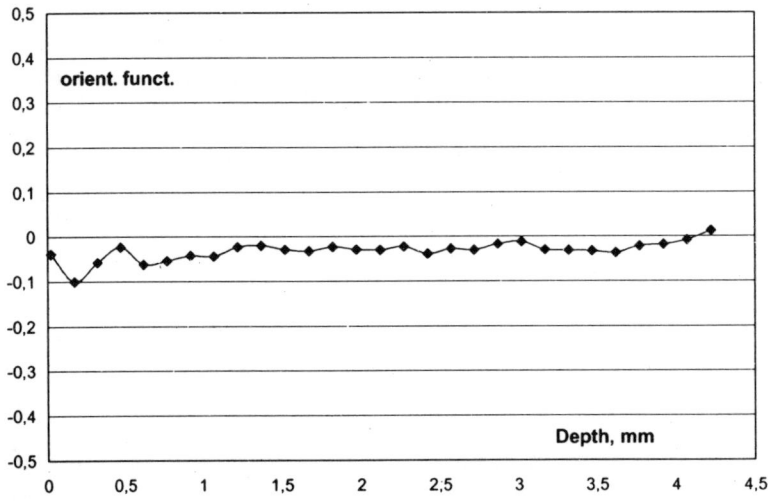

Figure 14. *PA6 orientation distribution (by infrared) versus depth*

In Figure 13 crystallinity measurements based on Infrared data for PA6 samples are reported, for solidification pressures of 0.1 and 8 MPa. From an accurate analysis of the figure three different zones can be distinguished: a low crystallinity zone (low crystallinity plateau), a "transition zone" and a high crystallinity zone (high crystallinity plateau). The thickness of the initial layer near the cooled wall characterized by low IR crystallinity values is around 700 •m, which matches the thickness of the low density region already mentioned and discussed above (Figure 6). As for the high crystallinity region, its starting point is located around 1500 •m, in the same place as the starting point of the high microhardness plateau is located (Figure 8). To sum up, the scale of sensitivity to structural changes of crystallinity calculated by means of the Infrared technique is of the same order as the scale of sensitivity of density in the high cooling rate region (depth less than 700–800 •m), while it is of the same scale of sensitivity of microhardness in the low cooling rate region (depth greater than 1500 •m). In other words IR crystallinity is a very efficient technique for evaluating structural modifications when the crystalline phase content is high, while it is a poor technique in the case of very low crystalline phase content. As for the influence of solidification pressure on the final crystallinity, Figure 13 shows that an increase of pressure from 0.1 MPa up to 8 MPa is not very effective in modifying the crystalline content. This outcome was already found in density results, showing that density does not vary significantly from 0.1 and 8 MPa, while remarkable differences may be observed between the lowest pressure (0.1 MPa) and the highest pressure (40 MPa). This result is also of deep interest as far as a comparison between iPP and PA6 behavior under pressure is concerned. As a matter of fact in the case of iPP, not only an increase of solidification pressure determines a decrease of the final density, but also the most significant changes in density are noticeable by varying the solidification pressure from 0.1 MPa to 8 MPa.

Finally, in the case of PA6 IR transmission spectra were also used to evaluate orientation. The orientation function (calculated according to Hermans' function) takes into account the average orientation of the macromolecules in the space. The orientation profile for a PA6 sample crystallized at ambient pressure 'is reported in Figure 14. It is reported the Hermann's orientation function versus depth. As it may be easily noticed, the orientation function has values near to zero for the whole sample depth, showing that the assumption of considering negligible the orientation induced by the injection into the cavity was absolutely correct. In other words, the use of a very low filling rate adopted for the present experiments allowed us to obtain samples that crystallized under a constant pressure field in quiescent conditions.

3.5. Comparison of the crystallization behavior of iPP and PA6

By looking at the density dependence upon cooling rate at a constant pressure for the two studied polymers, one may assert that an increase of cooling rate determines a continuous decrease of density. This result is an extension to higher pressures of the density dependence upon cooling rates already known (Brucato *et al.*, 1991; 1998, Piccarolo, 1992, Piccarolo *et al.*, 1992).

If one compares the density dependence upon solidification pressure of iPP and PA6 at a constant cooling rate, several observations arise. First of all, one may notice that the cooling rate range where the majority of the density drop takes place ("the transition region") falls between 20 and 100°C/s in the case of iPP (Figures 3 and 5), while it falls between 5 and 20°C/s in the case of PA6 (Figure 7). On the basis of these results one may define iPP as a faster crystallizing polymer than PA6, in the sense that the "transition region" of iPP falls at higher cooling rates than in the case of PA6 (*i.e.* the "crystallizability" of iPP is larger than the one of PA6 (Ziabicki, 1976). Secondly, in the case of iPP there is another region where a large density change takes places, located around 1°C/s, due to the occurrence of a secondary crystallization. From an accurate analysis of Figure 7, showing the density dependence of PA6 upon cooling rate for various pressure values, one may assert that PA6 does not show a secondary crystallization process, at least in the range of experimental conditions explored along this work. A comparative analysis of Figure 10 and Figure 11, showing the phase content after WAXS deconvolution, leads to the conclusion that, at a constant pressure, an increase of cooling rate determines a decrease of the content of the more stable phase (the •-phase), that is replaced by a less stable phase (mesomorphic phase for iPP and •-phase for PA6), while the amorphous phase content does not vary remarkably. As for the effect of pressure at a constant cooling rate, one should notice that, in the case of iPP, an increase of solidification pressure determines a decrease of density due to the decrease of alpha phase and the increase of mesomorphic phase content. In the case of PA6, the increase of solidification pressure is coupled to an increase of density due to a slight increase of alpha phase and a slight decrease of amorphous content.

A similar pressure dependence on density has also been observed by He and Zoller (He *et al.*, 1994) using a standard dilatometer by GNOMIX Inc. to measure sample specific volume during crystallization from the melt. A constant slow cooling rate (2–3°C/min) under constant pressure was applied bringing the sample back to a fixed pressure at the end of the test (10 MPa). He and Zoller observed an increase of specific volume with increasing crystallization pressure in the case of iPP while PA66 and PET show an opposite behavior (He *et al.*, 1994). Although they tried to explain the density decrease with the formation of • phase, which is less dense than • phase, in our samples there is no evidence of • phase formation (La Carrubba *et al.*, 2000, La Carrubba, 2001). Their results are however in agreement with our results, where final density (measured at atmospheric pressure) of iPP samples solidified under pressure turns out to be lower than the one of samples solidified at atmospheric pressure, while the opposite occurs in the case of PA66 samples.

In the case of microhardness measurements, few more considerations are needed. In the case of iPP, three major differences with respect to density and WAXS crystallinity results must be highlighted. Firstly, microhardness varies continuously with cooling rate, also in the high cooling rate region (above 20–30°C/s), where density and crystallinity tend to level off, this outcome being also true for PA6 microhardness results. Secondly, a very large difference in microhardness may be noticed by increasing solidification pressure from 0.1 to 8 MPa; this difference is almost constant with cooling rate, and does not tend to level off in the high cooling rate region (as observed for density and crystallinity). Thirdly, the curve of microhardness at 24 MPa (Figures 3 and 4) is nearly overlapping the one at 8 MPa, what was not seen to occur for density and crystallinity (Figures 1–3 and 5). These evidences show that microhardness, which is a mechanical property, exhibits a larger sensitivity to cooling rate since the latter in turn determines to a larger extent morphology, while density is a "bulk" property. Furthermore, microhardness sensitivity to pressure, although it seems to be very high while detecting structural changes from 0.1 to 8 MPa, turns out to be inadequate (in the case of iPP) for detecting any further transformation occurring at higher pressures. In the case of microhardness measurements on PA6, microhardness exhibits a higher sensitivity to pressure at low cooling rates, and a low sensitivity to pressure at high cooling rates. A possible interpretation of this experimental evidence could be attempted on the basis of some of the results recently found for PET (Piccarolo et al., 2000). As a matter of fact in the case of PET a continuous variation of microhardness with cooling rate (for samples crystallized at ambient pressure) was observed, even in the region where density turns out to be not affected by cooling rate, being almost constant. For PET the variations of MH were explained with the occurrence of a metastable phase forming during crystallization from the melt. This metastable phase (whose evidence was discussed (Piccarolo et al., 2000) strongly affects the microhardness values, while its influence on density could be considered negligible. A similar dependence should be assessed in the case of PA6, where the existence of a mesophase (called •*) was already proposed some years ago (Fu et al., 1994). The presence of this mesophase determines the observed increase of microhardness with decreasing cooling rate. On the opposite density is not very sensitive to these structural variations, as the density value of this mesophase is not far away from the density value of the crystalline phase. Finally, it should be pointed out that the amount of the formed mesophase does not vary with cooling rate, its formation relying only on constraints imposed to the stable phase (thermal motion, diffusion, mobility).

Last but not least, one may question why the crystallinity dependence of iPP upon pressure (with a lower density upon increasing solidification pressure) was observed only for this material, while PA6 and also PET (La Carrubba, 2001, La Carrubba et al., 2001) showed an opposite behavior (with an increase of crystallinity upon increasing solidification pressure). The answer to this open question may rely on the intrinsic nature of the •-crystalline structure of iPP. As a matter of fact, •-phase is characterized by a monoclinic unit cell, with layers made of isochiral helices parallel to the ac plane. Alternate layers are antichiral, and further they differ by the azimuthal settings of the chain (Lotz, 1998). If the alternate sequence of left-handed and right-handed helices is lost, then the mesomorphic phase may be formed. The perturbation necessary to generate the mesomorphic structure could be created by an increase of the hydrostatic pressure during the solidification process, which may limit the necessary free volume to give rise to the organization of helices in the crystalline register required by the •-form. The effect of solidification pressure observed in the case of PET may be different because of the different intrinsic nature of PET triclinic cell, where the crystalline register does not require such a complex helical organization necessary in the case of iPP. In the case of PET, the "kinetic" effect due to the pressure increase is almost negligible, and then the "thermodynamic" effect turns out to prevail, giving rise to a larger density as the solidification pressure increases. PA6 is situated somewhat halfway of these two extremely contrasting behaviors, exhibiting a small increase of crystallinity upon increasing solidification pressure.

An intensive study of solidification under pressure of polymers characterized by a structural organization similar to the one of iPP (such as isotactic polybutene iPB, polytrimethylentherephthalate PTT exhibiting helical structures) may address this open question to the correct interpretation in terms of structural prerequisites determining the response of a polymer during non-isothermal solidification with an applied hydrostatic pressure field.

Finally, a conformational analysis may also give a robust contribution towards a correct interpretation of the phenomena involved during crystallization in non-isothermal conditions under pressure. This analysis, together with the knowledge of the stability of the crystalline forms characterizing a given polymer, could lead to a reliable prediction of material behavior under processing conditions.

On the basis of the crystallization kinetics behavior of the polymers studied over the last few years (iPP, PA6, PET), one could even try to give a "rule of thumb" or zeroth approach for the description of polymer crystallization behavior under non-isothermal conditions: the faster the polymer crystallizes (as iPP), the more intensively it will be affected by the "kinetic effect" of pressure (a decrease of crystallinity with increasing solidification pressure). On the other hand, the slower the polymer crystallizes (as PET) the more intensively it will be affected by the "thermodynamic effect" of pressure (an increase of crystallinity when pressure increases). PA6, which has a crystallization behavior intermediate between iPP and PET, is expected to behave in an intermediate way (slight increase of crystallinity upon increasing pressure).

4. Conclusion

From a comparative study of the crystallization behavior of iPP and PA6 under pressure and high cooling rates, the following conclusions can be drawn:

IPP:

– An increase of solidification pressure is seen to determine a parallel decrease of density and alpha phase content, the greatest differences being located among ambient pressure and 8 MPa, *i.e.* at a pressure quite low with respect to the one attained in polymer processing. Microhardness results confirm the trend, with some differences in the high cooling rate region.

– For cooling rates above 20°C/s the effect of pressure could be considered negligible; in the rest of the explored cooling rate range (from 0.01 to 20°C/s) the effect of pressure on the final density is very remarkable.

– The observed decrease of the alpha phase content is mainly balanced by an increase of mesomorphic phase content, while the amorphous phase content remains nearly constant.

PA6:

– An increase of solidification pressure determines a larger density and a larger microhardness. The increase of density and MH is of the same order of magnitude than the decrease observed in iPP.

– A "critical cooling rate" around which the most significant structural changes occur was found to be around 10°C/s at all the investigated pressures.

– The increase of pressure determines an increase of alpha phase crystallinity leaving almost unaffected the gamma crystallinity.

– The increase of pressure is accompanied by a process of perfectioning of the crystalline domains, as witnessed by the increase of long period in the low cooling rate region (*i.e.* characterized by the highest level of crystallinity).

Acknowledgements

This work was supported by: Italian Ministry of Research (MURST) PRIN 1999 and UE BRITE Contract BRPRCT960147.

References

Balta Calleja F. J., Rueda D. R., Garcia Pena, J., Wolf F. P., Karl V. H., "Influence of pressure and crystallization rate on the surface microhardnessof high-density polyethylene", *J. Mater. Sci.*, Vol. 21, No. 4, 1986, pp. 1139–44.

Bhatt S. M., McCarthy S., "Study of pressure, volume and temperature (PVT) relations and high-pressure DTA", *Polym. Prepr. (Am. Chem. Soc., Div. Polym. Chem.)*, Vol. 31, No. 1, 1990, pp. 561–2.

Brucato V., Crippa G., Piccarolo S., Titomanlio G., "Crystallization of polymer melts under fast cooling. I: nucleated polyamide 6", *Polym. Eng. Sci.*, Vol. 31, No. 19, 1991.

Brucato V., Piccarolo S., La Carrubba V., Titomanlio G., "Polymer solidification under pressure and high cooling rates", *International Polymer Processing*, Vol. 15, No. 1, 2000, pp. 103–110.

Brucato V., Piccarolo S., Titomanlio G., "Influence of a nucleating agent on PA6 crystallisation kinetics", *Int. J. of Forming Processes*, Vol. 1, No. 1, 1998, pp. 35–52.

Chiu C., Liu K., Wei J., "A method for measuring PVT relationships of thermoplastics using an injection mo¹ding machine", *Polym. Eng. Sci.*, Vol. 35, No. 19, 1995, pp. 1505–10.

Choi C., White J.L., "Correlation and modelling of the occurrence of different crystalline forms of isotactic polypropylene as a function of cooling rate and uniaxial stress in thin and thick parts", *Polym. Eng. Sci.*, Vol. 40, No. 3, 2000, pp. 645–655.

Dee G., Sauer B., Haley B., "Thermodynamic properties of perfluorinated linear alkanes and poly(tetrafluoroethylenes) measured by PVT and thermal analysis", *Macromolecules*, Vol. 27, No. 21, 1994, pp. 6106–11.

Fleischmann E., Koppelmann J., "Effect of cooling rate and shear-induced crystallization on the pressure-specific volume-temperature diagram of isotactic polypropylene", *J. Appl. Polym. Sci.*, Vol 41, No. 5–6, 1990, pp. 1115–21.

Fu Y., Annis B., Boller A., Jin Y., Wunderlich B., *J. of Polym. Sci., Polym. Phys.*, Vol. 32, 1994, pp. 2289.

He J., Zoller P., "Crystallization of polypropylene, nylon-66 and poly(ethyleneterephthalate) at pressures to 200 MPa: kinetics and characterization of products", *J. Polym. Sci., Part B: Polym. Phys.*, Vol. 32, No. 6, 1994, pp. 1049–67.

La Carrubba V., "Polymer Solidification under pressure and high cooling rate", PhD Thesis, CUES, Salerno, 2001, ISBN : 88-87030-27-8.

La Carrubba V., Brucato V., Piccarolo S., "Isotactic Polypropylene solidification under pressure and high cooling rates. A master curve approach", *Polym. Eng. Sci.*, Vol. 40, No. 11, 2000, pp. 2430–2441.

La Carrubba V., Brucato V., Piccarolo S., "Influence of 'controlled processing conditions' on the solidification of iPP, PET and PA6", *Macrom. Chem. Macrom. Symp*, 2001, accepted.

Lotz B., "Alpha and Beta phases of isotactic polypropylene: a case of growth kinetics phase reentrency in polymer crystallization", *Polymer*, Vol. 39, No. 18, 1998, pp. 4561–4567.

Martorana A., Piccarolo S., Scichilone F., *Macromol Chem. Phys.*, Vol. 198 , 1997, pp. 597.

Mears D.R., Pae K.D., Sauer J.A., "Effect of hydrostatic pressure on the mechanical behaviour of polyethylene and polypropylene", *J. Appl. Phys.*, Vol. 40, 1969, pp. 4229.

Piccarolo S., "Morphological changes in Isotactic Polypropylene as a function of cooling rate", *J. Macromol. Sci. – Phys.*, Vol. B31 No. 4, 1992, pp. 501.

Piccarolo S., Brucato V., Kiflie Z., "Non-isothermal crystallization kinetics of PET", *Polym. Eng. Sci.*, Vol. 40, No. 6, 2000, pp. 1263–1272.

Piccarolo S., Saiu M., Brucato V., Titomanlio G., "Crystallization of polymer melts under fast cooling. II high-purity iPP", *J. Appl. Poly. Sci.*, Vol. 46, 1992, pp. 625.

Rodriguez E. L., Filisko F. E., *J. of Mater. Sci.*, 22, 1987, pp. 1934–1940.

Shlykova T. S., Tchalaya N. M., Abramov V. V., Kuleznev V. N., "Effect of high-pressure injection molding on the properties of high density polyethylene (HDPE)", *J. Polym. Eng.*, Vol. 12, No. 3, 1993, pp. 219–27.

Silano A.A., Pae K.D., Sauer J.A., "Effects of hydrostatic pressure on shear deformation of polymers", *J. Appl. Phys.*, Vol. 48, 1975, pp. 4076.

Strobl G., *The physics of polymers*, 2nd edition, Chapter 4, Springer, Berlin, 1997.

Thermodynamik, Kenndaten fur die Verarbeitung thermoplastischer Kunststoffe, Carl Hanser Verlag, Munchen, Wien, 1979.

Ziabicki A., *Fundamentals of fibre formation*, Wiley, London, 1976

Zihlif A.M., "Mechanical properties of isotropic polypropylene under hydrostatic pressure", *J. Phys.*, Vol. 6, 1975, pp. 97.

Zoller P., Fakhreddine Y., "Pressure-volume-temperature studies of semi-crystalline polymers", *Thermochim. Acta*, Vol. 238, No. 1–2, 1994, pp. 397–415.

Chapter 21

Arbitrary Lagrangian-Eulerian Simulation of Powder Compaction Processes

A. Pérez-Foguet, A. Rodríguez-Ferran and A. Huerta
Dept. Matemàtica Aplicada III, ETS d'Enginyers de Camins, Canals i Ports, Universitat Politècnica de Catalunya, Barcelona, Spain

1. Introduction

A key process in powder forming is cold compaction. It consists in the vertical compaction of a fine powder material through the movement of a set of punches at room temperature. The process transforms the loose powder into a compacted sample with a volume reduction (and therefore a density increase) of about 2–2.5 times. The design of these processes includes the definition of the initial dimensions of the sample and the movements of the punches that lead to compacted samples with uniform density distributions. In this context, efficient and reliable numerical simulations can play an important role as a complement to experimental tests.

Two ingredients are crucial for the numerical simulation of powder compaction processes: the constitutive model and the kinematics formulation of the problem. Several constitutive models have been proposed, including microscopic models, flow formulations and solid mechanics models; see [OLI 96] and [LEW 98] for a general overview and additional references. One of the most common approaches is the use of elastoplastic models based on porous or frictional materials. Here, an elliptic yield function expressed in terms of the relative density and the Kirchhoff stresses is used [OLI 96]. The plastic model is originally formulated within the framework of isotropic finite strain multiplicative hyperelastoplasticity [SIM 92], with some simplifications derived from the assumption of small elastic strains [OLI 96]. In this work, large elastic strains are included in the formulation [PER 01]. As shown in the following, this does not represent any drawback from a modeling point of view. On the contrary, it allows application of numerical techniques and material models developed for the general kinematics framework in a straightforward manner.

Up to date, a common feature of powder compaction simulations with solid mechanics constitutive models is the use of a Lagrangian kinematics formulation. This approach has been shown to be adequate for problems that do not exhibit large mass fluxes among different parts of the sample *(i.e.* homogeneous tests). But in practical problems, such as those that appear in realistic design processes, the Lagrangian approach leads to highly distorted and usually useless meshes [LEW 98, KHO 99]. In order to solve these problems, different *h*-adaptive procedures have been presented recently [KHO 99, CAN 99]. However, *h*-refinement is computationally expensive and information must be interpolated from the old mesh to the new mesh. Moreover, in Lagrangian formulations, the curvature of contact surfaces is a major problem because of the piecewise boundary discretization. A limit situation is found in sharp corners. There, it is extremely difficult to model the mass flux between the two sides of the corner because the elements do not slide smoothly from one side to the other. In previous powder compaction simulations, this problem has been avoided by allowing powder-tool overlapping [OLI 96] (thus violating the boundary conditions), using interface elements [LEW 98] (thus neglecting the possible powder-tool separations), or with remeshing techniques [KHO 99, CAN 99]. In other forming processes, such as rolling or extrusion, fluid-based rbitrary Lagrangian-Eulerian formulations have been widely applied to model piece-tool interaction. ALE formulations were first proposed for fluid problems with moving boundaries [DON 77]. Since then, ALE has been successfully extended to nonlinear solid and structural mechanics [LIU 86, BEN 89, GHO 91, ROD 98, ROD 01]. The ALE formulation for hyperelastoplasticity recently presented in [ROD 01] is used here.

The main features of the constitutive model are presented in Section 2. After that, in Section 3, the ALE formulation is summarized. The proposed approach is applied to three representative examples in Section 4. The present results are compared with experimental data and numerical simulations presented previously in the literature. Section 5 contains some concluding remarks.·

2. Constitutive model

The powder material is modelled with density-dependent finite strain multiplicative hyperelastoplasticity [SIM 92, PER 01]. The elastic behaviour is assumed to obey the standard Hencky's hyperelastic law. The dependence on the density is incorporated in the plastic equations through the relative density, η. An elliptic yield function is considered [OLI 96, PER 01]:

$$f_{ellip}(\mathbf{T},\eta) = 2J_2(\mathbf{T}) + a_1(\eta)(I_1(\mathbf{T})/3)^2 - 2a_2(\eta)(\sigma_y)^2/3 \qquad [1]$$

with τ the Kirchhoff stress tensor, $I_1(\tau)$ the first invariant of τ, $J_2(\tau)$ the second invariant of the deviatoric part of τ, σ_y the yield limit for fully compacted material and the density-dependent parameters

$$a_1(\eta) = \begin{cases} \left(\dfrac{1-\eta^2}{2+\eta^2}\right)^{n_1} & \eta \leq 1 \\ 0 & \eta > 1 \end{cases} \qquad [2]$$

and

$$a_2(\eta) = \begin{cases} \left(\dfrac{0.02\,\eta_0}{1-0.98\,\eta_0}\right)^{n_2} & \eta \leq \eta_0 \\ \left(\dfrac{\eta-0.98\,\eta_0}{1-0.98\,\eta_0}\right)^{n_2} & \eta > \eta_0 \end{cases} \qquad [3]$$

The dependence of a_1 and a_2 on η for the material parameters presented in Table 1 is depicted in Figure 1. The traces of the yield function on the meridian plane $p\tau$-$q\tau$, with $p\tau = I_1(\tau)/3$ and $q\tau = 3\,J_2(\tau)$, for different relative densities are depicted in Figure 2. Note that f_{ellip} becomes the von Mises yield function for $\eta \geq 1$ (fully compacted material). Associate plasticity is considered; therefore the flow direction is given by the partial derivatives of f_{ellip} with respect to τ.

Figure 1. *Dependence of parameters $a_1(\eta)$ and $a_2(\eta)$ on the relative density, η*

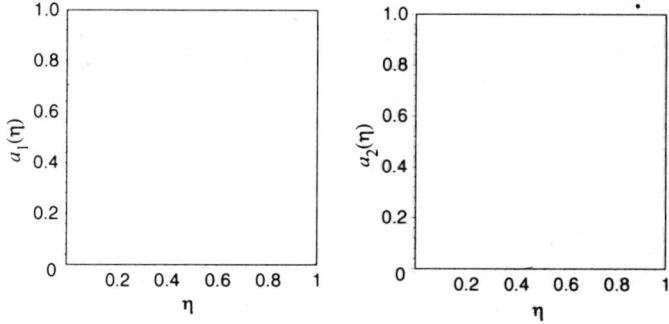

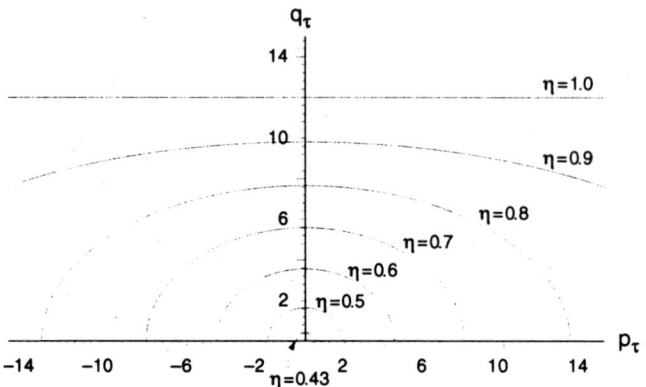

Figure 2. *Traces of the elliptic yield function on the meridian plane* q_τ-p_τ *for different relative densities,* η

Table 1. *Material parameters*

E	50 000 [Mpa]
v	0.37
σ_y	12 [MPa]
η_0	0.41
n_1	0.5
n_2	2.2

3. ALE formulation

The ALE description in nonlinear solid mechanics is nowadays standard for hypoelastic-plastic models [ROD 98]. An extension to hyperelastic-plastic models is used here [ROD 01]. In contrast to previous ALE formulations of hyperelasticity [YAM 93] or hyperelastoplasticity [ARM 00], the deformed configuration at the beginning of the time-step, not the initial undeformed configuration, is chosen as the reference configuration. As a consequence, convecting variables are required in the description of the elastic response. This is not the case in previous formulations, where only the plastic response contains convection terms. In exchange for the extra convective terms, however, the proposed ALE approach has a major advantage: the ALE remeshing strategy must ensure the quality of the mesh only in the spatial domain. In previous formulations, it is also necessary to keep the distortion of the mesh in the material domain under control. Thus, the full potential of the ALE description as an adaptive technique can be exploited here.

The evolution of the elastic left Cauchy-Green tensor, $\mathbf{b}^e$, is defined in the ALE setting as:

$$\left.\frac{\partial \mathbf{b}^e}{\partial t}\right|_\chi + \mathbf{c}\nabla_x \mathbf{b}^e - \mathbf{l}\,\mathbf{b}^e - \mathbf{b}^e\,\mathbf{l}^T = -2\dot{\gamma}\,\frac{\partial f_{\text{ellip}}}{\partial \tau'}\,\mathbf{b}^e \qquad [4]$$

where $|_\chi$ means *holding grid points fixed*, $\mathbf{c}$ is the convective velocity, ∇_x is the gradient operator with respect to spatial coordinates, $\mathbf{l}$ is the velocity gradient tensor, and $\dot{\gamma}$ is the plastic multiplier, which is determined with the classical Kuhn-Tucker conditions.

The evolution of the relative density η is given by the mass conservation principle, which in ALE formulation reads

$$\left.\frac{\partial \eta}{\partial t}\right|_\chi + \mathbf{c}\nabla_x \eta + \eta \nabla_x \cdot \mathbf{v} = 0 \qquad [5]$$

where $\nabla_x \cdot \mathbf{v}$ is the divergence of the particle velocity.

The numerical time-integration of equations [4] and [5] is carried out by means of a fractional-step method [ROD 98]. Every time-step is divided into two phases: the Lagrangian phase and the convection phase. During the Lagrangian phase, convection is neglected and the increment of particle displacements is computed in the usual Lagrangian fashion [PER 01]. After that, an ALE remeshing algorithm is employed to compute the increment of mesh displacements. During the convection phase, the convective term is taken into account. An explicit Godunov-like technique is used for that purpose [ROD 98].

4. Numerical simulations: compaction of a rotational flanged component

In the following, the proposed approach is applied to the simulation of the compaction of a rotational flanged component [LEW 98, KHO 99]. The geometry of the sample is depicted in Figure 3. This test is a challenging example for the proposed approach because it involves sharp geometry boundaries and large mass fluxes between different parts of the sample. It has been used to illustrate the applicability of a dynamic approach with a cone-cap hypoelastic-plastic model [LEW 98] and to show the utility of an adaptive remeshing technique to reduce mesh distortion [KHO 99]. Some experimental results are available [LEW 94].

Three different compaction tests are simulated: 1) a vertical movement of the top punch (6.06 mm); 2) a vertical movement of the bottom punch (5.10 mm); and 3) a simultaneous movement of both punches (6.06 mm for the top punch and 7.70 mm for the bottom punch). Eight-noded elements with reduced integration (four Gauss points per element) are used in the three tests. The die wall friction is simulated with a Coulomb friction coefficient of 0.08 and the radial displacement at the punches is restrained [LEW 98]. The analysis is performed with the material parameters calibrated in [CAN 98] for the compaction of a plain bush component (see Table 1).

In the ALE simulations of the three tests, the ALE remeshing technique consists in prescribing equal height elements in the upper and lower part of the domain, and a constant width for each element. Therefore, the mesh is always in contact with the boundary and the mass flux around the corner is allowed. The improvement of this approach over the Lagrangian one is analysed with the top compaction test and the bottom compaction test, those that involve larger mass fluxes.

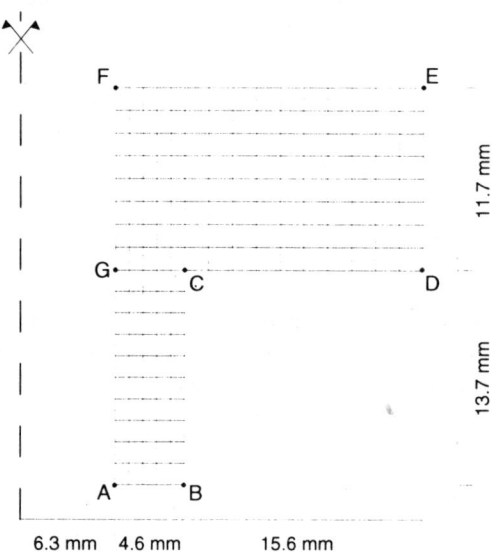

Figure 3. *Flanged component. Problem definition and computational mesh [LEW 98]*

4.1. *Top punch compaction test*

ALE formulation

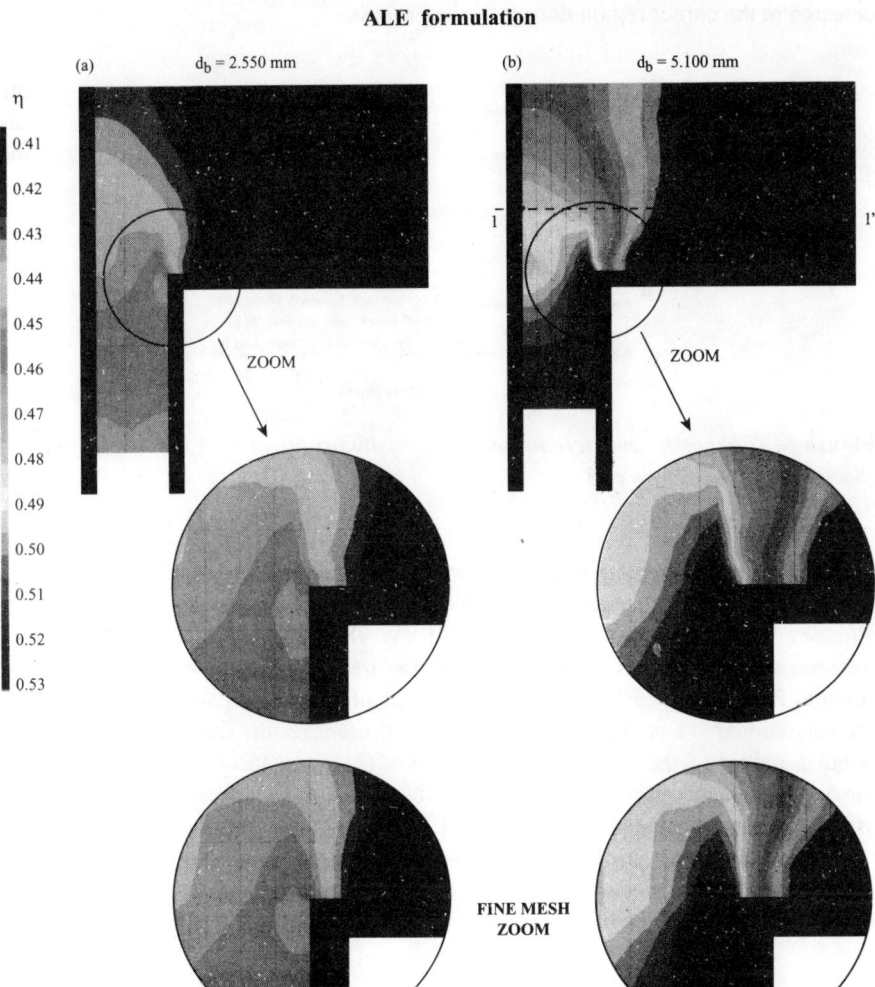

Figure 4. *Top punch compaction. Relative density distribution at the middle and the end of the test (Lagrangian and ALE solution)*

The evolution of the relative density fields computed with Lagrangian and ALE formulations are summarized in Figure 4. The two results are in general agreement. However, in the Lagrangian simulation, the elements close to the corner, point C, are so distorted that the results are not quantitatively reliable. The compaction process leads to a clearly non-homogenous density distribution. As expected, higher values are found in the outer region of the sample and lower ones close to the bottom punch. A smooth transition from higher to lower densities is found. A dense zone is detected in the corner region during all the process.

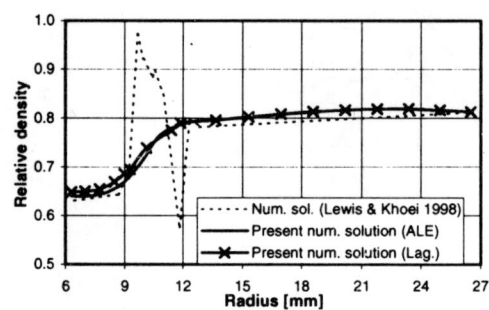

Figure 5. *Top punch compaction. Relative density profile at 1.88 mm above line GD (Section 1–1', see Figure 4)*

Final relative density profiles are depicted in Figure 5, together with the results presented in [LEW 98]. Two zones with a quasi-uniform relative density are found in all cases. The density profile obtained in [LEW 98] exhibits a significant oscillation between these two zones. On the contrary, the transitions obtained in this work are smooth. Remarkably, in this profile, the results of Lagrangian and ALE simulations are very similar. Thus, the difference between present results and those of [LEW 98] is not a matter of the kinematical formulation (Lagrangian or ALE). On the other hand, in compaction problems, the use of elliptic or cone-cap plastic models does not affect significantly the results [PER 01]. Therefore, the difference between both results is probably caused by the use of different time integration schemes (quasistatic problem solved with a mix explicit-implicit scheme in this work versus dynamic problem solved with a generalized Newmark scheme in [LEW 98]).

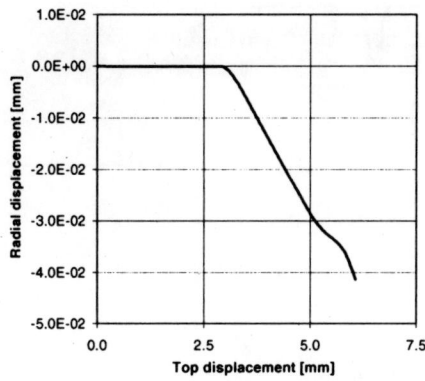

Figure 6. *Top punch compaction. Radial displacement of point C (Lagrangian solution)*

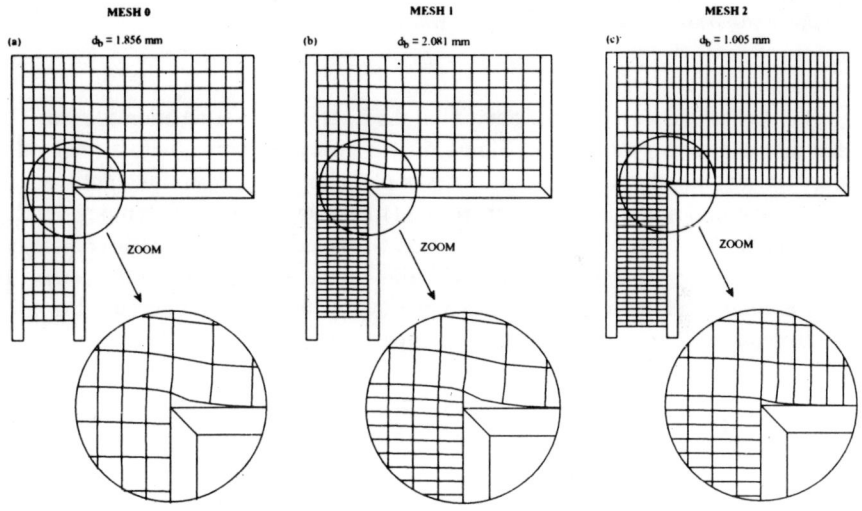

Figure 7. *Bottom punch compaction. Last converged configuration with three different meshes (Lagrangian solution)*

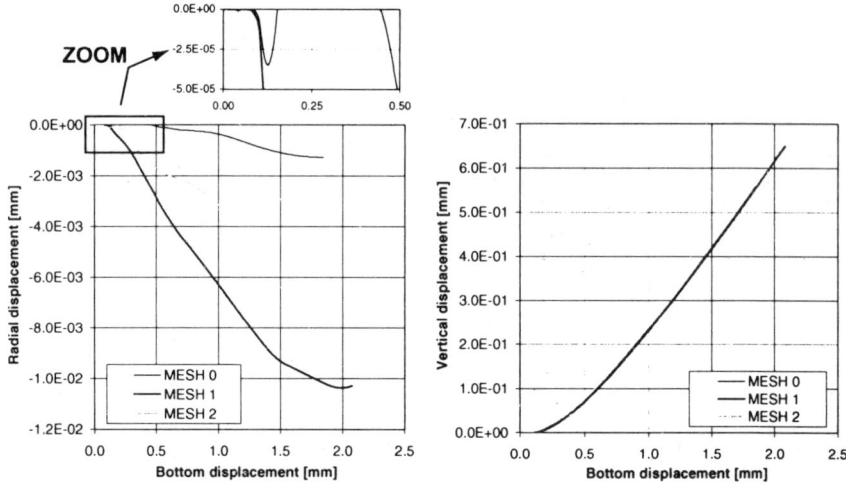

Figure 8. *Bottom punch compaction. Radial and vertical displacements of point C with three different meshes (Lagrangian solution)*

The vertical displacement of point C in the Lagrangian formulation is zero. The radial displacement is depicted in Figure 6. It is always negative (with a final value about 1% of the thinner part of the sample and not perceptible until a top displacement of 3 mm). Therefore, there is a (small) separation between powder and the die on the left of point C. This separation is caused by the spatial discretization, which, in a Lagrangian approach, does not allow the mass flux around the corner. In the ALE simulation, the remeshing technique avoids this situation because the point C of the mesh remains fixed but the material particles located there do not.

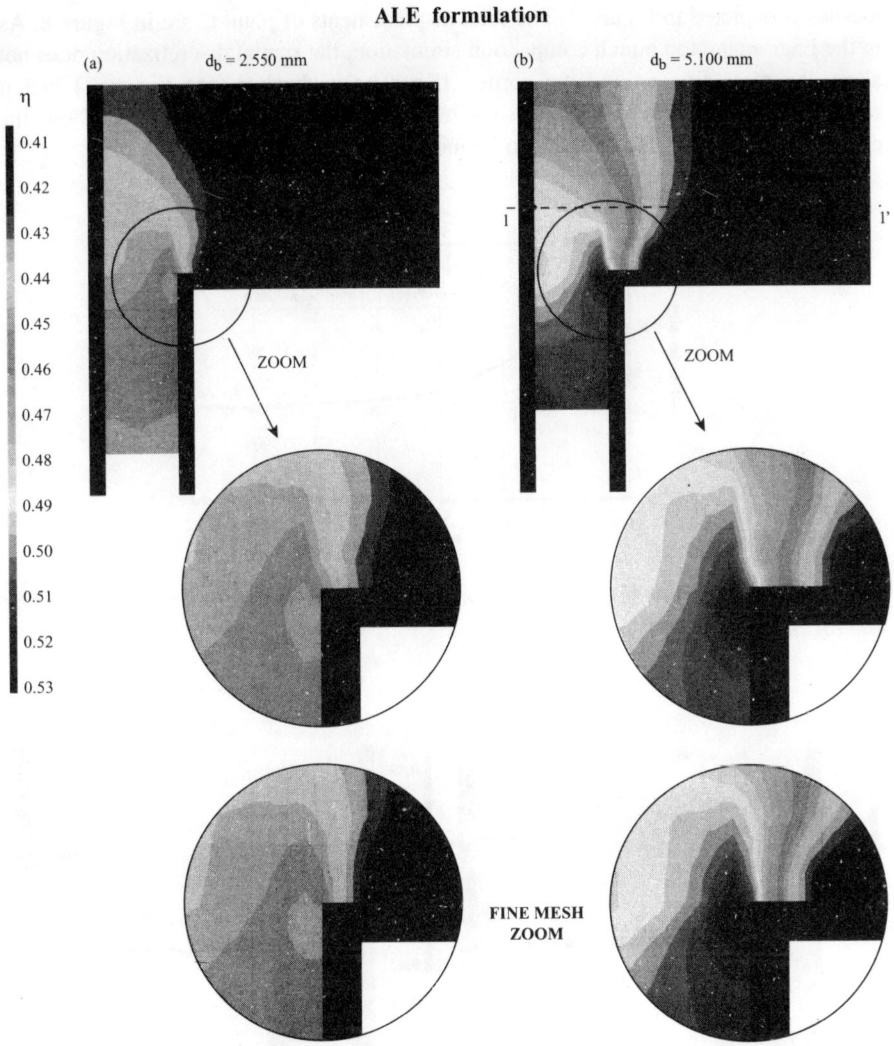

ALE formulation

Figure 9. *Bottom punch compaction. Relative density fields at the middle and the end of the test (ALE solution)*

4.2. *Bottom punch compaction test*

The bottom punch compaction test cannot be simulated entirely with the Lagrangian approach. At a certain bottom punch displacement the iterative scheme, which includes the consistent tangent moduli for density-dependent plastic models [PER 01], diverges, independently of the size of the time-step. This limit value depends on the element size. The final configuration obtained with three different meshes is depicted in Figure 7. The total displacements of point C are in Figure 8. As in the Lagrangian top punch compaction simulation, the spatial discretization does not allow the mass flux around the corner. It has been checked (see Figure 7) that it cannot be solved with a finer mesh. The deformation pattern does not allow the elements of boundary BC to move to boundary CD.

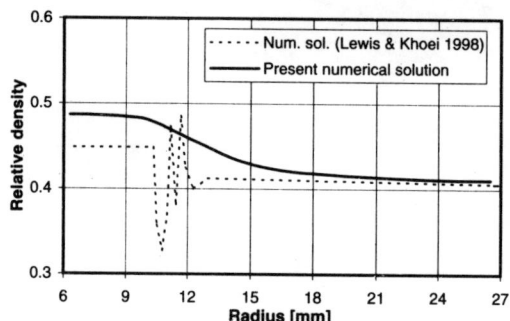

Figure 10. *Bottom punch compaction. Relative density profile at 3.9 mm above line GD (Section 1–1', see Figure 9)*

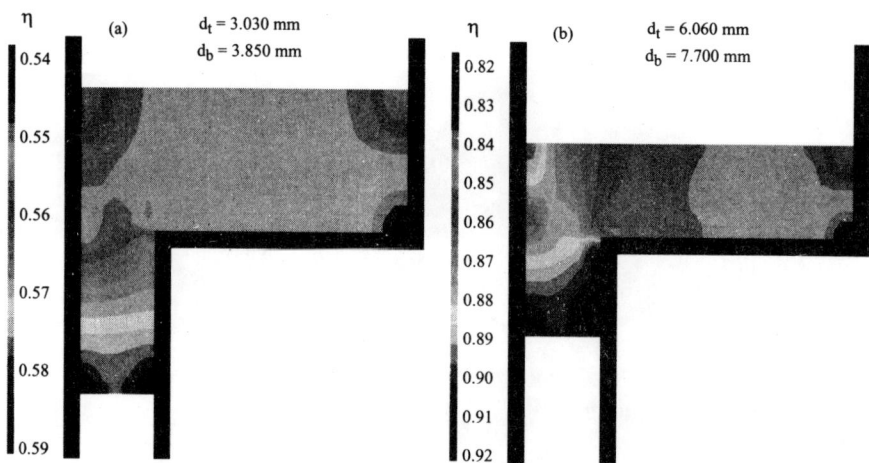

Figure 11. *Double punch compaction. Relative density fields at the middle and the end of the test*

With the ALE approach it has been possible to simulate the entire test. The evolution of the relative density distribution is summarized in Figure 9. The expected values are obtained: higher densities in the lower part of the sample and lower densities in the outer part. A dense zone is detected in the corner region during all the process. This result agrees with the simulation of the top punch compaction (note that in this case the mass flux has the opposite direction). However, contrary to the top punch compaction, the dense zone extends in the mass flux direction. This can be related with the higher convective behaviour of the problem and the different mass flow pattern.

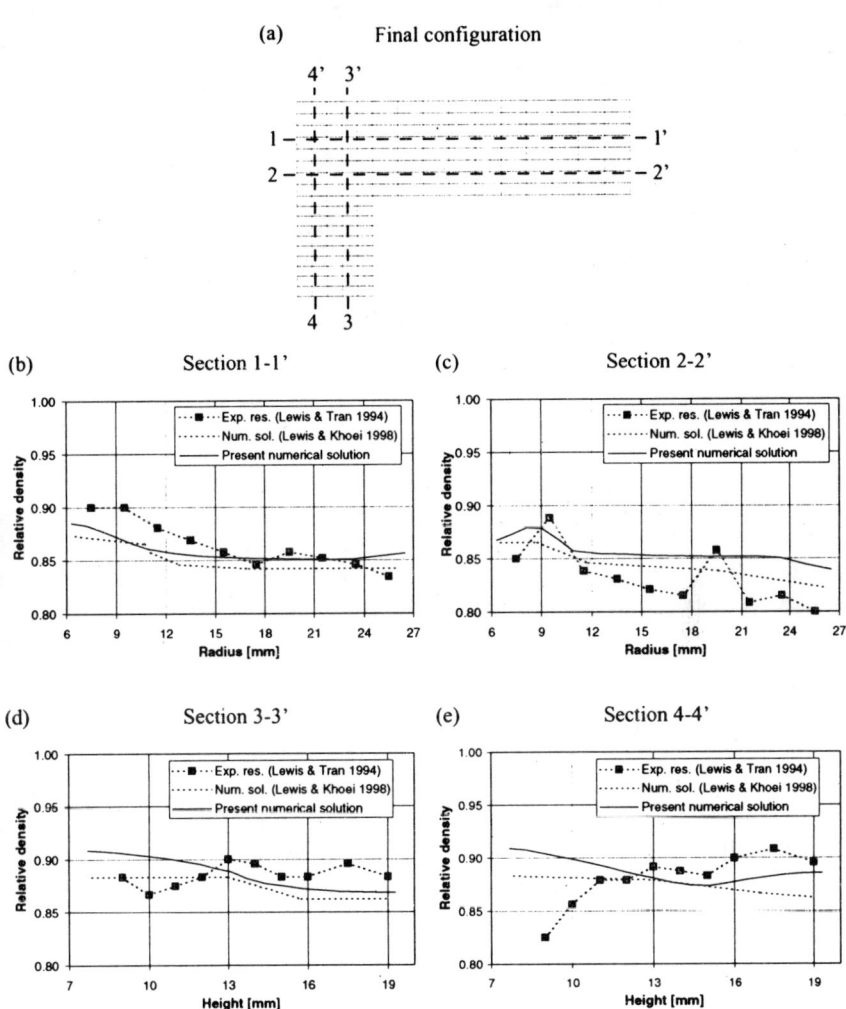

Figure 12. *Double punch compaction. Relative density profiles at four locations*

A final relative density profile is depicted in Figure 10. The numerical results presented in [LEW 98] are included in the same figure. Both simulations lead to similar values in outer and inner parts of the sample. The main difference is the presence of oscillations in the results of [LEW 98]. As in the top compaction test, this difference is probably due to the use of different time integration schemes.

4.3. Double punch compaction test

In the last test, a simultaneous top and bottom punch movement performs the compaction of the flanged component. The results obtained with the ALE approach are presented. The evolution of the relative density distribution is summarized in Figure 11. The compaction process leads to a quasi-homogeneous density distribution during the entire test, with differences less than 10%. Higher values are found in the bottom of the sample and lower values in the external part. Four relative density profiles are depicted in Figure 12. The numerical results presented in [LEW 98] and the experimental data available [LEW 94] are included in the same figure. Note that all the results are in good agreement.

4.4. Mass balance

The variations of the mass of the sample during the ALE simulations of the three tests are depicted in Figure 13. These variations correspond to the truncation error in the discretization of the convective term of the ALE formulation (both temporal and spatial discretization). Note that higher variations are found as larger mass fluxes are simulated. However, in all cases the error in the global mass balance is less than 1%.

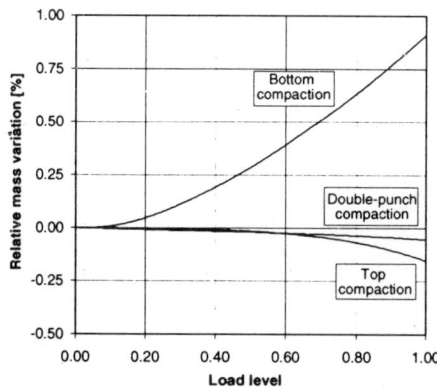

Figure 13. *Relative mass variation during the three compaction processes of the fanged component. The load levels are referred to the punch displacements imposed at the end of each test*

5. Conclusion

An ALE formulation for finite strain density-dependent hyperelastoplasticity has been applied to the simulation of the compaction of a rotational flanged component. It has shown to be very useful to model the mass flux around sharp corners. The uncoupling between mesh and particle displacements allows avoiding the excessive distortion of Lagrangian meshes and simplifies the control of contact between tools and powder sample.

The use of simple ALE remeshing schemes and the explicit character of the Godunov-like technique has guaranteed a low computational overhead of the ALE approach with respect to the standard Lagrangian formulation.

The efficiency of the proposed approach is complemented with the high accuracy of the results. The mass conservation principle has been verified with very low relative errors in all examples. This represents a significant improvement with respect to previous results based on non-adaptive and h-adaptive Lagrangian approaches found in the literature.

The present results are in general agreement with the experimental data available and the previous results presented in the literature. Moreover, the relative density distributions do not present spurious oscillations, even in compaction tests involving high convective effects.

Acknowledgements

The partial financial support of the Ministerio de Ciencia y Tecnología (grant numbers TAP98-0421 and 2FD97-1206) is gratefully acknowledged.

References

[ARM 00] ARMERO F. and LOVE E., "An Arbitrary Lagrangian-Eulerian (ALE) finite element method for finite strain elasto-plasticity", *Technical Report UCB/SEMM-2000/04*, U.C. Berkeley, 2000.

[BEN 89] BENSON D. J., "An efficient, accurate, simple ALE method for nonlinear finite element programs", *Comp. Meth. Appl. Mech. Engrg*, 72, 305–350, 1986.

[CAN 98]. CANTE J. C, OLIVER J. and OLLER S., "Simulación numérica de procesos de compactación de pulvimateriales. Parte 2: Validación y Aplicaciones Industriales", *Rev. Int. de Métodos Numéricos para Cálculo y Diseño en Ingeniería*, 14, 101–116, 1998.

[CAN 99] CANTE J. C., OLIVER J. and HERNÁNDEZ R., "Simulación numérica de las etapas de transferencia y prensado del proceso de compactación de pulvimateriales", in *IV Congreso de Métodos Numericos en Ingeniería*, Eds. R. Abascal, J. Domínguez and G. Bugeda, SEMNI, Spain, 1999.

[DON 77] DONÉA J., FASOLI-STELLA P. and GIULIANI S., "Lagrangian and Eulerian finite element techniques for transient fluid structure interaction problems", in *4th Int. Conf. on Struc. Mech. in Reactor Tech.*, paper B1/2, San Francisco, 1977.

[GHO 91] GHOSH S. and KIKUCHI N., "An Arbitrary Lagrangian-Eulerian finite element method for large deformation analysis of elastic-viscoplastic solids", *Comp. Meth. Appl. Mech. Engrg*, 86, 127–188, 1991.

[KHO 99] KHOEI A. R. and LEWIS R. W. "Adaptive finite element remeshing in a large deformation analysis of metal powder forming", *Int. J. Num. Meth. Engrg*, 45, 801–820, 1999.

[LEW 94] LEWIS R. W. and TRAN D. V., "Finite element approach to problems in particulate media with special reference to powder metal forming", *Bull. Tech. Univ. Istambul*, 47, 295–310, 1994.

[LEW 98] LEWIS R. W. and KHOEI A. R., "Numerical modelling of large deformation in metal powder forming", *Comp. Meth. Appl. Mech. Engrg*, 159, 291–328, 1998.

[LIU 86] LIU W. K., BELYTSCHKO T. and Chang H., "An Arbitrary Lagrangian-Eulerian finite element method for path-dependent materials", *Comp. Meth. Appl. Mech. Engrg*, 58, 227–245, 1986.

[OLI 96] OLIVER J., OLLER S. and CANTE J. C., "A plasticity model for simulation of industrial powder compaction processes", *Int. J. Sol. & Struc.*, 33, 3161–3178, 1996.

[PER 01] PÉREZ-FOGUET A., RODRÍGUEZ-FERRAN A. and. HUERTA A, "Consistent tangent matrices for density-dependent finite plasticity models", *Int. J. Num. Anal. Meth. Geomech.*, 25, 1045–1075, 2001.

[ROD 98] RODRÍGUEZ-FERRAN A., CASADEI F. and HUERTA A., "ALE stress update for transient and quasistatic processes", *Int. J. Num. Meth. Engrg.*, 43, 241–262, 1998.

[ROD 01] RODRÍGUEZ-FERRAN A., PÉREZ-FOGUET A. and HUERTA A., "Arbitrary Lagrangian-Eulerian (ALE) formulation for hyperelastoplasticity", *Int. J. Num. Meth. Engrg.*, in press, 2001.

[SIM 92] SIMO J. C., "Algorithms for static and dynamic multiplicative plasticity that preserve the classical return mapping schemes of the infinitesimal theory", *Comp. Meth. Appl. Mech. Engrg.*, 99, 61–112, 1992.

[YAM 93] YAMADA T. and KIKUCHI F., "An Arbitrary Lagrangian-Eulerian finite element method for incompressible hyperelasticity", *Comp. Meth. Appl. Mech. Engrg.*, 102, 149–177, 1993.

Chapter 22

Austenite-to-Ferrite Phase Transformation During Continuous Casting of Steels

Ernst Gamsjager and Franz Dieter Fischer
Christian Doppler Laboratory – Functionally Oriented Material Design, Institut für Mechanik, Montanuniversität Leoben, Austria

Christian M. Chimani
VOEST-Alpine Industrieanlagenbau GmbH, Linz, Austria

Jiri Svoboda
Institute of Physics of Materials, Academy of Sciences of the Czech Republic, Brno, Czech Republic

1. Introduction

During continuous casting of carbon steel the as-cast material is transported from the mold through the strand support rolls to the end of the casting machine. In consequence of a temperature reduced ductility, surface cracking may occur. From the point or region of solidification to a temperature of about 1500K (depending on the steel quality) hot ductility increases during cooling. For several steel grades a minimum in hot ductility has been observed at temperatures between 1400K and 1000K, which is generally associated with the precipitation of non-metallic inclusions and the γ-α phase transformation, see *e.g.* [SUZ 84], [MAE 90], [LEW 98], [NAG 99] and [CHI 99]. The properties of the strand depend not only on its chemical composition. Physical quantities and parameters like the austenite grain size before transformation, the cooling regime and global as well as local strains and strain rates determine the quality of the steel slab. The quantities mentioned above are interrelated with the kinetics of the solidification as well as the γ-α phase transformation.

The Fe-C diagram shows that between 996K and 1184K a binary low-carbon alloy can exist in two phases, the γ-phase characterized by an fcc-crystal structure and the α-phase, where the iron atoms are arranged in a bcc-structure. Proeutectoid ferrite nucleates preferably at the austenite grain boundaries. Metallographical investigations show that the ferrite tends to grow to a thin film, which is able to surround the austenite grains completely at a ferrite volume fraction of about 5 to 10% [LEW 98]. The bright zones in Figure 1 indicate the formation of ferrite allotriomorphs along the former austenite grain boundaries. The soft ferrite film appears as a network and can be distinguished from the comparably hard austenite prisms. Additionally, an inter-crystalline crack has been formed.

Figure 1. *Grain structure parallel to the strand surface*

The stress/strain state has been investigated in two different austenite/ferrite composites *via* two-dimensional micromechanical models. The mechanical loading condition the strand has to sustain during unbending has been simulated by a uniaxially applied tension test. These models provide accurate results, if the growth of the ferrite film is negligible during the loading period. Since the diffusional phase transformation is a time-consuming process, the condition mentioned above is a realistic approach in some cases. However, it is of great interest to investigate the transformation kinetics as a function of the grain size, temperature, the mobility of the interface and the diffusion coefficient. A numerical routine has been developed to simulate the transformation kinetics. This procedure has been incorporated in a finite element analysis in order to study the influence of a certain loading condition on the transformation kinetics.

2. Austenite/ferrite phase arrangements

2.1. *Unit cell models*

Periodic unit cell type models are used to describe the two-phase microstructure, see Figure 2a and 2b. The austenite grains in the models are assumed to be infinitely long, equally sized prisms with a hexagonally shaped cross-sectional area. Therefore, generalized plain strain elements are used in both finite element models. The constraints are indicated by symbols in both models.

Symmetrical boundary conditions

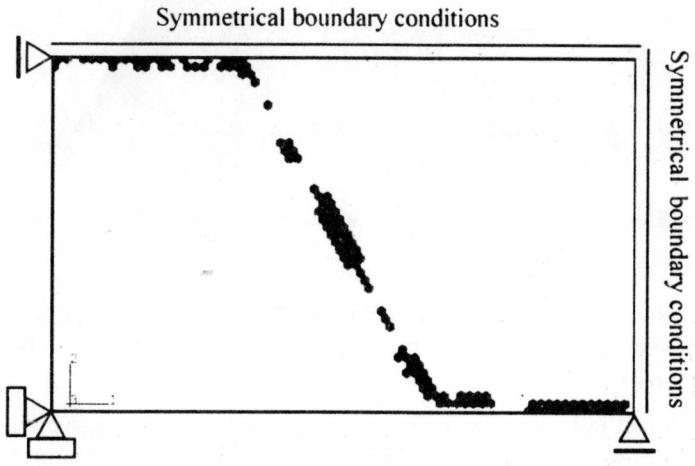

Figure 2a. *Model 1: Phase arrangement with a volume fraction of 3.5% ferrite*

The unit cell model 1 (Figure 2a) consists of two quarters of a hexagon that are put together into a rectangle. Boundary conditions are applied at the edges of the rectangle in order to take the symmetries of these lines into account. The grain boundary is situated at the edges of the hexagons. The dark areas inside the rectangle indicate the ferrite (volume fraction of 3.5%) having been formed. Ferrite plays the role of an inclusion in an austenite matrix. The phase pattern has been created with the finite element preprocessor HEXGRAIN [BOE 97] allowing creation of micromechanical unit cells for two-phase composite materials. The mesh is built up by triangular, second order generalized plain strain elements. 6 triangular elements form a hexagon. The microstructure is modelled by assigning ferrite or austenite material properties to each hexagon.

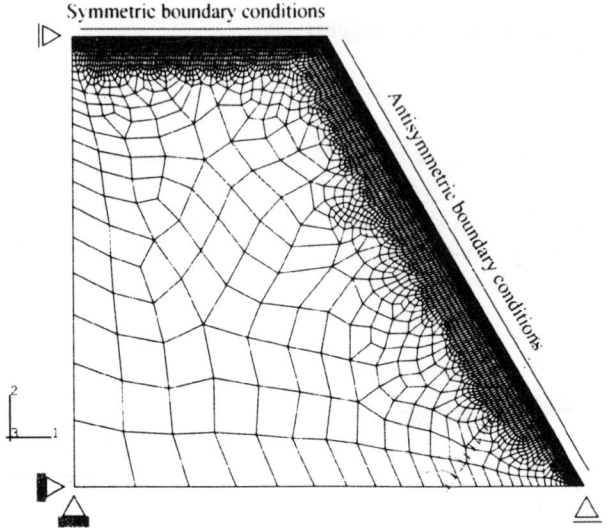

Figure 2b. *Model 2: Phase arrangement with a volume fraction of 5% ferrite*

As already mentioned above, a volume fraction of 5% can lead to a thin, but already continuous soft ferrite film along the grain boundaries. This case is considered in the second unit cell model (Figure 2b).

The unit cell in model 2 is the smallest valid tiling of a hexagonally tessellated space and consists of a quarter of a hexagon. Proper boundary conditions are applied to this model, symmetrical boundary conditions at the top and the bottom of the hexagons, antisymmetrical ones at the other edges of the hexagons. The finite element mesh of model 2 has been generated by the pre-processor PATRAN [PAT] and consists of 6-node elements, where 2 extra nodes simulate generalized plane strain conditions.

Since the ferrite regions have grown together to a thin, but continuous film, austenite inclusions are embedded in a ferrite matrix. The ferrite film appears darker in Figure 2b. In order to avoid artificial strain concentrations, the interface between austenite and ferrite has been smoothed. This means that the sharp corners are approximated by arcs, which indeed is a better model of reality. It is remarkable that the value of the strain concentration located at the "corner" is nearly independent of the radius of the applied smoothing curve (Figure 3b).

2.2. Material model

Generally speaking, steel shows a non-linear visco-elasto-plastic behaviour in the temperature region mentioned above. The material behaviour of the two phases is formulated in creep laws [HAR 96]. A creep routine has been implemented in ABAQUS [ABA] and has been used to calculate the equivalent creep strains.

2.3. *Equivalent creep strain*

An isothermal uniaxial tensile test in horizontal (1)-direction was applied to model 1 (Figure 2a). The maximum global strain in the strain controlled simulation was set to 1%.

In Figure 3a the contour plot for the equivalent creep strain is presented. The global strain rate has a constant value of 10^{-4} s^{-1}. Since the strain concentration in the ferrite drives the crack formation, only a very low fraction of the ferrite phase needs to be considered.

Figures 2a and 3a demonstrate that the phase arrangement in connection with the applied strain plays an important role. A high level of straining in the ferrite phase occurs at the grain boundary in a zone which is inclined with respect to the direction of the tension test. Ferrite zones at this grain boundary that are going to coalesce suffer the highest creep strains. Furthermore, it can be seen that the equivalent creep strain is relatively high at the grain vertices compared to the neighbouring regions. The ferrite regions in the lower corner of the unit cell and half way up the unit cell carry comparably high strain values. In these regions ferrite tends to grow to a thin film in an unfavourable orientation with respect to the externally applied load. The local strain can be twenty times higher than the global strain.

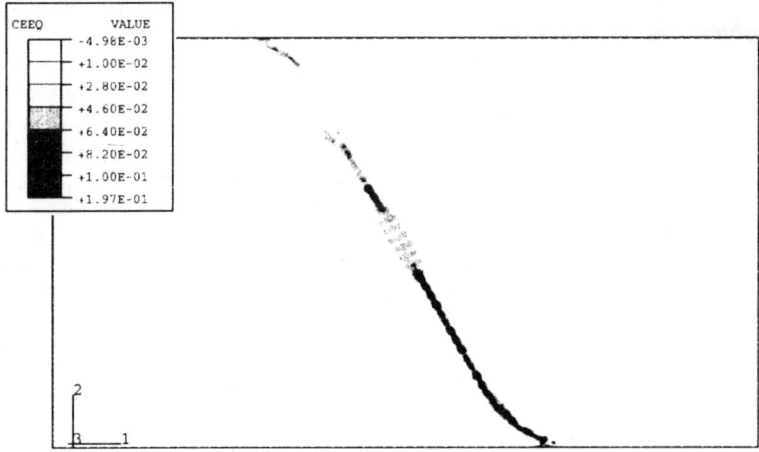

Figure 3a. *Contour plot of the equivalent creep strain in model 1*

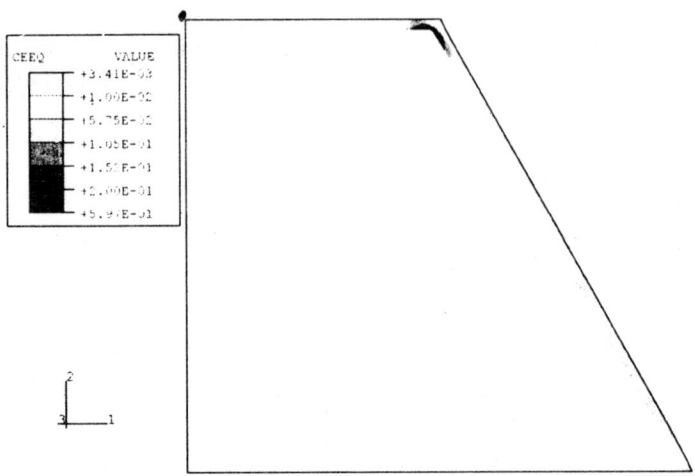

Figure 3b. *Distribution of the equivalent creep strain in model 2 (continuous ferrite film)*

In Figure 3b a contour plot shows the distribution of the equivalent creep strain in model 2. Again a uniaxial tension test in direction 1 has been simulated, and the global strain rate has been set to 10^{-4} s^{-1}. The creep strain in the ferrite zone, on the right edge of the unit cell, is about 8 to 10 times larger than the global strain. The maximum equivalent creep strain is observed in the region near the "corner" and is up to sixty times higher than the global strain.

2.4. Influence of the strain rate

Experimental investigations have shown that increasing the strain rate improves the hot ductility in the transformation temperature range (750–850°C) at a certain volume fraction of ferrite [LEW 98]. This holds true over the entire volume range. Creep simulations at various creep rates have been carried out. The global creep rates were set to $(10^{-1}, 10^{-2}, 10^{-3}$ and $10^{-4})$s^{-1}.

The equivalent creep strains $\varepsilon_{eq,\ elem}$ have been examined in a ferrite element with a comparably high strain concentration observed in the previous test (Figure 3b).

The strain ratio $\varepsilon_{eq,\ elem}/\varepsilon_{eq,\ g}$ was plotted versus the equivalent global strain $\varepsilon_{eq,\ g}$ (see Figure 4) at the various creep rates. $\varepsilon_{eq,\ g}$ includes the elastic strain, while $\varepsilon_{eq,\ elem}$ includes the creep strain only. For low $\varepsilon_{eq,\ g}$ both phases deform elastically, therefore the ratio is zero.

At low global strains the strain ratio $\varepsilon_{eq,\ elem}/\varepsilon_{eq,\ g}$ increases rapidly to an almost stationary value. At higher strain rates the stationary value is reached at higher strains. The strain ratio decreases with an increasing strain rate. As far as local strain concentrations can be associated with the hot ductility behaviour these results are in good agreement with experimental investigations found in [LEW 98]. It is stated that high strain rates improve the hot ductility behaviour.

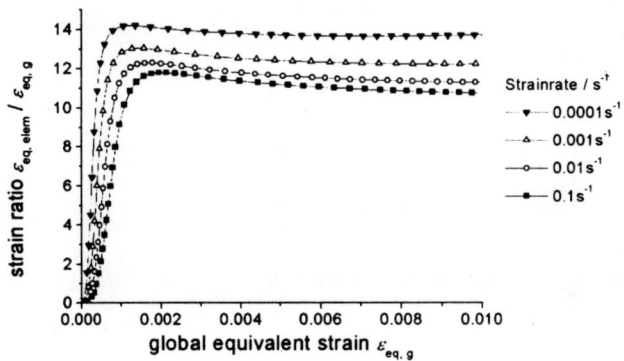

Figure 4. *Strain ratio as a function of the global equivalent strain at various strain rates*

2.5. Damage indicator

In [MAE 90] one possible mechanism of crack initiation and growth in low-alloy steels is described schematically. Due to precipitation of non-metallic inclusions at the grain boundaries, micro-voids are formed under a certain deformation state and eventually lead to the decohesion of precipitate/matrix interfaces.

The growth of pores in a ductile metallic matrix is not only a function of plastic or creep deformation but is also strongly correlated to the hydrostatic part of the stress state. To account for this behaviour a damage indicator [FIS 95] has been implemented in the FE-model. This damage indicator can be used to determine critical regions of crack initiation (Figure 5).

The zones where the indicator approaches the value 1 are critical with respect to damage.

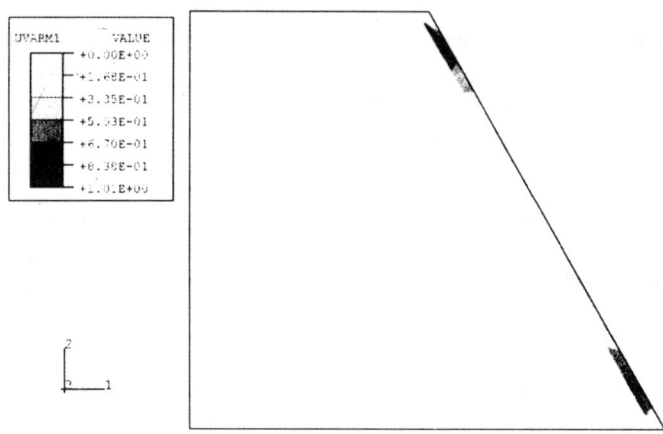

Figure 5. *Damage indicator in the ferrite film*

3. Transformation kinetics

The previous models are based on the assumption that the transformation takes a long time in comparison to the loading time, and the crucial cases with respect to the austenite/ferrite arrangement have been investigated. This section treats the γ/α transformation kinetics and provides the basis for implementing the transformation process into a unit cell model. In order to optimize the conditions during continuous casting, it is essential to predict the time period in which the volume ratio has a critical value.

When a ferrite film already exists, the γ/α–phase transformation can be described by two processes under the assumption of a sharp interface (the thickness of the interface approaches zero).

– Firstly, the ferrite layer grows and the interface migrates simultaneously, while carbon accumulates at the austenite side of the interface. Therefore migration of the interface means a gradual rearrangement of the lattice of the parent phase γ into the lattice of the product phase α.

– Secondly, carbon diffuses from the interface into the austenite grain. A similar behaviour will be observed whenever the solute is less soluble in the newly formed phase α and its initial mole fraction in austenite is higher than the solubility in ferrite.

3.1. *Mathematical description of the diffusive γ/α phase transformation*

The mathematical description of the diffusional phase transformation requires three steps [HIL 99], [SVO 01]. The mass balance is required at the interface. The

C-diffusion from the interface into the γ-bulk obeys the diffusion equation. Therefore, the following equations have to be solved simultaneously:

– Mass balance at the interface

$$J_V = (V_m)^{-1} \cdot X_{int} \cdot v,$$ [1]

$$J_{diff} = -D \cdot (V_m)^{-1} \cdot \frac{\partial X_{int}}{\partial x},$$ [2]

$$J_{diff} = J_v,$$ [3]

where J_v is the flux of the C-atoms due to the movement of the interface, J_{diff} denotes the flux of C-atoms due to diffusion, v is the interface velocity, X_{int} the C-mole fraction at the interface, V_m the molar volume and D denotes the diffusion coefficient. V_m equals approximately $7.194 \cdot 10^{-6}$ m^3 mol^{-1} for steel in the temperature region investigated.

– Kinetics

The transformation kinetics usually applied in these cases is a linear relation between the driving force and the interface velocity:

$$v(t) = (V_m)^{-1} \cdot M \cdot \Delta F.$$ [4]

The interface velocity is a linear function of the driving force ΔF. The mobility M is the linear multiplier.

– Diffusion equation

$$\frac{\partial X}{\partial t} = D \frac{\partial X^2}{\partial x^2};$$ [5]

X is the mole fraction of carbon, t denotes the time.
A transformation routine has been developed to solve these equations simultaneously.

3.1. Driving force

The driving force ΔF is defined as the energy that is spent on the motion of the interface. A certain amount of undercooling is given. Then, ΔF can be either positive (γ–α transformation) or negative (reaustenitization) or becomes zero after equilibration. ΔF is composed of two contributions:

– the decrease of Gibbs energy available for interface migration or the chemical driving force ΔF_{chem},

– and the decrease of the mechanical energy ΔF_{mech} in the system provided for the migration,

$$\Delta F = \Delta F_{chem} + \Delta F_{mech}.$$ [6]

3.1.1. Chemical driving force

The following terms are valid for the binary Fe-C system. The partial molar Gibbs energy (chemical potential) of iron μ_{Fe} in a phase and the partial molar Gibbs energy (chemical potential) of carbon μ_C in the same phase are given by:

$$\mu_{Fe} = g - X \cdot \frac{\partial g}{\partial X}$$ [7]

and

$$\mu_C = g + (1 - X) \cdot \frac{\partial g}{\partial X}.$$ [8]

g denotes the molar Gibbs energy of the phase considered.

The total chemical energy $\Delta F_{chem}^{\gamma \to \alpha}$ that is available due to $\gamma \to \alpha$ transformation per mole of newly produced α-phase can be derived from the difference of the chemical potentials related to the corresponding component in the following way:

$$\Delta F_{chem}^{\gamma \to \alpha} = (1 - X^{\alpha}) \cdot (\mu_{Fe}^{\gamma} - \mu_{Fe}^{\alpha}) + X^{\alpha} \cdot (\mu_C^{\gamma} - \mu_C^{\alpha}).$$ [9]

Hillert [HIL 99] argues that due to the diffusion of solute atoms (C) from the new phase α into the parent phase γ, a certain amount of Gibbs energy ΔF_{chem}^t is dissipated,

$$\Delta F_{chem} = \Delta F_{chem}^{\gamma \to \alpha} - \Delta F_{chem}^t.$$ [10]

In [SVO 01] it is shown that the "transinterphase"-diffusion vanishes ($\Delta F_{chem}^t = 0$) under the approximation of a sharp interface due to the continuity of the chemical potential across the interface,

$$\mu_C^{\gamma} = \mu_C^{\alpha}.$$ [11]

The Gibbs energy that is provided by the phase transformation is spent on the migration of the interface.

The second term in [9] vanishes, when condition [11] is applicable. Since the solubility of C in α is very low, X_α can be neglected and ΔF_{chem} follows as:

$$\Delta F_{chem} = \mu_{Fe}^{\gamma} - \mu_{Fe}^{\alpha}$$ [12]

The Gibbs energies g^γ and g^α in a given phase (γ or α) can be calculated as a function of the component (e.g. C) with the "Gibbs energy-minimizer" ChemSage

[ERI 90]. Combining [7] and [11] the chemical driving force ΔF_{chem} can be expressed by:

$$\Delta F_{\text{chem}} = g^{\gamma} - g^{\alpha} - X^{\gamma} \cdot \frac{\partial g^{\gamma}}{\partial X} \qquad [13]$$

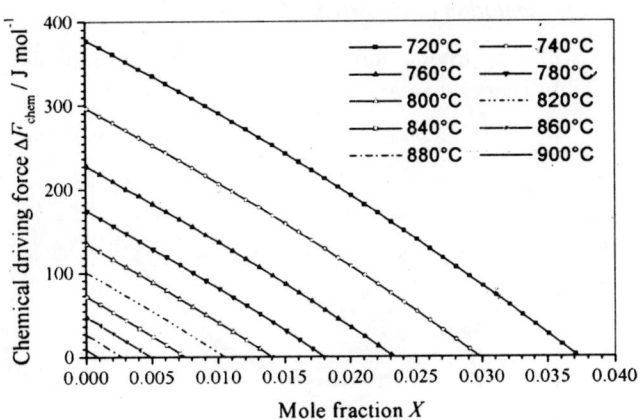

Figure 6. *Chemical driving force depending on the mole fraction X at various temperatures*

ΔF_{chem} is presented in Figure 6 as a function of the mole fraction X at various temperatures.

3.1.2. Mechanical driving force

The mechanical driving force ΔF_{mech} can be found from the energy dissipation due to the growth of a transformed microregion. The derivation is equivalent to that for a microregion developed by a martensitic transformation, for details see Fischer *et al.* FIS 98]. Due to the amorphous microstructure of the interface layer sliding may be possible along the interface and has to be taken into account by an additional term. However, if one assumes rather small tangential traction components the sliding term disappears. Here we refer to a forthcoming paper by Fischer et al. [FIS 01]. Finally the following relation can be written:

$$\Delta F_{\text{mech}} = V_{\text{M}} (<p> \cdot \delta - \Delta \widetilde{W}_{\text{el}}^{\tau} - \Delta \widetilde{W}_{\text{pl}}^{\tau}) \qquad [14]$$

$<p>$ is the average of the mean stresses in both phases near the interface and δ denotes the volume strain due to transformation. $\Delta \widetilde{W}_{\text{el}}^{\tau}$ is the additional strain energy per transformed volume, which is stored in the whole specimen and is not available for transformation. $\Delta \tilde{W}_{pl}^{\tau}$ is the corresponding additional plastic work finally dissipated into heat.

In the observed cases $<p>\cdot\delta$ exceeds the other terms in [14] by several orders of magnitude. A plane interface is assumed, and therefore, a surface term due to a curved interface does not occur in [14].

3.1.3. Graphical representation of the driving forces

The evaluation of the driving forces can be explained with the help of a schematic molar Gibbs energy diagram (Figure 7).

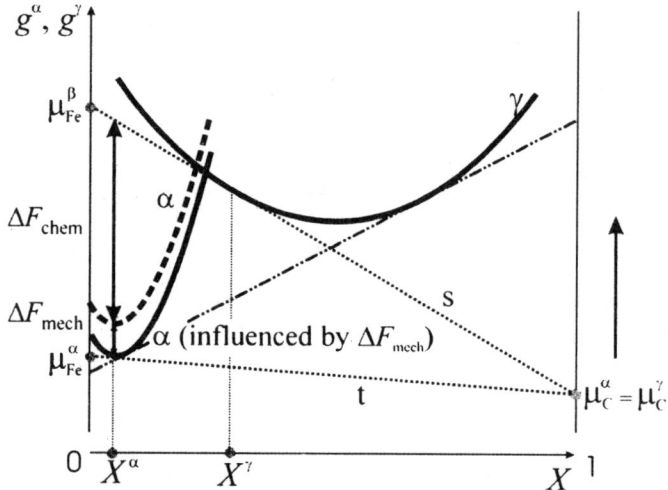

Figure 7. *Schematic molar Gibbs energy diagram*

The two energy curves (γ = solid curve, α = dashed curve) have been calculated as a function of the C mole fraction. The relative position has to be shifted by the value of ΔF_{mech}. We start from the two thick solid curves, the α- and the γ-energy curves. A tangent s is drawn at the initial mole fraction of γ in austenite. The intersections of a tangent with X = 0 and X = 1 provide the chemical potentials μ_{Fe} and μ_C. The chemical potentials μ_{Fe}^{γ} and μ_C^{γ} in the γ-phase are given by the intersection of tangent s with the vertical lines X = 0 and X = 1. Since μ_C^{γ} equals μ_C^{α} in the case of a sharp interface, this is the starting point of the tangent t on the α-phase. The total driving force ΔF is given as the distance between the two tangents s and t at the C mole fraction X_{α} in ferrite. The molar Gibbs energy diagram shows that ΔF is the difference of the chemical potentials of iron in both phases as X_{α} is approaching zero; compare [12]. During phase transformation the two tangents are shifted and twisted until the common tangent is reached.

3.2. *Modeling of the kinetics*

Firstly, the kinetics of the transformation has been calculated without taking the mechanical driving force ΔF_{mech} into account. The calculations have been performed with a numerical routine, the "transformation routine" . Therefore, it was possible to treat the rearrangement of the lattice during phase transformation and the time-consuming carbon diffusion simultaneously. Secondly, the numerical procedure has been implemented in an FE-model in order to investigate the influence of ΔF_{mech} on the phase transformation.

3.2.1. *Growth of a ferrite layer*

Figure 8 shows the two-phase arrangement and the growth of the ferrite layer. For sake of simplicity a plane interface has been assumed. The rectangle in Figure 8 represents the unit cell. The height of the rectangle may become zero. Therefore, a one-dimensional process is considered. The arrow from the grain boundary to the centre indicates the growth direction.

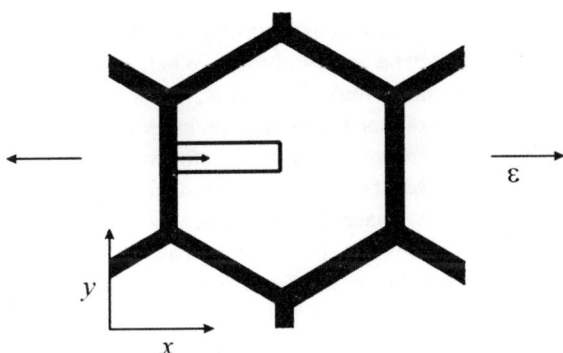

Figure 8. *Phase arrangement, growth of ferrite and uniaxially applied tension*

The dark regions in Figure 8 represent the ferrite phase, which tends to grow towards the grain centre for positive values of the driving force.

3.2.2. *Mobility and diffusion coefficient*

The thermally activated quantities are the mobility M and diffusion coefficient D,

$$M = M_0 \exp(-Q_M/RT), \ D = D_0 \exp(-Q_D/RT), \qquad\qquad [15]$$

$M_0 = 5.8 \cdot 10^{-6}$ m^2 s kg^{-1}; $Q_M = 147$ kJ mol^{-1},
$D_0 = 2.343 \cdot 10^{-5}$ m^{2}s^{-1}, $Q_D = 148$ kJ mol^{-1} are used in the calculations.

These values can be found in [LEE 99].

3.2.3. *Computational results of the transformation kinetics*

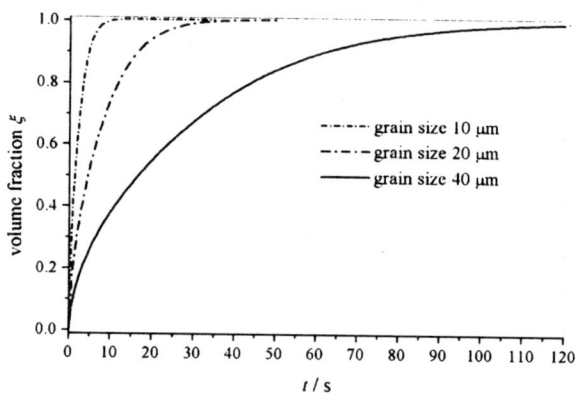

Figure 9. *Transformation kinetics in dependence of the initial grain size*

The volume fraction ξ is defined as the volume of ferrite already transformed in relation to the volume of ferrite at equilibrium. When ξ is plotted versus time at a constant temperature and initial mole fraction X, the influence of the grain size on the transformation can be examined. Larger grain sizes implicate larger distances for carbon diffusion and therefore much longer transformation times (Figure 9). Longer transformation times mean that the specimens are in a critical state over a longer time period with respect to the phase arrangement. As already said, thin films of allotriomorphic ferrite are known to be the worst case. Experimental investigations, for example [MIN 91], have shown that fine grained specimen are better with respect to hot ductility behaviour than coarse ones. This is in good accordance with this numerical result.

In order to optimize the conditions during continuous casting, it is essential to predict the time period in which the volume ratio has a critical value. Figures 10, 11 and 12 show the transformation kinetics for an initial mole fraction X_C of carbon of 0.004 , 0.002 and without any carbon. The kinetics are observed at different degrees of undercooling. The transformation time is defined as the period between the beginning of growth and at 99% completion (indicated by the arrows). At high temperatures (860°C) the extended period of equilibration is caused by the low value of the chemical driving force. At 850°C and 780°C the chemical driving force becomes higher the lower the temperature and, therefore, equilibrium is reached faster. At an even higher degree of undercooling (730°C) the thermally activated quantities mobility and diffusion coefficent "overbalance" the influence of the driving force. Therefore, equilibration takes longer again.

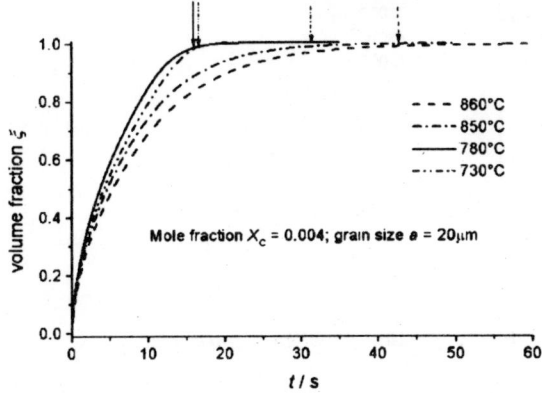

Figure 10. *Temperature-dependent kinetics for $X_C = 0.004$*

The same tendency of the transformation kinetics with respect to temperature has been obtained for a lower initial mole fraction of carbon (Figure 11).

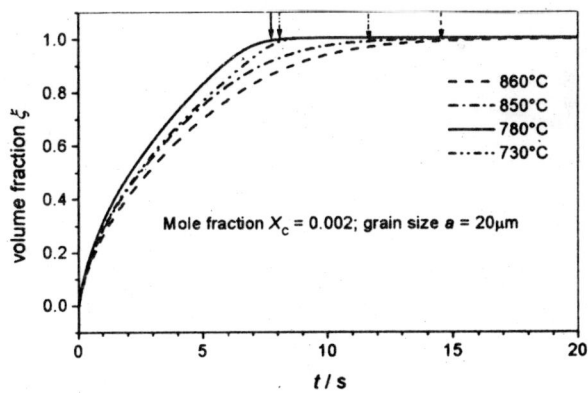

Figure 11. *Temperature-dependent kinetics for $X_C = 0.002$*

Additionally, it can be stated that a lower initial carbon content X_C speeds the transformation kinetics.

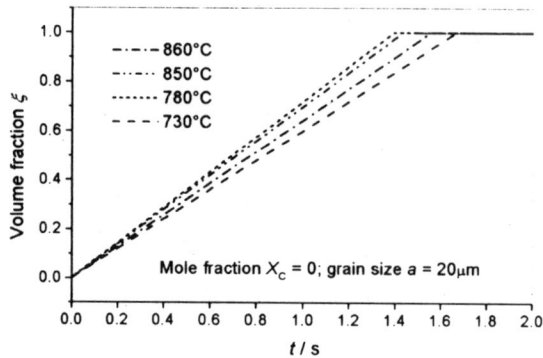

Figure 12. *Temperature-dependent kinetics in pure iron*

Even if the transformation of pure iron (initial mole fraction of carbon X_C being zero) is investigated, the interface velocity has a maximum in the temperature range where transformation occurs (Figure 12).

3.2.4. *TTT-diagrams*

Our model predicts a similar behaviour for the transformation periods as is usually documented in TTT-diagrams.

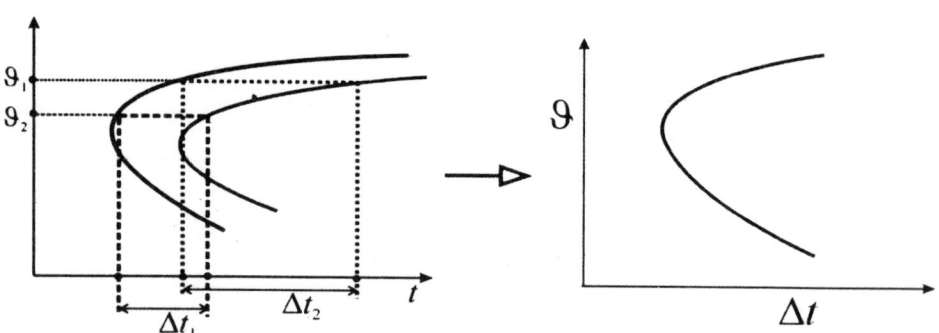

Figure 13a. *Typical TTT-diagram*   **Figure 13b.** *Growth kinetics*

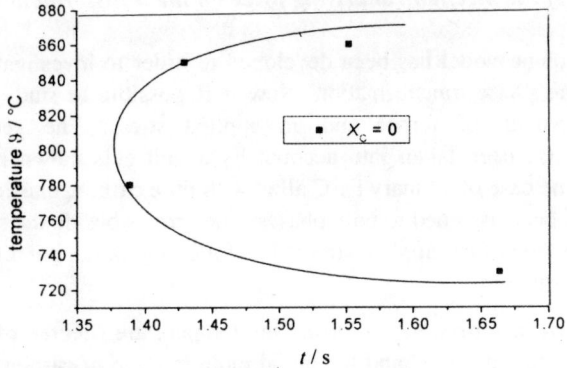

Figure 13c. *Growth kinetics in pure iron*

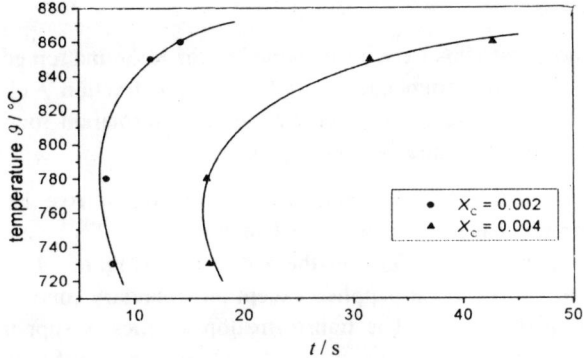

Figure 13d. *Growth kinetics for $X_C = 0.002$ and 0.004*

In typical TTT-diagrams the transformation time (as a sum of nucleation time and growth time) is plotted versus temperature ϑ, Figure 13a. Usually, two nose-like curves are plotted versus time, the first after 1%, the second after 99% has transformed. If nucleation occurs at the beginning of the transformation, the period Δt is consumed for growth. If Δt is plotted versus time, again a characteristic "nose-like" curve will be the result (Figure 13b).

The results of the numerical investigations show the same tendency when being plotted in TTT-like curves. This evaluation has been done at the three different initial compositions ($X_C = 0$, 0.002, 0.004), see Figures 13c and d.

•
3.3. *Influence of the mechanical driving force on the transformation*

A finite element model has been developed in order to investigate the influence of ΔF_{mech} on the phase transformation. Now it is possible to study the interaction between the growth of ferrite and an applied stress. The geometry of the microstructure has been taken into account by a unit cell shown in Figure 8. As appropriate in the case of a binary Fe-C alloy with little carbon, elastic-ideally plastic properties have been assigned to both phases. The arrows beside the sketch in Figure 8 show the direction of the applied stress. The following initial conditions have been specified to the model:

The grain size has been set to 20μm, the temperature (degree of undercooling) has a constant value of 800°C and the initial mole fraction of carbon in austenite X_C is assumed to be 0.004.

A strain-controlled tension test in x-direction (see Figure 8) has been simulated. The volume fraction f of ferrite (volume of ferrite divided by the total volume) has been plotted versus time.

First of all no constraints have been applied to the right and top edges of the unit cell (Figure 8). The time dependence of the volume fraction f obtained by this analysis coincides with the curve given by the "transformation routine", when the same conditions prevail (Figure 14: free expansion).

Tension tests with three different loading conditions have been performed. Firstly, the strain rate has been set to a constant value of $5 \cdot 10^{-4} s^{-1}$, and the total strain reaches a maximum value of 0.01 in the x-direction (Figure 14: $\varepsilon_x = 0.01$). No constraints in the y-direction are applied except the boundary conditions required to assure geometrical symmetry. The transformation kinetics is supported by tensile stresses. The high hydrostatic pressure induced by the transformational volume change works against the phase transformation and slows it down. However, an external tension stress state drives the transformation process. This effect is in agreement with experimental investigations [LEW 98], where it is stated that straining accelerates the austenite-to-ferrite transformation.

Secondly, the displacement in the x-direction has been inhibited (Figure 14, fixed in the x-direction). This constraint retards the transformation kinetics. In a third case transformation kinetics is investigated under the assumption that no displacement in x and y-direction is possible (Figure 14, fixed in the x-direction and the y-direction). These strict constraints and the blow-up of the volume due to transformation operate against each other, and less ferrite than in the previous cases will be formed.

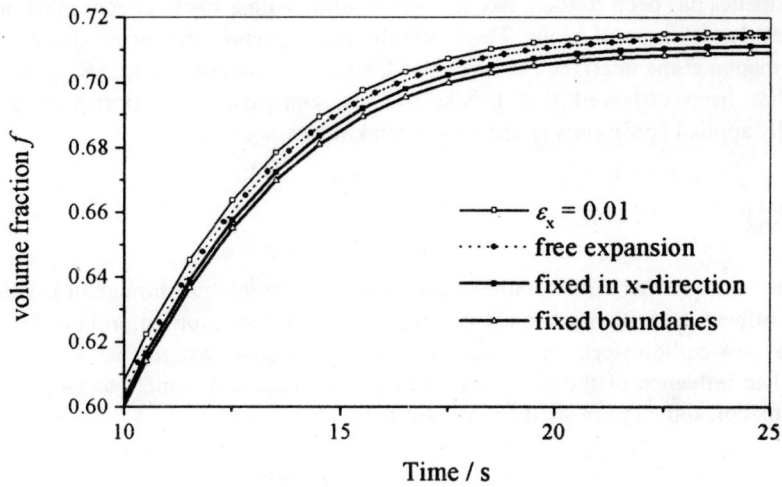

Figure 14. *Influence of a mechanical driving force on the kinetics*

Since the chemical driving force ΔF_{chem} exceeds the mechanical driving force ΔF_{mech} at almost all stages of transformation, the influence of ΔF_{mech} is small at least at the temperature and initial carbon content investigated above. It is known [LEW 98] that the influence of tensile stresses at temperatures with a small amount of undercooling (some degrees below Ae_3) is more pronounced. This can be explained by the fact that ΔF_{mech} gains in importance at a small amount of undercooling.

The magnitude of ΔF_{mech} strongly depends on the mechanical propertie of the specimens. The volume change would lead to comparably high values of ΔF_{mech} in a purely elastic composite or in an elastic-plastic material with a high capability for plastic hardening.

4. Conclusion

Large strain concentrations in the austenite-ferrite microstructure can be predicted by unit cell calculations. These analyses provide for explanation of the ductile damage qualitatively.

The coupled problem of an interface migration and carbon diffusion during γ–α–transformation in the binary Fe-C-system has been solved by a numerical routine. The transformation kinetics have been evaluated and the results have been compared with schematic TTT-diagrams. When the growth of a ferrite layer is investigated, the variation of the temperature (amount of undercooling) leads to the expected "nose-like" characteristics. The influence of the mechanical driving force on the kinetics has been studied, too. In an Fe-C-alloy with a small amount of carbon almost no hardening will occur. Therefore, the mean stresses that occur due to the transformation at the interface can be reduced, and ΔF_{chem} overbalances ΔF_{mech}. Even so, it has been observed that tensile stresses support the transformation and externally applied constraints retard the growth kinetics.

5. Outlook

In the future further efforts will be made to investigate the transformation kinetics and the influence of the mechanical driving force on the transformation in different low-alloy low-carbon steels. It is planned to work on ternary systems like Fe-C-Mn-alloys. The influence of the very small AlN-precipitates with respect to the ferrite transformation and damage shall be investigated in detail, too.

References

[ABA] www.abacom.de

[BÖH 97] Bohm H.J., "Hexgrain V.A-2.", Institute for Light Weight Structures and Aerospace Engineering, TU – Vienna, Austria, 1997.

[CHI 99] Chimani C.M., Mörwald K., "Micromechanical investigation of the hot ductility behavior of steel", ISIJ International, 39, 1194–1197, 1999.

[ERI 90] Eriksson G., Hack K., "ChemSage - A computer program for calculating of complex chemical equilibria", Metallurgical Transactions B, 21B, 1013–1023, 1990.

[FIS 95] Fischer F.D., Kolednik O., Shan G. X., Rammerstorfer F.G., "A note on calibration of ductile failure damage indicators", International Journal of Fracture, 73, 1995, 345–357, 1995.

[FIS 98] Fischer F. D., Reisner G., "A criterion for the martensitic transformation of a microregion in an elastic-plastic material", Acta Materialia, 46, 2095–2102, 1998.

[FIS 01] Fischer F.D., Simha N. K., Svoboda J., "A micromechanical study of the kinetics of diffusional transformations in elastic-plastic materials", submitted to Int. J. Mech. Phys. Solids, 2001.

[HAR 96] Harste K., Schwerdtfeger K., "Thermomechanical properties of iron: viscoplasticity of ferrite and of austenite-ferrite mixtures", Materials Science and Technology, 12, 1996, 378.

[HIL 99] Hillert M., "Overview No. 135: Solute drag, solute trapping and diffusional dissipation of Gibbs energy", Acta Materialia, 47, 4481–4505, 1999.

[LEE 99] VAN LEEUWEN Y., KOP T. A., SIETSMA J., VAN DER ZWAAG S., "Phase transformation in low-carbon steels; modelling the kinetics in terms of the interface mobility", *Journal de Physique IV*, 9, 401–409, 1999.

[LEW 98] LEWIS J., JONAS J.J., MINTZ B., "The formation of deformation induced ferrite during mechanical testing", *ISIJ International*, 38, 300–309, 1998.

[MAE 90] MAEHARA Y., YASUMOTO K., TOMONO H., NAGAMICHI T., OHMORI Y., "Surface cracking mechanism of continuously cast low carbon low alloy steel slabs", *Materials Science and Technology*, 6, 793–806, 1990.

[MIN 91] MINTZ B., YUE S., JONAS J. J., "Hot ductility of steels and its relationship to the problem of transverse cracking during continuous casting", *International Materials Reviews*, 36, 187–217, 1991.

[NAG 99] NAGASAKI C., KIHARA J., "Evaluation of intergranular embrittlement of a low carbon steel in austenite temperature range", *ISIJ International*, 39, 75–83,1999.

[PAT] PATRAN, www.macsch.com

[SUZ 84] SUZUKI H. G., NISHIMURA S., IMAMURA J., NAKAMURA Y., "Embrittlement of steels occurring in the temperature range from 1000 to 600°C", *Transactions ISIJ*, 24, 169–177, 1984.

[SVO 01] SVOBODA J., FISCHER F. D., FRATZL P., GAMSJÄGER E., SIMHA N.K., "Kinetics of interfaces during diffusional transformation", *Acta Materialia*, 49, 1249–1259, 2001.

Index

adaptive FEM simulation, iterative
and
hydroforming processes, tubular
parts for 179 et seq
method 184
adiabatic softening 306
advection–diffusion equations 2
filling state 4
AlMg sheet, modeling of
forming, elevated temperatures at
165 et seq
ALE formulation 344
aluminium extrusion, finite element
simulations 139 et seq
anisotropic sheet-metals, bend-
reverse bend experiments in 197
et seq
non-linear kinematic hardening
197 et seq
arbitrary Lagrangian-Eulerian
simulation 341 et seq
powder compaction processes of
341 et seq
area normalization 75.
austenite/ferrite phase arrangements
359
-to-ferrite phase
continuous casting of steels 357
et seq
autqmomotive transmission tunnel,
validation for 98
axis tilt effect 294

back cutting detection 295
backward global solution scheme 8
bearing, simulating 141

bending-unbending test 202
bend-reverse bend experiments,
anisotropic sheet-metals in 197 et
seq
non-linear kinematic hardening
197 et seq
Bergstrom model, the 169
bias extension test 76
blanking
device 125
experimental and numerical
analysis
thin sheet metal parts for 123
et seq
material behaviour in 127
process, simulation of 131
bottom punch compaction 349, 351,
352
test 351
boundary conditions modeling 297

calibration step 179
camshaft cover 297
coating film, thermal resistance of
245
interfaces 245
Cockcroft and Latham criterion 249
et seq, 263, 264
fracture
steel fasteners, cold heading
of 249 et seq
validation of 264
cold 249 et seq, 341
compaction 341
heading, steel fasteners of

Cockcroft and Latham
    fracture criterion 249 et
    seq
collar test, the 113
combined extension, shear and 83
  deformed unit cell under 83
  test
      reaction force, comparison
      of 83
compaction
  cold 341
  powder 341
      processes
          arbitrary Langrangian-
          Eulerian simulation
          341 et seq
  rotational flanged component 345
  punch 347, 348, 349, 350, 354
      bottom 349, 350, 352
          test 351
      double 352, 353
          test 354
      top 347, 348, 349
complex shape, mold filling 13
computation 288, 296
  force 288
  form error 296
continuous casting of steels 357 et
  seq
  austenite-to-ferrite phase 357 et
  seq
convergent die, flow in 11
crystallinity, infrared 330, 331
  iPP 330
      measurements 331
  PA6
      measurements 331
crystallization
  behaviour
      iPP, PA6 and 333
  iPP and PA6 under pressure 315 et
  seq
cup, deep drawing of 157
cutting force model 288

damage
  distribution, prediction of 134
  indicator 363
  modeling 133
  models 110
      ductile 105
deep drawing
  cup, of 157
      experiments, simulations
      and 172
deformed unit cell, combined
  extension and shear 83
dilational plastic strain rate 109, 110
displacement field, recording of 129
double punch compaction 352, 353
  test 354
draw-bend test 152, 155
  results of 155
drop weight
  compression testing (DWCT) 250
  results 250
  test machine 250
ductile damage models, comparison
  of 105
dynamic TCR, model of 241

ejection, handling of parts and 66
elevated temperatures, modeling of
  AIMg sheet forming at 165 et seq
engineering alloys, semi-solid
  processing 39 et seq
  twin-screw rheomolding process
  39 et seq
equations 2, 4, 218
  advection
      -diffusion 2
      filling state 4
  rate, plastic flow and hardening
  218
equivalent creep 261
exhaust manifold 299
experimental 92, 123 et seq, 128
  analysis of blanking, numerical
      and
      thin sheet metal parts 123 et
      seq

load-penetration, punch 128
predicted, and
    shear compliance
        woven fabric 92
extrusion 139 et seq, 140
    aluminium, finite element
        simulations 139 et seq
    process, numerical model 140

fabric 87 et seq, 91, 92, 93, 94, 95
    deformation, characterization and
        modeling
            textile composites, forming of
                87 et seq
    forming simulation 94
    non-crimp 95
            shear compliance curves for
                93
    reinforcements, woven and non-
        crimp
            shear 91
    woven, shear compliance
            experimental and predicted
                92
face milling, turning and
    simulation of
            finite element method 285
                et seq
FEM simulation, adaptive and
    iterative 179 et seq
    tubular parts, hydroforming
        processes for 179 et seq
filling 4, 16, 13, 32
    mold
            complex shape 13
            two branch process 32
    state, advection equations for 4
    time, mold filled ratio vs. 16
finite element method
    face milling and turning,
        simulation of 285 et seq
    simulations, aluminium extrusion
        of 139 et seq
    software 132
flow
    convergent die, in a 11

curves, material 309
fluid-particle, metal injection
    molding during
        phase segregation effects,
            analysis of 1 et seq
plastic, hardening and
    rate equations 218
simulation
        fluid 27, 29
            Newtonian 29
            short fiber molten composite
                31
fluid 1 et seq, 15, 27, 29
    flow simulation 27
        Newtonian 29
    -particle flows
            metal injection molding,
                during 1 et seq
    volume fractions, contours of 15
force computation 288
forging simulation 235 et seq
    thermal contact resistance 235 et
        seq
form error computation 296
forming 87 et seq, 94, 95, 165 et seq
    AlMg sheet, modeling of
            elevated temperatures 165
                et seq
    processes
            numerical simulation of 77
    simulation
            fabric 94
            mechanical 95
    textile composites 87 et seq
            fabric deformation,
                characterization and
                modeling of 87 et seq

gear box cover 300
Gelin model 119
    yield function 106
geometrical tolerances 182
geometric/kinematic models 94
geometry, mesh and 13
    material parameters 13

gradient theory of plasticity 215 et
    seq
  microhardness simulation, large
      strains 215 et seq
grain refinement 48
Gurson model 118

handling of parts, ejection and 66
hardening 218
  plastic flow and 218
        rate equations for 218
high strain 303 et seq, 305
  rate testing device, SHPB 305
  rates, material behaviour at 303 et
    seq
    modeling machining processes
        303 et seq
hydroforming
  processes, tubular parts for
        FEM simulation, adaptive
            and iterative 179 et seq

imposed curvature, strip at 201
indentation 215 et seq, 220
  surface, model problem 220
  tests 215 et seq
industrial forging design
  inverse technique, using 267 et
    seq
infrared crystallinity 330, 331
  iPP 330
        measurements 331
  PA6
        measurements 331
injection
  molding
        machine 58
        metal, fluid-particle flows
            during
            phase segregation effect,
                analysis of 1 et seq
        micro 55 et seq, 58, 65
        process 63
        simulation of 64
  processes, short fiber molten
      composites 19 et seq

unit
        micromelt'- 60
        screw/piston 61
intensively sheared semisolid slurry,
    secondary solidification of 49
intercept length, particle density and
    48
  shear rate 48
interfaces 236, 245
  thermal resistance of
        coating film 245
        workpiece-die, TCR 236
intra-ply shear 88
inverse problem 270
inverse technique, industrial forging
    design using 267 et seq
investigation of springback 151 et
    seq
iPP 315 et seq
  density 321, 323
        depth profile 324
  infrared crystallinity 330
        measurements and PA6 331
  microhardness 320, 322
  PA6 and
        crystallization of
            behaviour 333
            under pressure 315 et seq
  phase distribution 329
iterative method 185

kinematic 94, 197 et seq
  geometric models 94
  hardening, non-linear
        bend-reverse bend
            experiments, anisotropic
            sheet-metals in 197 et seq

large strains, microhardness at 215 et
    seq
  gradient theory of plasticity 215 et
    seq
  simulation of 215 et seq
LIGA-techniques 62
load-penetration curve 127, 128
  experimental

punch 128

machine processes, modeling 303 et
   seq
material 13, 71 et seq, 78, 303 et seq,
   309
   behaviour
      blanking, in 127
      high strain rates
         modeling machining
            process 303 et seq
   characterization, numerical
      approach for 78
   flow curves 309
   models, stamping of plain woven
      composites
         materials characterization
            methods and 71 et seq
         parameters, geometry and mesh
            13
materials characterization methods
   71 et seq
   material models, stamping of plain
      woven composites for 71 et seq
mechanical conditioning,
   repeatability through 74
   forming simulation 95
mesh, geometry and 13
   material parameters 13
meshless techniques, natural element
   method 23
metal injection molding (MIM) 1 et
   seq
   fluid-particle flows, during 1 et
      seq
microhardness 215 et seq, 320, 322,
   326, 327
   iPP 320, 322
   large strains, simulation of
      gradient theory of plasticity
         215 et seq
   PA6 326, 327
micro injection molding 55 et seq,
   58, 65
   parts, definition of 55 et seq
micromelt'-injection unit 60

microstructural characteristics,
   rheomolded alloys of 43
   evolution, primary phase 44
mixed natural element formulation
   28
mixture front, contours 14
model 108, 118, 119, 140, 220, 235
   et seq, 241
   Gelin 119
   Gurson 118
   numerical, extrusion process 140
   problem, surface indentation 220
   Rice and Tracey 108
   simplified, thermal contact
      resistance of 235 et seq
   TCR, dynamic 241
   void growth 108
modeling 87 et seq, 133, 303 et seq,
   165 et seq
   AIMg sheet of, forming at elevated
      temperatures 165 et seq
   damage 133
   fabric deformation,
      characterization and
         textile composites, forming of
            87 et seq
   machining processes
      material behaviour, high
         strain rates for 303 et seq
models 71 et seq, 94, 105, 110
   damage 110
      ductile 105
   geometric/kinematic 94
   material, materials characterization
      methods and
         stamping of plain woven
            composites 71 et seq
mold 16, 32
   filled ratio vs filling time 16
      complex shape 13
      filling 13
      process, two-branch 32
molding 1 et seq, 55 et seq, 58, 63,
   64, 65
   injection
      machine 58

metal, fluid-particle flows
  during
    phase segregation effects,
      analysis of 1 et seq
  micro 55 et seq, 58, 65
  process 63
    simulation of 64
molds 62
moment-curvature
  (aluminium 2) 207
  curve (aluminium 1) 203, 207
  (mild-steel) 209
  theoretical 205, 209

Nadai model, extended 168
natural element method 25
  meshless techniques 23
  mixed formulation 28
Newtonian fluid flow simulation 29
nodal variables, recovery of 28
non-crimp fabrics 95
  reinforcements, woven and
    shear, in 91
    shear compliance curves 93
non-linear kinematic hardening,
  bend-reverse bend experiments
  with 197 et seq
  anisotropic sheet-metals in 197 et
  seq
nose cone, validation for 97
numerical analysis of blanking,
  experimental and
  approach, material characterization
  for 78
  model, extrusion process 140
  simulation, forming processes of
  71
  thin sheet metal parts for 123 et
  seq

optimization process 183

PA6 density 325
  infrared crysallinity
    measurements, iPP and 331
  iPP and

crystallization of
  behaviour 333
    under pressure 315 et seq
  microhardness 326, 327
  phase distribution 329
packaging 67
particle density, intercept length and
  48
  shear rate 48
parts, ejection and handling of 66
phase 1 et seq, 44, 329
  distribution
    iPP 329
    PA6 329
  primary, microstructural evolution
    of 44
  segregation effects, analysis of
    fluid-particle flows, metal
      injection molding during 1
    et seq
picture-frame shear rig 89
plain woven composites, stamping of
  71 et seq
  materials characterization methods,
    material models for 71 et seq
plastic flow, hardening and 218
  rate equations 218
plasticity, gradient theory of 215 et
  seq
  microhardness at large strains,
    simulation of 215 et seq
p-norm method 287
powder compaction processes
  arbitrary Lagrangian-Eulerian
    simulation 341 et seq
primary phase, microstructural
  evolution 44
  solidification 49
process limits, product performance
  or 181
  optimization 183
product performance 182
  process limits or 181
punch 127, 128, 347, 348, 349, 350,
  351, 352, 353, 354
  compaction

bottom 349, 350, 352
  test 351
double 352, 353
  test 354
top 347, 348, 349
load-penetration curve for 127
  experimental 128

quality control 65

reaction force combined extension,
  shear test and 83
repeatability
  mechanical conditioning through
    74
  shear response of 75
rheomolded alloys, microstructural
  characteristics 43
rheomolding process, twin-screw 39
  et seq, 40, 51
  semi-solid processing of
    engineering alloys by 39 et seq
Rice and Tracey model 108
rotational flanged component,
  compaction of 345

screw/piston injection unit 61
secondary solidification 49
  intensively sheared semisolid
    slurry 49
segregation effect 2
  phase, analysis of
    fluid-particle flows, metal
      injection molding during 1
      et seq
  predictions, numerical schemes for
    7
self-regularization method 10
semi-solid metal (SSM) 39
  processing, engineering alloys of
    twin-screw rheomolding
    process 39 et seq
shape optmization technique 276
shear 72, 75, 83, 88, 89, 91, 92, 93
  combined extension and
    deformed unit cell 83

compliance
  curves, non-crimp fabrics
    93
  experimental and predicted
    woven fabrics 92
fabric reinforcements, woven and
  non-crimp 91
frame 72
intra-ply 88
rate, intercept length and particle
  density 48
response, repeatability of 75
rig, picture-frame 89
test, combined extension
  reaction force 83
short fiber molten composites 19 et
  seq, 31
  flow simulation 31
  injection process 19 et seq
SHPB 303
  high strain rate testing device 305
simple forward explicit method 7
simulating the bearing 141
simulation 27, 29, 31, 64, 71, 94, 95,
  131, 139 et seq, 172, 179 et seq,
  215 et seq, 235 et seq, 285 et seq,
  341 et seq
  arbitrary Lagrangian-Eulerian
    powder compaction
    processes 341 et seq
  blanking process 131
  deep drawing experiments, and
    172
  face milling, turning and
    finite element method 285
    et seq
  FEM, adaptive and iterative
    hydroforming processes for
    tubular parts 179 et seq
  finite element, aluminium
    extrusion of 139 et seq
  fluid flow 27
    Newtonian 29
  forging
    thermal contact resistance,
      adapted to 235 et seq

forming
    fabric 94
    mechanical 95
injection molding process 64
microhardness, large strains
    gradient theory of plasticity
        215 et seq
numerical, forming processes 71
short fiber molten composite flow
    31
solidification 49
    primary 49
    secondary
        intensively sheared
            semisolid slurry 49
solid volume fractions, contours of
    15
space-time distribution, TCR 238
specific yield function 107
springback
    analysis 210
    investigation of 151 et seq
stamping, plain woven composites of
    71 et seq
    materials characterization methods,
        material models for 71 et seq
steel fasteners, cold heading of 249
    et seq
    Cockcroft and Latham fracture
        criterion 249 et seq
stress
    history
        yield surface, along 261
    trixiality, equivalent strain versus
        261
strip, imposed curvature with 201
structural integrity 182
superelement method (SEM) 292
    performance 293
surface 111, 183, 220, 261
    characteristics 183
    indentation, model problem 220
    yield 111, 261
        comparison of 111
        stress history 261
suspension support 298

Taylor-Galerkin method 8
TCR 236, 238, 241
    model of dynamic 241
    space-time distribution 238
    workpiece-die interface 236
tensile test curves (mild-steel) 209
textile composites, forming of fabric
    deformation, characterization and
    modeling 87 et seq
theoretical moment-curvature
    (aluminium 1) 205
thermal
    contact resistance (TCR) 235
        dynamic, model of 235 et
            seq, 241
        forging simulation, adapted
            to
        simplified model 235 et
            seq
        space-time distribution 238
        workpiece-die interface 236
    resistance
        coating film 245
        interfaces 245
thin sheet metal parts 123 et seq
    blanking, experimental and
        numerical analysis of 123 et
        seq
tool finite element mesh 277
top punch compaction 347, 348, 349
trellising results 81
T-shape test 190
tube-expansion test 186
tubular parts, hydroforming processes
    for 179 et seq
    FEM simulation, adaptive and
        iterative 179 et seq
twin-screw reomolding process 39 et
    seq, 40, 51
    semi-solid processing, engineering
        alloys of 39 et seq
two-branch mold filling process 32

void growth model 108
volume fractions, contours of 15
    fluid 15

solid 15

wave synchronization 308
workability test 116
workpiece-die interface, TCR 236
woven composites, stamping of plain
fabric
        reinforcements, non-crimp
        and
          shear, in 91
        shear compliance,
          experimental and predicted
        92

materials characterization methods,
    material models and 71 et seq

yield 106, 111, 261
    function, Gelin 106
    surfaces
            comparison of 111
            stress history 261